Climate Change and Global Energy Security

Climate Change and Global Energy Security

Technology and Policy Options

Marilyn A. Brown and Benjamin K. Sovacool

The MIT Press
Cambridge, Massachusetts
London, England

For information on quantity discounts, email special_sales@mitpress.mit.edu.

Set in Sabon by Toppan Best-set Premedia Limited. Printed and bound in the United States of America.

Library of Congress Cataloging-in-Publication Data

Brown, Marilyn A.
Climate change and global energy security : technology and policy options / Marilyn A. Brown and Benjamin K. Sovacool.
p. cm.
Includes bibliographical references and index.
ISBN 978-0-262-01625-4 (hardcover : alk. paper)—ISBN 978-0-262-51631-0 (pbk. : alk. paper)
1. Renewable energy sources. 2. Power resources. 3. Energy policy. 4. Environmental policy.
5. Greenhouse gas mitigation—International cooperation. I. Sovacool, Benjamin K. II. Title.
TJ808.B76 2011
363.738'746—dc22

2011001927

10 9 8 7 6 5 4 3 2 1

A fundamental principle of a civilized society is to protect people from harm done by others. If we are to call ourselves a civilized species, there is nothing for it but to change the way we produce and consume energy.

David Archer and Stefan Rahmstorf, *The Climate Crisis: An Introductory Guide to Climate Change* (Cambridge University Press, 2010), p. 228

Contents

Acknowledgments

Many people and organizations helped make this book possible. Marilyn's husband Frank Southworth (a scientist at Oak Ridge National Laboratory and the Georgia Institute of Technology) offered encouragement and technical insights, particularly on transportation issues. Numerous other scholars and practitioners provided valuable background information. Paul Barter of the National University of Singapore and Kian Keong Chin of the Singaporean Land Transport Authority imparted helpful comments in revising our case study on Singapore. Dipal C. Barua of Grameen Shakti provided excellent assistance and comments for our case study on Bangladesh. Laurenz Pinder of the Fundação O Boticário de Proteção à Natureza and Thais Cercal Dalmina Losso of Losso, Tomasetti & Leonardo Advogados Associados kindly offered to review our case study on the Oasis Project. All who participated in our research interviews have our deepest gratitude for taking time out of their busy schedules to meet and speak with us. Also helpful were the staff of the Newman Library at the Virginia Polytechnic Institute and State University—especially the wonderful people in the interlibrary loan department who helped us collect hundreds of hard-to-find articles and research materials.

Financial support for the book was provided by several organizations, including the Ivan Allen College of Liberal Arts at the Georgia Institute of Technology and the Centre on Asia and Globalisation of the Lee Kuan Yew School of Public Policy at the National University of Singapore. Financial support was also provided by the Bipartisan Policy Center, by the National Security Coordination Centre and the Centre for Strategic Futures at the Singaporean Prime Minister's Office, by the U.S. National Science Foundation (grants SES-0522653, ECS-0323344, and SES-0522653), by the Singaporean Ministry of Education (grant T208A4109), by the Rockefeller Foundation (grant RES2009315), by the National University of Singapore (faculty start-up grant 09-273), and by the MacArthur Foundation Asia Security Initiative (grant 08-92777-000-GSS). The financial assistance allowed us to acquire the services of a veteran copy editor, Charlotte Franchuk, who deserves special credit for formatting all nine chapters. Valuable assistance with fact checking was provided by Kelly

Siman of the National University of Singapore and by Matt Cox of Georgia Tech. Beth Daihl and Lizzie Noll of Georgia Tech helped with computer graphics, and Marty Sung with GIS and computer cartography.

Any opinions, findings, and conclusions or recommendations expressed in the book are those of the authors and do not necessarily reflect the views of any of the sponsoring organizations.

1

Motivation and Organization of the Book

Marion King Hubbert, the geophysicist who famously predicted when oil production would hit its peak in the United States, once commented that all of industrial civilization was threatened by the incompatibility of two elemental systems: the age-old system of matter and energy (with which humans and other animals have evolved symbiotically for thousands of years) and the more recent money-based culture (which has operated without respect for limits and constraints on growth).[1] Hubbert believed that the two systems could coexist only so long as the matter-energy system had sufficient reserves to enable the industrial-monetary system to grow. He worried that the "disparity between a monetary system which continues to grow exponentially and a physical system which is unable to do so" would be the great challenge of the twenty-first century.

Although the merits of Hubbert's prophecy can be debated, his classification of the two systems as prehistoric is apt. We rely on dwindling reserves of fossilized fuels that have existed for millions of years to provide a majority of our energy needs and services. The belief that civilization has a limitless opportunity to grow is also prehistoric in a sense, for it contravenes even rudimentary lessons from physics, thermodynamics, ecology, and biology. Much of what we can do to address climate and energy challenges we already know. But for business, social, cultural, and political reasons, societies all over the world allow their consumption of resources to damage the capacity of the planet. We agree with Hubbert that "our ignorance is not so vast as our failure to use what we know."

Imagine if present-day policy makers were given the assignment of designing the global economy, human civilization, and the energy systems underpinning it from the start. Would they have wished for the following?

• global mobility that depends almost exclusively on the refined products of a single resource—crude oil—with known reserves that are concentrated in a handful of largely volatile places—notably the Middle East, Russia, Nigeria, and Venezuela

• a production system that puts billions of tons of toxic materials into the air, water, and soil every year, requiring vigilance by future generations

• thousands of complex health and safety regulations that are needed not to keep people safe, but to ensure that they are not poisoned passed "acceptable" thresholds

• more than 2 billion people who live on less than the equivalent of 2 US dollars per day; 1.6 billion people without access to electricity, 900 million people who do not have access to motorized transport, 1.8 million deaths per year due to unsafe drinking water, and 2.5 billion people without access to hygienic sanitation[2]

• at the same time, millions of people driving to and from work alone in powerful "sport utility vehicles" and later relaxing in spacious homes with more televisions than people, each television consuming 30 watts of electricity when turned off to enable an instantaneous "on"

• valuable materials placed in landfills and waste sites all over the planet, from which they can rarely be retrieved

• prosperity and wealth created and maintained by digging, cutting, mining, and burning natural resources

• the diversity of species, the beauty of tropical forests, and coastal habitats threatened by a warming climate and rising sea levels[3]

Clearly, no single person is responsible for oil dependence, waste, pollution, environmental destruction, climate change, and electricity and transport poverty. The modern economy was not designed from the ground up; it emerged in fits and starts through an iterative process that has spanned thousands of years. This exercise should, however, make readers wonder if there might be better technology and policy options than the ones on which we have relied to date.

1.1   Socio-Technical Approach

This book explicitly recognizes that the world's energy-security and climate-change challenges occur at the intersection of technology, policy, and society—that is, they are socio-technical. Technology and policy options must be considered alongside the social and institutional context in which decisions about energy and climate are made.

Specifically, this book explores the commercially available technologies and practices that can enhance the secure provision of energy services and address global climate change, the barriers to their wider adoption, and the public policies that can overcome those barriers. In doing so, it informs readers about not only the causes and consequences of insecure energy and climate change, including water scarcity, deforestation, and environmental degradation, but also effective solutions to them.

(*Energy security* is broadly defined as the equitable provision of available, affordable, reliable, efficient, environmentally benign, properly governed, socially acceptable energy services to citizens.)

The book is motivated by the premise that tackling climate change and improving energy security are two of the most significant challenges to prosperity in the twenty-first century, and that success will require the provision of energy services (such as heating, mobility, computing, and lighting); improved standards of living; and the preservation of the natural environment without forcing tradeoffs among them.[4] It argues that society has all the technologies it needs to meet these challenges, but that better public policy is crucial to ensuring their adoption while continuing to improve their performance and affordability. Without smart policies, "socio-technical gaps" will grow, with society continuing to invest in products, infrastructure, and practices that compromise their energy security and accelerate climate change. The book offers one of the most comprehensive assessments to date of potential policies that could shrink the gaps between technically possible solutions and actual social choices, evaluated within the context of behavior and values, market and policy failures, and catalysts for change in a global context.

The book refuses to approach the issues of energy security and climate change within disciplinary boundaries. It therefore differs from previous scholarship on climate and energy policy in four fundamental ways.

1.1.1 Interdisciplinary Interactions

Rather than focusing separately on technology, barriers, and policies, the book investigates how these components interact. Most books and academic articles treat climate and energy problems narrowly from either a technological approach, a public-policy approach, or a barriers-and-impediments approach. Rarely do these works combine them to look at how technologies, policies, values, and impediments relate to each other. Our book explores not only cutting-edge technologies but also simple ones (such as light bulbs), the broader social and economic factors that drive both types of technology, their limitations and challenges, and the government policies that can promote them. It simultaneously analyzes technology, people, and solutions, and the factors that intertwine them. This is because the human origins of energy insecurity and climate change have both proximate and indirect causes. The proximate causes relate deeply to such technical processes as converting coal into electricity, discharging heavy metals into rivers, and harvesting a forest. The indirect causes are influenced by broader social forces, including population growth, economic development, and changes in social institutions and in human values.[5] Focusing on the proximate causes without understanding the indirect ones is like presenting *Hamlet* without the prince—it's missing a central component of the story.

1.1.2   Global Scope

Our book is global in its coverage of technologies, barriers, and policies. Instead of discussing the climate-change challenges in the United States or in other developed economies, it draws from case studies and experiences in Europe, North America, Asia, the Middle East, and South America. It therefore discusses "high-tech" and "large-scale" energy systems and technologies such as fuel cells, nuclear reactors, and carbon sequestration alongside more "mundane," "appropriate," and "small-scale" technologies that are seen as empowering citizens, such as improved cook-stoves, mopeds, and white roofs.[6] These examples are drawn from industrialized countries (including Germany, Japan, and the United States) and also from developing countries (e.g., Bangladesh, China, and Brazil).

1.1.3   Broad Coverage of Challenges

Rather than limiting the discussion about climate change to its principal determinants (electricity and transport, or "coal and cars"), we also address agriculture and forestry, waste and water, and other economic sectors and systems that emit greenhouse gases and threaten energy security. The book also strives to include the most up-to-date information on these different sectors and their technologies. Recent technological advances and improvements have accelerated the deployment of wind, solar, and energy-efficient technologies. Conversely, perpetually volatile oil prices, a worldwide economic downturn, the Fukushima nuclear accident, and the threat of cost overruns have slowed the momentum toward clean coal, nuclear, and hydrogen energy systems. The book assesses technology as it exists today and recent developments that suggest future trends.

1.1.4   Case Studies

We endeavored to conduct original research and to collect primary data that has either never been published before or is difficult to find. Our methodology is based on previous research that included interviews with more than 200 energy experts—interviews spanning more than 90 institutions in 11 countries that took 4 years to complete.[7] Likewise, we conducted 106 additional interviews in 12 countries over the course of 2 years exclusively for the case studies presented in chapter 8. These interviews were supplemented by an exhaustive review of the contemporary scientific and technical literature on greenhouse-gas-reducing technologies, technological and social impediments faced by those systems, and public policies that have facilitated their acceptance among users and communities.

1.2   Climate Change and Energy Security

The primary motivation behind writing this book concerns the steady degradation of energy security and the global climate that has occurred in the past few decades.

Not only does the security theme connote a concern that most citizens of the world can relate to, but it also provides a platform for constructing a multi-faceted discussion of related environmental and security issues. As we will show, the unchecked growth in the consumption of fossil fuels, the acceleration of global climate change, and related water, waste, agriculture, and deforestation challenges act as "threat multipliers," impinging on security around the world.[8]

How secure are today's energy supplies and infrastructure, and what do trends portend for the future? To answer this question, we first must consider what is meant by "energy security."[9] The classic conception of energy security addresses the relative availability, affordability, and safety of energy fuels and services.[10] The World Bank, for example, tells us that energy security is based on the three pillars of energy efficiency, diversification of supply, and minimization of price volatility.[11] Consumer advocates and users tend to view energy security as reasonably priced energy services without disruption. Major oil and gas producers focus on the "security" of their access to new reserves, while electric utility companies emphasize the integrity of the electricity grid. Politicians dwell on securing energy resources and infrastructure from terrorism and war.[12] From a distinct vantage point, scientists, engineers, and entrepreneurs characterize energy security as a function of strong energy R&D, innovation, and technology-transfer systems.[13]

These diffuse conceptions of energy security map onto distinct national energy-security concerns. In the United States, energy security has generally meant the availability of sufficient energy resources and services at affordable prices. The oil-security policy of the United States was formalized by the Carter Doctrine, which stated that any effort by a hostile power to block the flow of oil from the Persian Gulf would be viewed as an assault on the vital interests of the United States and/or would be repelled by "any means necessary, including military force."[14] Under various presidents, oil security has meant ending all oil imports, eliminating imports only from the Middle East, merely reducing dependence on foreign imports, and entirely weaning the country off oil.[15] US energy-security policy has historically also included maintaining a strategic petroleum reserve, reducing physical threats to energy infrastructure, and preventing the proliferation of nuclear weapons in "non-nuclear-weapons states" and non-signatories to the Nuclear Non-Proliferation Treaty such as Iran and North Korea.[16] More recently, concern about an increasingly fragile US electricity grid has become more evident because of the expanded electrification of US military operations.[17]

Other countries with limited energy resources have deployed different strategies to achieve security. Japan has pursued an energy-security strategy of diversification, trade, and investment, as well as selective engagement with neighboring Asian countries to jointly develop energy resources and offset Japan's stark scarcity of domestic reserves.[18] In the United Kingdom, energy security tends to be associated with promoting open and competitive energy markets that will provide fair access to energy supplies, foster investment, and deliver diverse and reliable energy at competitive prices.[19]

Similarly, the focus on energy security in countries that are struggling to meet their energy requirements is quite distinct. China, for example, has viewed energy security as an ability to rapidly adjust to their new dependence on global markets and engage in energy diplomacy, shifting from its former commitments to self-reliance and sufficiency (*zi li geng sheng*) to a new desire to build a well-off society (*xiaokang shehui*). China's current approach to energy security entails buying stakes in foreign oil fields, militarily protecting vulnerable shipping lanes, and an all-out "energy scramble" for resources.[20]

Among the countries with excess supplies of oil and natural gas, the focus on energy security takes on yet another emphasis. As one example, Russia appears to pursue an energy-security strategy of asserting state influence over strategic resources to gain primary control over the infrastructure through which it ships its hydrocarbons to international markets. Restricting foreign investment in domestic oil and gas fields is an important element of this strategy.[21] Buoyed by this strategy, Russia was recently able to triple the price of natural gas exported to Belarus and Ukraine because those countries were completely dependent on Russian supply.

Other countries have enacted strategies shaped by their substantial endowments of energy resources. For example, Saudi Arabia pursues energy security by maintaining a "security of demand" for its oil and gas exports.[22] In contrast, Australia's strategy involves cultivating a strong demand for uranium, natural gas, and coal trading.[23] Venezuela and Colombia focus on minimizing attacks on oil, gas, and electric infrastructure.[24]

International comparisons of energy security highlight the interdependence of countries enmeshed in larger relationships between and within producers and consumers of energy fuels and services. Globally, trade in energy commodities amounted to 900 billion US dollars (696.5 billion euros) in 2006, including almost two-thirds of the oil produced in the world, and much of it was in natural gas and uranium.[25] As a result, few countries are truly energy independent. As figure 1.1 shows, the world's known oil reserves (1.2 trillion barrels) are concentrated in volatile regions, as are the largest petroleum companies. The three biggest petroleum companies—the Saudi Arabian Oil Company, the National Iranian Oil Company, and Qatar Petroleum—own more crude oil than the next 40 largest oil companies combined. The 12 largest oil companies control roughly 80 percent of petroleum reserves and are all state owned. Therefore, although oil and gas are internationally traded in what superficially resembles a free market, most supplies are controlled by a handful of government-dominated firms. The distribution of other conventional energy resources, including coal, natural gas, and uranium, is equally consolidated. Eighty percent of the world's oil can be found in nine countries that have only 5

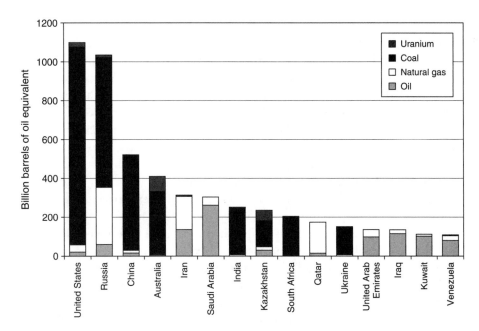

Figure 1.1
Global distribution of energy reserves (million barrels of oil equivalent) of the top 15 countries. Oil, gas, and coal data from Energy Information Administration, *International Energy Annual Review 2006*, 2006). Uranium data from International Atomic Energy Agency, *Uranium 2005—Resources, Production, and Demand*, 2006; Luis Echávarri and Yuri Sokolov, *Uranium Resource* (Nuclear Energy Agency and IAEA, 2006): 13, converted to million tons of oil equivalent using the rate of 65,000 tons of uranium ore = 24 quads of energy. (A quad is equivalent to 10^{15} British thermal units or 1.06×10^{18} joules.)

percent of the world's population, 80 percent of the world's natural gas is in 13 countries, and 80 percent of the world's coal is in six countries. Many of the same countries are among the six that control more than 80 percent of the world's uranium resources.[26]

As a result, threats to energy security take distinct forms. Japan and Chile have essentially no domestic fossil fuels and thus are completely dependent on foreign supplies. Saudi Arabia is the largest exporter of crude oil but must import refined gasoline. Russia exports natural gas but must import uranium. The United States is a net exporter of coal but imports oil and natural gas. This interdependence explains why any discussion of energy security must consider the interactions between countries as much as it considers the resources of individual countries, serving as a useful reminder that energy security does not stand abstractly by itself; rather, it is most meaningful in a geographic context.

The deterioration of energy security has also become increasingly multi-dimensional as its links to other challenges have become clearer.

1.2.1   Growing Worldwide Demand for Energy

The growing worldwide demands for electricity and for mobility compound issues of energy security. The world is in transition from a position of abundant fossil energy supplies to a largely resource-constrained supply future. The demand for energy is expected to increase by 45 percent between now and 2030, and by more than 300 percent by the end of the century. Coal without carbon capture and sequestration is projected to account for the largest share of this overall rise, with oil and natural gas consumption also expanding rapidly. (See figure 1.2.)

1.2.2   Growing Imbalance between Supply and Demand

The growing imbalance of oil production and consumption exacerbates the risk of fuel shortages and interruptions in supply, which will take a fairly rapid turn for the worse for many countries if alternative fuels such as ethanol and biodiesel are not widely deployed. The likely geographic pattern of expected oil production and consumption over the next two decades suggests that oil dependence in Europe, China, India, and other Asian countries could grow rapidly, each importing 75 percent or more of its oil by 2030.[27] All of the growth in oil demand is forecast by the International Energy Agency to come from non-OECD countries, with China contributing 43 percent and the Middle East and India each about 20 percent. As figure 1.3 depicts, the increase in oil dependence in India is expected to be particularly dramatic, exceeding 90 percent by 2030.[28]

1.2.3   The Link to Global Climate Change

The destabilization of the world's climate (or, to be more precise, of certain climatic zones), driven by relentless emissions of greenhouse gases, has the potential to exacerbate food and water shortages, advance the spread of infectious disease, induce mass migration, damage trillions of dollars of property, and precipitate extreme weather events—all of which could lead to increased conflict worldwide.[29]

This broad range of threats to energy security necessitates a holistic treatment of causes and effects, including energy and climate issues as well as water and waste, and agriculture and forestry. Assembling all these pieces into a single book highlights important interactions that often go unnoticed. Without a fully articulated appreciation of these complexities, different strands of energy and climate policy run the risk of competing with each other or, worse, trading off so that the net result is continued emissions, higher prices, greater energy poverty, and degraded security.

1.3   Preview of Chapters

The book begins by extending our discussion of energy security and climate change to include five challenges that threaten the prosperity of future generations.

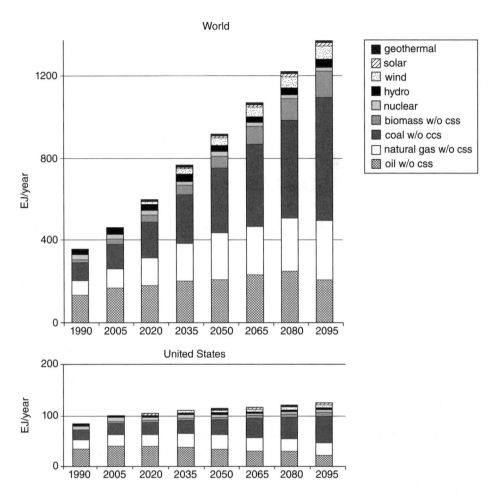

Figure 1.2
Projections of world and US energy use, 1990–2095. Redrawn from data published in L. Clarke et al., *CO₂ Emissions Mitigation and Technological Advance*, Pacific Northwest National Laboratory Report PNNL-18075, 2009.

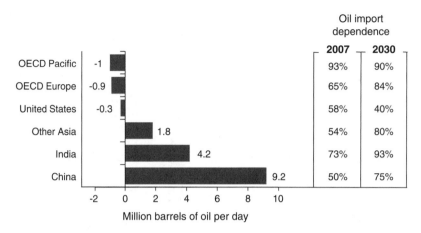

Figure 1.3
Change in oil demand in selected countries and regions, 2007–2030. Drawn from data published in International Energy Agency, *World Energy Outlook 2008* and in Energy Information Administration, *Annual Energy Outlook 2009*.

Specifically, chapter 2 describes four resource-intensive sectors of the global economy—electricity supply, transport, agriculture and forestry, and waste and water—that are responsible for a majority of the world's emissions of greenhouse gases, then takes up the subject of climate change. The trends in each of these interrelated sectors are described from a global perspective, drawing on the experiences of developing and developed countries. By ending with a description of the science and consequences of climate change, the chapter underscores the need for transformational energy and climate technologies and policies. The chapter also describes the policy conflicts and complementarities that confound simultaneous solutions to climate-change and energy-security challenges.

Since greenhouse gases originate from almost every human activity, technologies that limit emissions of greenhouse gases are both numerous and diverse. Chapter 3 focuses on the array of plausible technology solutions to enhance energy security and mitigate climate change. Since it is difficult to speculate about which of the competing solutions currently in laboratory research will emerge victorious, we examine the best current practices and the best-performing technologies currently being prototyped and demonstrated.

Most of the scientific and policy dialog surrounding global climate change has focused on reducing emissions of greenhouse gases. Chapter 4 expands the discussion to include "geo-engineering" approaches—ways to remove CO_2 from the atmosphere and ways of reflecting sunlight to cool the Earth—as well as "adaptation" approaches. In contrast with the mitigation technologies and approaches described in chapter 3, geo-engineering interventions tend to be more speculative and uncertain, and they are fraught with ethical and scientific complexities. Nevertheless, in view of the lack of evidence that societies around the world will successfully curb their emissions of greenhouse gases in the next several decades, it can be argued that geo-engineering is needed as a fallback option. Similarly, actions should be taken to reduce the vulnerability of humans and ecosystems to the effects of global climate change. Such "adaptation" actions can be either anticipatory or reactive. Stabilizing and then reducing the emissions of greenhouse gases is possible, but it looks very unlikely to occur before the end of the century. Attention is beginning to turn to ways to alter the climate and ways to adapt to the consequences of climate change.

Chapter 5 enumerates the tenacious barriers that prevent climate-friendly technologies from being rapidly adopted in the global marketplace. A thorough understanding of these impediments provides a basis for developing effective strategies to shrink the socio-technical gap between the cost-effective technologically feasible and the socially achievable. The chapter begins by briefly introducing the concepts of market failure, public goods, and policy failure. It then discusses a typology of 20 barriers and obstacles. The chapter ends by elaborating on the notion of "carbon lock-in" and what the presence of such barriers means for the promotion of climate-friendly technologies and programs.

Chapter 6 begins by explicating one final persuasive rationale for public-policy intervention, the precautionary principle, which is compared and contrasted with the more common "risk paradigm" in environmental policy making. The chapter offers a typology of public-policy mechanisms and summarizes different methods of evaluating policies, including cost-benefit analysis, cost-effectiveness analysis, and various hybrids. The chapter finishes by exploring the potential and pitfalls of putting a price on carbon, the dynamics of carbon cap-and-trade schemes, and how pricing carbon can be complemented with other policies in the electricity supply, transport, agriculture and forestry, and waste and water sectors.

Because of the diverse spatial dimensions connected to energy and climate problems, chapter 7 argues that similarly multi-dimensional scales must be utilized to implement policies that respond to them. In addition, success requires combining multiple stakeholders (such as government regulators, business leaders, and civil society). Polycentric approaches—those that blend scales and engage multiple stakeholder groups—have the potential to capture all the benefits of local, regional, and global action, and to reduce, or in some cases eliminate, their costs. The chapter begins by discussing five benefits to global action: consistency, economies of scale, equity, mitigation of spillovers, and minimization of transaction costs. It then discusses five strengths of local action (diversity, flexibility, accountability, simplicity, and positive contagion) and explains the benefits behind polycentrism, or how properly designed policies can capture most of the advantages of both global and local action while avoiding their disadvantages. The chapter concludes by summarizing the challenges to multi-scalar governance.

Chapter 8 presents eight case studies that exemplify empirically successful approaches to improving energy security and reducing emissions of greenhouse gases: Denmark's approach to energy policy and wind power, Germany's feed-in tariff, Brazil's ethanol program, Singapore's congestion road pricing and vehicle moratoriums, Grameen Shakti's efforts to distribute small-scale renewable energy technologies in Bangladesh, China's improved cookstoves program, the Oasis Project in Brazil (which prevents deforestation and improves water quality), and the Toxics Release Inventory in the United States (which tracks hazardous pollutants). These approaches addressed social and technical barriers simultaneously, relied on polycentric scales of action, and rapidly achieved their goals. They illustrate the types of initiatives that are needed for a secure and sustainable energy future.

Chapter 9 presents our conclusions. It reemphasizes the socio-technical aspects of energy and climate challenges, and it argues that efforts to improve technologies and alter human behavior must work together to produce meaningful change. The chapter summarizes some of the changes in energy and climate policy that must be implemented. It furthermore explains why polycentric approaches tend to be the most successful at initiating such changes, implying that governments, individuals, corporations, and institutions at a variety of scales must all play mutually supportive roles.

2

A Tale of Five Challenges

In the 1830s, the American philosopher and poet Ralph Waldo Emerson traveled to Europe on a sailing ship and returned on a steamship. He crossed the Atlantic on a recyclable vessel powered by solar and wind energy and controlled by craftsmen practicing ancient arts, but returned on what would become a steel rust bucket spewing oil into the water and smoke into the sky, operated by men shoveling coal into boilers in the dark. Emerson ruefully contemplated the symbolic implications of this shift, noting that the more "advanced" vessel cut passengers off from the forces of wind and nature as well as the open sea.[1] A few decades later, Emerson lamented that the new age of technology epitomized by steamships and railroads "had an engine, but no engineer,"[2] suggesting that advances in the use of energy were accelerating without direction and guidance.

It may not seem so, but we are now experiencing an equally significant shift in the use of energy—an experience with equally unpredictable consequences. From 1900 to 2000, Earth's population nearly quadrupled (going from 1.6 billion to 6.1 billion), but the annual average supply of energy per capita grew at an even greater rate, from 14 gigajoules in 1900 to roughly 60 gigajoules in 2000. Over this period, energy consumption more than tripled in the United States, quadrupled in Japan, and increased by a factor of 13 in China. Individual examples of energy-intensive lifestyles and practices are even more striking. In 1900, an affluent farmer in the American midwest holding the reins of six large horses plowing a field would generate about 5 kilowatts of animate power. A century later, that same type of farmer could sit in a large tractor, with an air-conditioned and stereo-equipped cabin, consuming 250 kilowatts. In 1900, an engineer operating a locomotive could reach speeds close to 100 kilometers per hour with about 1 megawatt of steam power, yet in 2000 the pilot of a Boeing 747-400 flying 11 kilometers above the Earth's surface can exceed 900 kilometers per hour and a discharge of 120 megawatts.[3]

These modern patterns of energy use reflect a fundamental transition from principal sources of energy derived directly from the sun (such as human and animal muscle power, wood, flowing water, and wind) to those dependent on fossil fuels.

Globally, use of hydrocarbons as a fuel by humans, for example, increased 800-fold in the years 1750–2000 and 12-fold again in the years 1900–2000.[4] Indeed, we have seen four major changes in human energy use in the past century:

• Use of electricity has increased significantly because of instant and effortless access, easily adjustable flow, facilitation of high precision speed and process controls, and cleanliness and silence at the point of use. Electricity now powers not only lights, refrigerators, televisions, and radios but also vehicles, electric fireplaces, electric arc furnaces, and movable sidewalks. In 1900, less than 2 percent of the world's electricity came from fossil fuels, but by 1950 the number had jumped to 10 percent, and in 2000 it passed 30 percent.[5]

• The introduction of mobile engines and inexpensive liquid fuels has enhanced personal mobility and facilitated new modes of transport. The number of mass-produced motorized vehicles in the world jumped from a few thousand in 1900 to more than 700 million in 2000, accompanied by a notable increase in vacation travel and other non-essential trips.

• The mobility of people has been matched by the increasing movement of goods and services as trade and commerce have accelerated. In 2000, trade alone accounted for 15 percent of the world's economic activity.

• Humans have become more dependent on information and media technologies such as the Internet and television. This has brought dramatic increases in the manufacturing of information storage, telecommunications, and electronics and in the amounts of energy needed to operate such devices.[6]

Such trends are expected to continue in the next few decades as hundreds of millions of people are born and developing economies continue to industrialize.

This chapter, however, warns that a continuation of today's trends could threaten to cause serious alterations in Earth's climate as a result of increased emissions of greenhouse gases and a barrage of social, environmental, economic, and political consequences. According to the Intergovernmental Panel on Climate Change (IPCC), human sources emitted 49 billion metric tons of carbon dioxide equivalent into the atmosphere in 2004.[7] Globally, GHG emissions increased 70 percent from 1970 to 2004 and, if trends continue, could increase 130 percent by 2040. Yet the climate-related impacts of these emissions could last longer than Stonehenge, time capsules, and perhaps even high-level nuclear waste. One-fourth of the carbon dioxide we leave in the atmosphere today will still be affecting the atmosphere 1,000 years from now.[8,9] Put another way, the climate system is like a bathtub with a very large tap and a small drain.[10] As figure 2.1 shows, four resource-intensive sectors of the economy—electricity supply, transport, agriculture and forestry, and waste and water—are responsible for most of these dangerous emissions. This chapter discusses

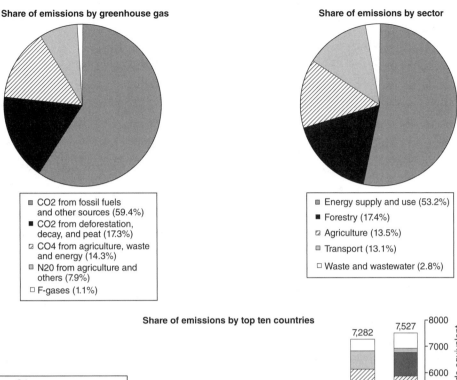

Share of emissions by greenhouse gas

- ▣ CO2 from fossil fuels and other sources (59.4%)
- ■ CO2 from deforestation, decay, and peat (17.3%)
- ▨ CO4 from agriculture, waste and energy (14.3%)
- ▨ N20 from agriculture and others (7.9%)
- ▢ F-gases (1.1%)

Share of emissions by sector

- ▣ Energy supply and use (53.2%)
- ■ Forestry (17.4%)
- ▨ Agriculture (13.5%)
- ▨ Transport (13.1%)
- ▢ Waste and wastewater (2.8%)

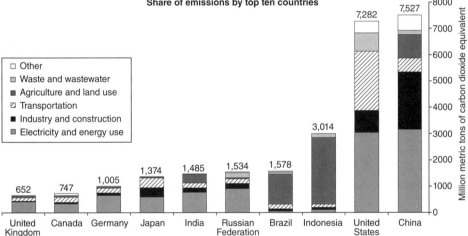

Share of emissions by top ten countries

Legend:
- ▢ Other
- ▨ Waste and wastewater
- ■ Agriculture and land use
- ▨ Transportation
- ■ Industry and construction
- ▣ Electricity and energy use

Million metric tons of carbon dioxide equivalent

United Kingdom: 652
Canada: 747
Germany: 1,005
Japan: 1,374
India: 1,485
Russian Federation: 1,534
Brazil: 1,578
Indonesia: 3,014
United States: 7,282
China: 7,527

Figure 2.1

Global anthropogenic greenhouse-gas emissions by gas, sector, and country. Data for GHG emissions by source and sector from *Climate Change 2007: Synthesis Report* (Intergovernmental Panel on Climate Change, 2008): 36; data for country emissions from *Greenhouse Gas Inventory Data* (UN, 2009); World Resources Institute, *Earth Trends Climate and Atmosphere Country Profiles* (WRI, 2009); Environment Canada, *Canada's 2007 Greenhouse Gas Inventory: A Summary of Trends* (Environment Canada: February, 2009); Ministry of the Environment, *National Greenhouse Gas Inventory Report of Japan* (Center for Global Environmental Research, April, 2009); European Environment Agency, *Greenhouse Gas Country Profiles* (Copenhagen, 2008); Larry Parker and John Blodgett, Greenhouse Gas Emissions, report RL32721, US Congressional Research Service, 2008; Jane A. Leggett and Jeffrey Logan, *China's Greenhouse Gas Emissions and Mitigation Policies*, report RL34659, Congressional Research Service, 2008. Consistent data on country emissions beyond 2000 are difficult to find publicly. Data for China and United States are for 2007, data for Indonesia for 2004, data for Brazil for 2004, data for Russia for 1999, data for Japan for 2007, data for India for 2003, data for Germany for 2006, data for Canada for 2007, and data for United Kingdom for 2006. Country emissions do not treat the European Union as a single entity and include emissions from changes in land use.

challenges and trends in each of these four interrelated sectors, and then explores the science and the consequences of climate change.

2.1 Electricity

Electricity has transformed industry and society by improving the productivity and the quality of life of populations and regions around the world. Nearly every aspect of daily life in a modern economy depends on it. Electricity's extraordinary versatility as a source of energy means it can be put to almost limitless uses to heat, light, transport, communicate, compute, and generally power the world. The backbone of modern industrial society is, and for the foreseeable future will be, the use of electricity.

The electricity industry is one of the most important sectors of the US economy. Expenditures on electricity in the US reached $355 billion (equivalent to €275 billion) in 2007 (3.2 percent of the country's gross domestic product in that year).[11] The US has more electric utilities and power providers than Burger King restaurants.[12] As a percentage of gross domestic product, the electricity industry is larger in capital investment than auto manufacturing and roughly the same size as the massive telecommunications sector.

As another measure of value, consider the lives of the 1.6 billion people in the world who have little to no access to electricity.[13] As a result of "electricity deprivation" or "energy poverty," millions of women and children spend significant amounts of time searching for firewood, and then burning either it, dung, or charcoal indoors to heat their homes and prepare their meals, emitting localized pollution into the living space. Worldwide, nearly 2.4 billion people use wood, charcoal, and other such "traditional" biomass fuels for cooking and heating, and 1.6 billion do not have access to electricity. (See table 2.1 and figure 2.2.) Even if we take into account significant increases in development assistance and rural electrification programs, by 2030 about 1.4 billion will still be at risk of having to live without modern energy services.[14] The indoor air pollution resulting from cookstoves shortens the lives of 2.8 million people every year, almost equal to the number dying per year from HIV/AIDS.[15] Of these deaths, 910,000 are children under the age of 5 who, in their final months, suffer from debilitating respiratory infections, chronic obstructive pulmonary disease, and lung cancer. Access to electricity, therefore, can raise populations out of poverty and significantly improve the health of millions of people, mainly women and children.

On the other hand, electricity is also a source of numerous detrimental effects. First, many of the fuels used to generate electricity worldwide degrade the natural environment through processes such as mountaintop removal and resulting toxic sludge to water and air pollution. Second, electricity prices are rising and are

Table 2.1
Number of people relying on traditional biomass in developing countries. Source: V. Modi et al., *Energy Services for the Millennium Development Goals* (Energy Sector Management Assistance Program, UN Development Programme, UN Millennium Project, and World Bank, 2005).

	Millions	Percentage of population
China	706	56
Indonesia	155	74
Rest of East Asia	137	37
India	585	58
Rest of South Asia	128	41
Latin America	96	23
North Africa/Middle East	8	0.05
Sub-Saharan Africa	575	89
Total	2,390	52

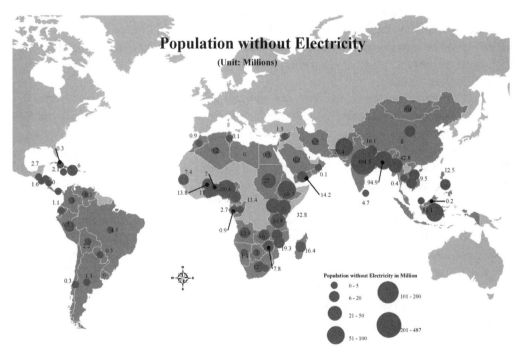

Figure 2.2
Population without electricity (millions). Drawn from data published in *World Energy Outlook 2008* (International Energy Agency, 2008).

predicted to increase substantially, which could significantly penalize those regions of the world that are highly dependent on inexpensive, carbon-intensive electricity. Historically, electricity has been so affordable that it has been used wastefully, leading to an electricity supply, distribution, and end-use infrastructure that is highly inefficient and will be expensive to transform.

2.1.1   The Environmental Impacts of Electricity Generation

Every kilowatt-watt hour of electricity generated from conventional fuels produces a multiplicity of environmental damages that may include radioactive waste and abandoned uranium mines and mills, acid rain and its damage to fisheries and crops, water degradation and excessive consumption, particulate pollution, and cumulative empoverishment of ecosystems and biodiversity through species loss and habitat destruction.[16]

Consider coal, one of the most polluting fuels. Coal plants account for approximately 39 percent of all US carbon dioxide (CO_2) emissions and one-third of all GHG emissions. These emissions are forecast to increase both absolutely and as a proportion of total emissions in the next 25 years and perhaps longer.[17] The demand for coal has been increasing faster than the demand for any other energy source, reflecting the rapid worldwide increase in electrification and electricity consumption. Coal is the largest source of electricity generation both worldwide and in the United States (figure 2.3).[18]

Yet in the United States, coal-burning power plants pose serious threats to human health. They release an average of 68 percent of their waste by volume directly into the environment, and more than one-third of the coal-fired power plants currently operating in the US (approximately 123 gigawatts, out of 300 gigawatts) do not have advanced pollution controls installed. Incidences of drinking wells and surface water being contaminated by leaching from coal waste ponds has swelled, with 137 coal or oil ash waste sites contaminated with heavy metals and other toxic materials.[19] State requirements for the handling of coal ash vary widely, and about three-fourths of the approximately 300 active surface impoundment sites are unlined. Using conservative estimates, the costs of the environmental damage from electricity in 2007 surpassed 420 billion US dollars —143 billion more than the industry's annual reported earnings in the US. A recent report by the National Research Council examined the damage caused by pollution from energy production and consumption in the US.[20] The study committee concluded that these damages amounted to $120 billion in 2005, excluding any costs of climate change, the effects of mercury, the impacts on ecosystems, and other external damages difficult to monetize. The total cost estimates were dominated by human sickness and suffering from air pollution associated with electricity generation and vehicle transportation.[21]

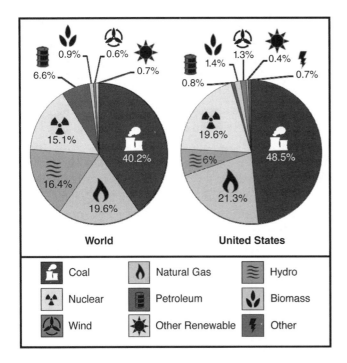

Figure 2.3
Sources of electricity in the world and in the United States. Drawn from data published in *World Energy Outlook 2008* (International Energy Agency, 2008) and in *Annual Energy Outlook, 2008* (Energy Information Administration).

Coal mining also removes mountaintops and causes acid drainage into river systems; toxic sludge is created while preparing coal to be burned, and it can spill out of its impoundment (as occurred in Kingston, Tennessee in December 2008).[22] The acid precipitation from coal combustion damages fisheries, crops, forests, and livestock, and when coal is burned, it releases sulfur dioxide, nitrogen oxide, mercury, particulates, and carbon dioxide that contribute to global climate change.

Globally, electricity generation is the largest contributor to climate change, producing more than 10 billion tons of carbon dioxide a year—the greatest contribution from any particular industry or sector. In monetary terms, the social and environmental damage from worldwide electricity generation amounts to roughly 13 US cents (10 euro cents) per kilowatt-hour, or 2.6 trillion US dollars (2 trillion euros) in extra costs every year.[23]

All of these environmental assaults must be addressed; however, the most immediate need is to ensure that the carbon contained in coal yet to be used is not released to the atmosphere. According to the National Energy Technology Laboratory, 43 coal plants are in various phases of progression in the United States (ten permitted, six near construction, and 27 under construction), representing a total new capacity of 22 gigawatts. At the same time, a worldwide push for new coal plants has begun, with China constructing a coal plant each week and other developing countries—led by India—building new coal plants about one-third as fast (figure 2.4). Over their roughly 60-year lives, the new generating facilities in operation by 2030 could collectively emit as much carbon dioxide as was released by all the coal burned since the dawn of the Industrial Revolution.[24]

If the electricity industry were to remain configured as it is today in the United States, engineers and architects would have to construct as many as 310,000 miles of new natural-gas pipelines, 10,000 natural-gas plants, 4,950 coal plants, 190 nuclear reactors, four large uranium-enrichment plants, five fuel-fabrication plants, and possibly three waste-disposal sites the size of Yucca Mountain by 2040. The costs of power outages would exceed $412 billion (€318.8 billion), the industry would consume and withdraw more water than the agricultural sector (threatening widespread shortages), more than 65,000 Americans would die prematurely from power plant pollution, and the country's electricity generators would dump 4.5 billion tons of CO_2 into the atmosphere every 12 months.[25]

2.1.2   Transmission and Distribution Reliability

Because electricity cannot be stored economically, the amount generated at every point in time must equal the amount consumed, yet the balkanized nature of most electricity grids prevents efficient transmission and distribution (T&D). In many countries, grids were designed and expanded slowly in an ad hoc fashion that distributes power sufficiently but not optimally. Evidence from transmission loading relief logs in the United States suggests increasing grid instabilities.[26] The operator of a system must coordinate schedules of generation, load, and power flow, and must balance deviations from expected supply or demand. With chronic under-investment in grid infrastructure, this has increased the instability of grids worldwide. As a result, electricity shortages and blackouts have disrupted life in the United States, in Europe, in Russia, and in many developing countries over the past 10 years.

One contributing factor to global blackouts is that the electricity industry has undergone dramatic structural changes in the last 15 years. Before 1990, in most countries the electricity system consisted of vertically integrated (generation, transmission, and distribution) monopolies that were highly regulated. The rationale for this system design was the assumption that electricity production is a natural monopoly, such that a single firm can provide complete supply at a lower cost than a collection of competing firms. For T&D, the case for natural monopoly and continued regulation remains relatively strong, since it doesn't make sense to build two or more sets of power lines to every home.

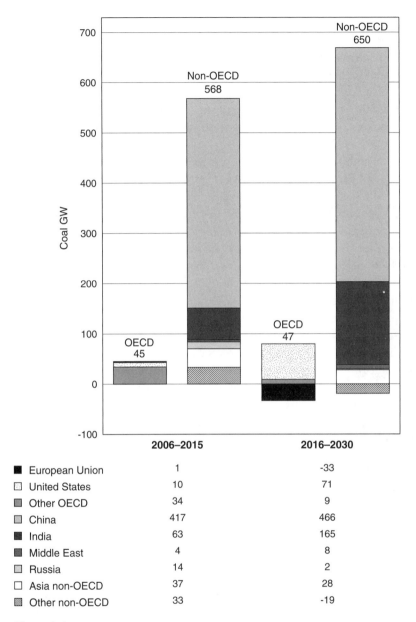

Figure 2.4
Incremental new coal capacity. Source of data: *World Energy Outlook 2008* (International Energy Agency, 2008).

Today, however, many countries operate according to a mixed system that now includes some elements of market competition. Privately owned electricity generators have been released from rate-of-return regulation, allowing them to earn market-based rates. Regulations make it possible for independent power producers and other utilities to compete in the merchant generation sector. Yet T&D systems were built for reliability of local service, not to support competitive regional wholesale electricity markets that require moving large quantities of power across long distances.

Moreover, renewable resources suitable for generating electricity tend to be located far from load centers and transmission infrastructure. For example, in the United States, wind resources are concentrated in the midwest and geothermal, solar thermal, and solar photovoltaic resources are concentrated in the west. The same is true in many countries—for instance, South Africa's greatest wind resources are along the remote west coast, and Argentina has a large wind resource in its Patagonia region, far from urban "load centers." The long distances between generation resources and metropolitan markets require expensive investments in transmission infrastructure that are prone to natural vulnerabilities. Risks are further compounded by al-Qaeda and other terrorist groups' commitment to attack the world's critical economic infrastructure, of which electricity is clearly a major component. The more expansive the energy infrastructure, the greater its vulnerability.

Another outcome of the shift from a regulated monopoly market to a restructured competitive market is that utilities have under-invested in T&D infrastructure. In the United States, one metric—"transmission loading relief" procedures ("TLRs")—determines which requests for transmission will be denied in order to prevent lines from becoming overloaded. The number of TLRs has grown substantially in recent years, indicating that transmission congestion is increasing. Experts appear to agree that the major blackout that affected a large part of the North American upper midwest and northeast on August 14, 2003 was not an anomaly and a blackout of similar magnitude will happen again.[27]

System theorists suggest that even if the infrastructure is expanded and "hardened," big outages will still happen. The northeast blackout affected 50 million people in eight US states and two Canadian provinces. Initially, 60,000–65,000 megawatts of load was interrupted (approximately 11 percent of the Eastern Interconnection), and eventually 531 generators were shut down at 263 plants. The economic cost of the blackout came to approximately $5 per forgone kilowatt-hour, a figure that is roughly 50 times the average retail cost of a kilowatt-hour in the United States. The economic loss is estimated at more than $6 billion, with a loss of $1 billion in New York City alone.[28] The "root cause" analysis of the blackout cited several deficiencies, including inadequate "situational awareness" at First Energy Corporation in Indiana (where the blackout began), inadequate tree trimming on

First Energy's transmission right-of way, and failure of reliability coordinators to promptly identify and deal with problems.

2.1.3   Electricity Prices and Inefficiencies of Use

Demand for electricity is highly price inelastic, in part because electricity does not have many short-run substitutes and in part because few consumers receive real-time prices that match supply costs. The lack of substitutes can force consumers to pay exorbitant prices for electricity with which to operate products such as portable electronic devices. For example, American consumers routinely pay $1.85 (€1.43) for a D-cell alkaline battery that can produce 0.017 kilowatt-hours, yielding an equivalent price of $108 (€83.58) per kilowatt-hour—1,000 times the cost of electricity sold on the grid.

Electrical energy is convenient as a carrier of energy that can be widely distributed and converted to provide a range of services, including light, heat, refrigeration, and air conditioning. How much energy one gets during the conversion of one form of energy to another is represented by the efficiency of the process. Most fossil fuels are converted to motive or electric power by burning the fuels to make heat and using a heat engine (e.g., a steam engine or an internal-combustion engine) to obtain the desired mechanical or electric power. Such conversions are not very efficient.

To produce electricity, for instance, coal is burned in a power station to turn a turbine, which then turns a generator to produce the electricity. A typical plant might have an efficiency of 38 percent, losing 62 units of the energy in the fuel during conversion to electricity. In the end, when you add the layers of inefficiency across the T&D network (2 units lost) to the consumer (where 34 units are lost to heat in incandescent lighting), only 2 percent of the energy embodied within the coal used to produce that electricity is converted into lumens inside an incandescent bulb.[29] Converting to a compact fluorescent bulb would improve this efficiency by a factor of about 4 but would do little to address remaining inefficiencies in grid transmission and distribution (figure 2.5). Similarly, a typical internal-combustion engine in an automobile utilizes only 15–25 percent of the energy of the gasoline to move the car; the rest of the energy, again, is lost as heat. These efficiencies are governed by the fundamental way in which the chemical energy of fuels is converted in heat engines.

Energy savings come in small pieces, not in large chunks. But when you add up the pieces, they can be substantial. For example, about 20 huge power plants operate around the clock only to energize US appliances and equipment that are turned off and in standby mode.[30]

Although electricity rates have not increased as rapidly as oil prices, they have been increasing steadily. From 2005 to 2009, residential electricity rates in the United States increased by 12 percent in real terms (that is, from 10.3 to 11.5 cents per

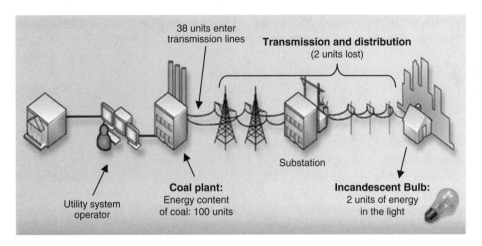

Figure 2.5
Layers of inefficiency. Source: Curt Suplee et al., *What You Need to Know About Energy* (National Academy of Sciences, 2008) (http://sites.nationalacademies.org).

kWh). In Europe, iron and steel manufacturers are experiencing increasing electricity and raw materials prices, which are damaging their already weakened industries.[31] The era of cheap energy has ended, making energy efficiency a "front-line" strategy for coping with rising energy prices and addressing global climate change.

Why are electricity prices rising? The answers are numerous. Here are three of them:

• China, India, and even Europe are consuming more coal for power production and steel mills than in the past, which has pushed coal prices up.

• The threat of new or tighter regulations on mercury, sulfur dioxide (SO_2), nitrogen oxides (abbreviated NOx), and carbon dioxide emissions has caused producers to invest in costly pollution-abatement equipment.

• Transportation bottlenecks are inhibiting the ability of the world's fuel suppliers to meet demand.

Official forecasts in the United States predict real increases in retail electricity prices over the next 20 years.[32] Several examples already point to this trend: Portland General Electric Company in Oregon increased its rates by 7.6 percent in 2009; in late 2010, two West Virginia utilities (Appalachian Power and Wheeling Power) received approval for a 13.8 percent electricity base rate hike; and in 2011, Georgia Power implemented one of the largest rate increases in the states' history. These increases reflect higher wholesale power and fuel costs, stricter environmental rules on coal-fired power plants, the construction of new plants, and other system investment costs.

In addition, patterns of electricity consumption are changing. Consumers have a growing appetite for new products and devices. The popular 42-inch plasma TV requires 250 watts—2.5 times as much as a 27-inch cathode-ray-tube television. Incandescent lighting is another "energy hog," costing 4 times as much to operate over its lifetime as fluorescent lighting (in this case, the newer technology), but incandescents are still purchased by consumers because they cost less "up front" than the alternatives. Air conditioning is another example. In the United States the presence of air conditioning in new single-family homes jumped from 49 percent in 1973 to 89 percent in 2006.[33] In hot and humid places, such as southern Florida, the percentage of houses with air conditioning went from 5 percent in 1950 to 95 percent in 1990. And there is a troubling trend toward air conditioning in developing countries in the tropics. The number of air conditioners in rural Thailand is expected to jump from 4 percent in 1995 to 100 percent by 2035 (when it will account for 40 percent of all electricity demand in the residential sector). We say "troubling" because people in tropical areas can keep cool effectively by wearing lightweight and light-colored clothing, designing homes with natural ventilation, and placing houses on stilts or near waterways.[34]

2.2   Transportation

The global proliferation of auto-dominated transportation systems and the monopoly of gasoline and diesel transportation fuels have caused severe social and environmental impacts in the developed countries that will soon be replicated worldwide unless current trends are altered.

The rise of the coveted automobile is often characterized as one of the great achievements of the twentieth century. During the first half of the century, the gasoline-powered vehicle evolved from a fragile, cantankerous, and faulty contraption to a streamlined, reliable, fast, luxurious, and widely affordable product.[35] The feats of automotive engineering were enhanced by the creation of interstate highway systems and urban infrastructures that have offered many people unprecedented mobility.

Despite its prominent status in modern society, however, the rise of the gasoline automobile was by no means inevitable. In 1899, urban residents had many transit options. A New Yorker could take an electric taxi to the subway, then catch an underground light train to the Grand Central Terminal, ride a locomotive to San Francisco, disembark and transit on a cable car or trolley, and then hail a taxi, take a horse, or walk to her final destination.[36]

Yet gasoline vehicles have become a transportation monoculture—cars and petroleum dominate. Public transport accounts for only about 3 percent of passenger travel in the United States,[37] and rail transport for less than 1 percent.[38] The 6.7

billion people on Earth in 2009 owned 850 million cars and trucks. If parked end to end, they would circle the planet almost 100 times. In the United States, 85 percent of all personal travel is by automobile, and Americans collectively drive 3.5 trillion miles per year on 4 million miles of roads, consuming 180 billion gallons of fuel bought from 170,000 gasoline stations. Petroleum fuels almost all of these vehicles. Non-petroleum fuels (including electricity, biofuels, and natural gas) account for only 4 percent of US transportation fuel consumption (up from 2 percent in 2007).[39] The motorization of America has resulted in more automobiles in the US than licensed drivers, and other countries are envious. Although 85 percent of the world's people lack access to a car, many aspire to car ownership—especially residents of the rapidly growing South and East Asia nations.

Globally, the use of automobiles is about to increase dramatically. In China, the number of conventional vehicles is expected to increase tenfold from the 2005 total of 37 million to 370 million by 2030.[40] As a result, per capita carbon dioxide emissions are expected to rise steeply. The negative effects of such a rapid increase in motorization (air pollution, GHG emissions, oil dependency, and potential social unrest) are being anticipated by the Chinese government.[41]

Within 20 years, the world is projected to have 2 billion gasoline-powered automobiles—twice the present number.[42] If China continues its car-centric development model, it could by itself add another billion cars by the end of the century. Such an increase in the use of automobiles, however, threatens to unleash a series of problems, including costly traffic congestion, accidents, deterioration of air quality, GHG emissions, susceptibility to interruptions in supply and price volatility of oil, and a growing reliance on petroleum produced in unstable and unfriendly regions of the world.

2.2.1 Traffic Congestion, Accidents, and Costs

Auto-centric transportation systems require the construction of extensive roads and highways, which threaten the vitality of urban centers, and depend on billions of dollars of government subsidies. In many countries that are not members of the Organisation for Economic Co-operation and Development (OECD), low oil prices and subsidies encourage excessive oil consumption. Most of the world's biggest petroleum-producing countries provide price subsidies so that their citizens can enjoy cheap oil (figure 2.6). In 2007, Iran's oil subsidies amounted to 35 billion US dollars (27.1 billion euros), and Saudi Arabia, Venezuela, and Egypt provided oil subsidies that ranged from 10 billion to 20 billion US dollars (7.7–15.5 billion euros). Oil subsidies are even offered in China ($24 billion; €18.6 billion) and India ($13 billion; €10.1 billion), both major oil-importing countries.[43] Cheap oil and subsidized construction of roads, highways, and parking structures result in degra-

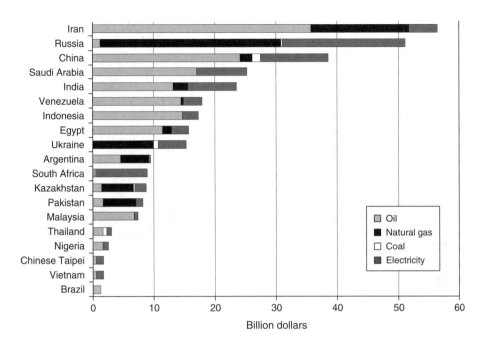

Figure 2.6
Energy subsidies in non-OECD countries, 2007. Source of data: *World Energy Outlook 2008* (International Energy Agency), figure 1.1.

dation of communities and in social exclusion of individuals who do not own vehicles. Automobile accidents kill more than 1.2 million people per year worldwide and injure 25–50 million more, which makes vehicles the third-largest contributor to death and injury in the world.[44]

In the United States, suburbanization and rising wealth transformed living and driving patterns dramatically after World War II. Not only did daily travel distances increase; so did the frequency with which households used their vehicles to get to work, to shop, and carry out personal business. Between 1970 and 2005, the average annual vehicle-miles traveled (VMT) per household—a telling measure of highway transportation demand—increased almost 50 percent, from 16,400 to 24,300.[45] At the same time, vehicle ownership per household increased even as average household size fell.[46] As figure 2.7 shows, total vehicle miles traveled in the United States jumped by almost a factor of 4 from 1960 to 2005, increasing from 780 billion to 3 trillion. Commercial truck travel increased even more rapidly (at an annual rate of 3.7 percent) compared to an annual rate of 2.8 percent for passenger travel.[47]

Increased travel is responsible for worsening traffic congestion, as well as for air pollution, wasted fuel, and rising carbon emissions. The congestion "invoice" for

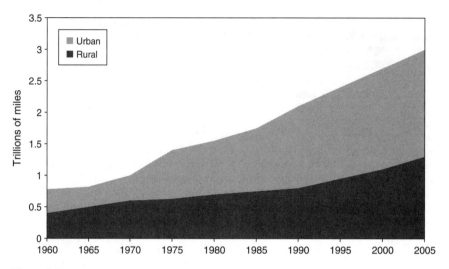

Figure 2.7
Vehicle-miles traveled in United States, 1960–2005. Source: William J. Mitchell et al., *Reinventing the Automobile* (MIT Press, 2010): 88.

the cost of extra time and fuel in 439 US urban areas in 2007 was estimated to be $87.2 billion —an increase of more than 50 percent in the past decade. This translates into a cost of $757 in delays and wasted fuel for each peak-period traveler, including the loss of 36 hours (the equivalent of almost one week's vacation), an increase from 12 hours in 1982.[48] In China, a country fast becoming infamous for painfully slow roads, one traffic jam near Beijing recently stretched for more than 100 kilometers and lasted 9 days.[49] In 2009, China also became the world's leader in manufacturing and sales of new gasoline-powered automobiles, yet less than 5 percent of the Chinese currently own a car. In Dhaka, Bangladesh, motorized travel is only slightly faster than walking—the city has few "flyovers," few traffic signals, no mass transit, and roads without marked lanes. The roads are clogged with cars, buses, scooters, motorcycles, baby taxis, rickshaws, bicycles, people, and the occasional horse or dog. In the Kathmandu Valley of Nepal, home to 6 million people and 4 million vehicles, police must still manage traffic without automated signals, and jams that often last half a day. Figure 2.8 shows a typical traffic jam on Sunday morning near the city center of Kathmandu.

Cars are also expensive. If we account for all of the personal costs of owning and operating a conventional automobile—that is, the time spent working to pay for it, driving it, getting it repaired, cleaning it, and so on—a car's owner spends 1,600 hours a year supporting it. If we divide all the miles traveled by the car in a year by the time spent giving such support, the result is an average speed of 5 miles per hour (8 kilometers per hour)—the speed of a toddler on a tricycle.[50]

Figure 2.8
Traffic jam in Kathmandu, Nepal, October 2010. Source: Benjamin K. Sovacool.

2.2.2 Air Pollution and Respiratory Diseases

Vehicles powered by fossil fuels spew a variety of unhealthy pollutants and particles into the air, from which they are ingested and inhaled, contributing to acid rain and ozone depletion. One local air pollutant from cars—smog—has been successfully addressed in the United States and in many other developed countries. Tailpipe standards have reduced smog emissions in the past several decades, enabled by improvements in emission-control technology. The United States and Japan have been leaders in the reduction of local pollutants.

Europe has a more lenient approach, allowing higher-efficiency, diesel-powered cars. (On a "well-to-wheels" life-cycle energy basis, diesel fuel is about 15 percent less energy and carbon intensive than gasoline.[51]) But Europe has suffered the consequences in local air pollution. Today, Athens, Milan, and other European cities are experiencing crumbling historic monuments and high rates of respiratory diseases. Also, cities in Mexico, India, and China are witnessing far more smog than those in the US and Europe, as a result of their looser of air-quality standards. Only 1 percent of China's 560 million city dwellers breathe air considered safe by the European Union; 297 of the largest 300 Chinese cities do not meet the minimal environmental standards for ambient air pollution set by the United States; and the World Bank estimates that the economic burden of premature mortality associated with air pollution in China exceeded 1.16 percent of the gross domestic product in 2007.[52]

Even in the United States, where clean-air regulations have been stricter than those of many other countries, local topography can create temporary climate inversions that trap pollution—especially in Denver and Los Angeles. The greater Los Angeles region is home to 16 million vehicles and the worst air quality in the United States. For the country as a whole, conventional automobiles contribute to 89 percent of the average cancer risk from air toxics.[53]

Consider emissions of particulate matter (PM), not a specific pollutant itself but a mixture of fine particles of harmful pollutants such as soot, acid droplets, and metals. Conventional automobiles are often the largest human-caused source of PM, and in places with stringent emissions requirements for vehicles, such as California or the European Union, PM makes up the second-largest human source after power plants. (Forest fires and dust storms are the leading non-human sources.) Thousands of medical studies have strongly associated inhalation of PM with heart disease, chronic lung disease, and some forms of cancer. Using some of the most recently available data, table 2.2 shows that in the United States deaths from PM pollution are comparable to those from Alzheimer's Disease and influenza and greater than the deaths from nephritis, septicemia, breast cancer, automobile accidents, prostate cancer, HIV/AIDS, and drunk driving combined. In France, the Agency for Health and Environmental Safety projects that normal automobile emissions kill 9,513 people per year and result in 6–11 percent of all lung cancer cases identified in people above 30 years of age.[54] Another report from the World Health Organization investigating

Table 2.2
Selected causes of death in the United States, 2008. Source: Benjamin K. Sovacool, "A transition to plug-in hybrid electric vehicles (PHEVs): Why public health professionals must care," *Journal of Epidemiology and Community Health* 64, 2010: 185–187.

	Estimated annual deaths
Heart disease	652,091
Cancer	559,312
Stroke	143,579
Alzheimer's Disease	71,599
Particulate-matter pollution	65,638
Influenza	63,001
Nephritis	43,901
Septicemia	34,136
Breast cancer	40,910
Automobile fatalities	36,710
Prostate cancer	30,142
HIV	18,017
Drunk driving	16,694

automobile emissions in Austria, Switzerland, and France calculated 40,000 deaths per year from PM emissions from automobiles.[55]

2.2.3 Emissions of Greenhouse Gases

Increasing use of conventional vehicles thwarts attempts to limit GHG emissions. Gasoline engines emit 20 pounds (74 liters) of CO_2 per gallon of fuel burned; diesel engines emit slightly less.[56] In many countries belonging to the OECD, gains in energy efficiency in the industrial and power sectors have been offset by energy use in the transportation sector. Motorized transport is already responsible for one-third of the "carbon footprint" of the United States, and emissions are increasing as drivers continue to switch from cars to light trucks, minivans, and "sport utility vehicles" (figure 2.9).[57] In 1970, cars accounted for more than 85 percent of US passenger vehicle sales. By 2006, that had dropped to less than 50 percent.[58] To

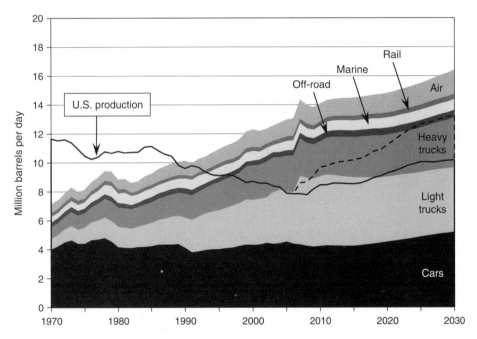

Figure 2.9
US petroleum production and consumption, 1970–2030. Sources of data: Stacy Cagle Davis et al., *Transportation Energy Data Book: Edition 28* (Oak Ridge National Laboratory), 2009; Energy Information Administration, *Annual Energy Outlook 2009*. US production has two lines after 2005. The solid line is conventional sources of petroleum. The dashed line adds in other inputs—ethanol, liquids from coal, and liquids from biomass. The sharp increase in values between 2006 and 2007 are caused by the data change from historical to projected values. Sources: *Transportation Energy Data Book: Edition 27*; EIA, *Annual Energy Outlook 2009*.

complicate matters, vehicle-miles traveled have been increasing at an average rate of 2.5 percent per year, outpacing population growth.

The situation in developing countries is quite different. Personal vehicles have been largely unaffordable for the masses, but this may be changing. Tata Motors, India's leading automobile manufacturer, recently introduced the Nano, a two-cylinder compact car that goes 55 miles on a gallon of gasoline (100 kilometers on 4.3 liters). This is one of the most efficient cars in the world. Priced at 100,000 rupees ($2,000; €1,548), it is listed in the *Guinness Book of World Records* as the world's cheapest car. The Nano emits an environmentally friendly 370 grams per kilometer (230 grams per mile) of carbon dioxide. If the Indian government were to encourage citizens who currently own vehicles to switch to the Nano, petroleum consumption and carbon emissions would plummet. Instead, the Nano is likely to lead to more car ownership and more widespread car use in India, resulting in sizable increases in oil consumption.

2.2.4   Oil Dependence and Transfers of Wealth

Reliance on crude oil to fuel most vehicles has transferred immense wealth to petroleum producers. Many terrorist groups receive funds indirectly from oil and gas revenues and then use those resources to plan attacks against oil and gas infrastructure. Saudi Arabia, for example, has used its oil wealth to offer more than $600 million (€464.3 million) in development and aid packages to al-Qaeda and the Taliban from 1993 to 2003, and has sent hundreds of thousands of barrels of oil per day to other groups in Afghanistan and Pakistan.[59]

Instead of reducing their reliance on imported oil, industrialized countries are becoming more dependent. Global dependence on Middle Eastern crude oil is expected to jump from 58 percent today to 70 percent by 2015. By 2030, the International Energy Agency predicts that the oil import dependence of Europe and Asia will exceed 75 percent, up 10–25 percent from today's levels.[60] At more than 19 million barrels per day, Asia's oil use already exceeds that of the United States, and India and China are projected to double their consumption of oil by 2025. As has already been noted, the number of cars in the US exceeds the number of licensed drivers. And as much as 40 percent of the US military budget can be attributed to protecting the oil trade.

The issue of oil dependence is just as stark for many developing countries. Increases in the costs of crude oil and gasoline mean that the foreign exchange required for oil imports create a heavy burden on the balance of trade for many developing countries. Although developed countries spend only 1 or 2 percent of their gross domestic product on imported crude oil, developing countries spend an average of 4.5–9 percent of their GDP on crude oil imports. Higher prices for oil also hit developing countries twice: once for costlier barrels of oil, and again for inflated transportation costs resulting from the increase in fuel prices to get the oil to those countries.[61]

2.2.5   Price Volatility and Peak Oil

To date, the world has consumed about a trillion barrels of oil. Although oil has been commercially produced since the nineteenth century, 99.5 percent of production has happened within the last 60 years. Today we consume 30 *billion* barrels of oil per year, and the demand for oil is projected to increase at more than twice the historic rate since 1980. It is estimated that there are remaining reserves of approximately a trillion barrels, enough to last only 35 years at today's consumption rate.

An accurate prediction of a peak in conventional oil supply is not possible because of gross uncertainties over cost and resource data. Nevertheless, reserve to production ratios can provide a range of plausible scenarios. Looking at all energy fuels, not just oil, if 2006 levels of production were to remain constant worldwide, known coal reserves would be exhausted within 137 years, and petroleum and natural-gas reserves would be exhausted in the next 50 years (table 2.3). If rates of production increase to keep up with demand, particularly in the rapidly developing "BRIC" countries (Brazil, Russia, India, China), known reserves of fossil fuels would be depleted much more rapidly. And if the world were to maintain its generation of nuclear power at 2004 levels, identified uranium resources would run out in 85 years.[62]

Thus, it is not clear how long the world can go on producing enough oil (and even enough coal, natural gas, and uranium) to meet the increasing demand. What is clear is that a growing percentage of the world's oil demand will be met using resources from the Middle East, since 45 percent of the world's proven reserves of conventional oil are located in Saudi Arabia, Iraq, and Iran. If transportation remains dependent on oil, and if the world's surplus oil production is limited to 1 million or 2 million barrels per day,[63] oil-importing countries will continue to be vulnerable to oil price shocks and volatility, as has been the case since the Yom Kippur War and oil embargo of 1973 (figure 2.10).

2.3   Forestry and Agriculture

We have changed our land use and our dietary patterns and attuned our crops to be dependent on fertilizers derived from fossil fuels, borrowing their energy to produce and transport food. Indeed, food now travels more than it did in 1980; with the average bite most people eat traveling 1,500–2,500 miles (2,414–4,023 kilometers) to reach their mouths. Even locally grown food is often taken somewhere nearby to be washed and packaged, then transported back home. One study looking at cans of

Table 2.3
Proven reserves of fossil fuels and identified uranium resources and rates of depletion. Life-expectancy calculations by the authors. Source for reserves and production data for coal, natural gas, and petroleum : *International Energy Annual 2006* (Energy Information Administration, 2006). Sources for uranium: *Uranium 2005* (International Atomic Energy Agency, 2006); Luis Echávarri and Yuri Sokolov, *Uranium Resources* (Nuclear Energy Agency and International Atomic Energy Agency, 2006).

			Life expectancy		
	Proven reserves (2005)	Current production (2006)	0% annual production growth rate	2.5% production growth rate	5% production growth rate
Coal	930,400 million short tons	6,807 million short tons	137	60	42
Natural gas	6,189 trillion cubic feet	104.0 trillion cubic feet	60	37	28
Petroleum	1,317 billion barrels	30.560 billion barrels	43	29	23
Uranium	Identified resources at $130 (€99) per kg U: 4,743,000 tons	Current production: 40,260 tons	With identified uranium resources: 85	With total conventional resources: 270	With total conventional resources and phosphates: 675

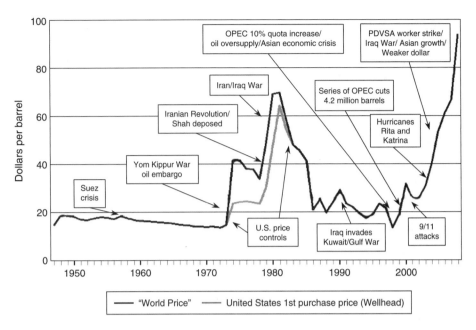

Figure 2.10
Prices of crude oil, 1945–2008. Source of data: *Energy Economist Newsletter* (http://www.wtrg.com).

strawberry yogurt produced in Germany found that the average carton was transported 8,000 kilometers (4,970 miles) by road for production and distribution.[64]

Classic agricultural practices attempted to match the growing patterns of food staples to the land; now we conform the land to produce crops that are in demand and highly profitable. Historically, for example, sound farming practices perfected through centuries of local knowledge avoided growing crops without also raising livestock. Farmers raised mixed crops and not monocultures, took care to preserve the soil and prevent erosion, and minimized waste by mixing vegetable and animal wastes into humus and fertilizer. Growth and decay balanced each other, care was taken to store rainwater, and both plants and animals were well equipped to protect themselves from pests and disease.[65]

Modern industrial farming inverts or ignores these practices. Farming is no longer based on natural energy flows and instead transforms ecosystems through the use of fertilizer, pesticides, and herbicides. The use of fossil fuels and chemicals deeply alters the land functionally, and it makes the entire infrastructure of commercial agriculture dependent on fossil-energy flows to ensure speed of transportation and communication and large-scale production and storage. The industrial food system depends on the same sources of fossil fuels as the electricity and transport sector, so farmers around the world have come to rely more on imported materials.[66] People in industrialized countries also consume so many goods, resources, and services that land in other parts of the world must be altered to support their habits. The average "ecological footprint" for someone in the United States is about 24 acres (9.7 hectares).[67] Yet the US has only about 13 acres (5.3 hectares) per person (and not all of it is productive), so there is a deficit of at least 11 acres (4.5 hectares) per person. Quite simply, the physical territory of the United States (as well as Japan, Germany, and many other industrialized countries) is insufficient to meet the demands of high-consumption lifestyles.[68]

How we use land and eat food have resulted in three significant challenging changes: increasing energy inputs but declining yields for food, a shift to a meat-centered and carbon-intensive diet, and rampant deforestation.

2.3.1 Increasing Energy Inputs but Declining Yields

Although cropping and agriculture have always taken different forms in different regions, global food production has become more energy intensive and has been transformed by the availability of fossil fuels and electricity. These high-energy inputs are used directly by farm equipment and irrigation systems, indirectly to produce machinery and agricultural chemicals, and as feedstocks in the synthesis of nitrogen fertilizers. In 1900, the world's farm machinery had a capacity of about 10 megawatts, and nitrogen applied to inorganic fertilizers (mostly in Chile) amounted to only 360,000 tons. In 2000, the total capacity of tractors and harvesters was about 500,000 megawatts, the Haber-Bosch synthesis fixed about 80 million tons of fertilized nitrogen, pumped irrigation served more than 100 million hectares (247 million acres) of farmland, and cropping was highly dependent on energy-intensive fertilizers and pesticides. In other words, the twentieth century saw a 150-fold increase in fossil fuels and electricity used in global cropping, yet only a sixfold increase in yields and a fourfold increase in productivity.[69] Larger farms are now worked with larger machines, need more fertilizers and pesticides, rely on centralized and energy-intensive delivery patterns, and depend more on packaging and processing.[70]

The overall energy efficiency of the agricultural system, moreover, is very poor. More than half of the synthetic nitrogen manufactured in 2006 was applied to corn, meaning that every bushel required the equivalent of one-fourth to one-third of a gallon (0.95–1.3 liter) of oil to grow it—inclusive of the natural gas needed for the fertilizer, diesel fuel needed for the tractor, and energy required for pesticides, drying, and transportation. Thus, it takes more than a calorie of fossil-fuel energy to produce a calorie from corn. (In pre-industrial times, it was the opposite: farms produced about two calories of food energy per calorie invested.) In some "primitive" societies in Africa, every calorie expended by a farmer to plant and tend crops produced 60 calories of food. The ratio is almost exactly the reverse in modern factory farming of pigs and other livestock, in which 65 calories of energy are expended to produce a single calorie of meat.[71] Also, it takes about 5 gallons (19 liters) of fresh water to process a bushel of corn, and for wet milling where feed material is steeped in water to soften the seed kernel, every calorie of processed food requires 10 calories of fossil-fuel energy.

Rice—like corn, a global staple—is also inefficient to produce. Global rice production has tripled in the past 50 years and is one of the world's largest users of fresh water, accounting for 34–43 percent of all irrigated water. Yet more energy is used for the growing of seedlings, transplanting, puddling, and irrigation of rice than we get out of eating it.[72] Even if the use of fossil fuels did not pose grave dangers to the environment and climate, the process would be unsustainable. As Michael Pollan has noted, "from the standpoint of industrial efficiency, it's too bad we can't simply drink the petroleum directly."[73]

2.3.2   More Carbon-Intensive Diets

Humans are not only becoming more reliant on fossil fuels to grow crops; they are also becoming more dependent on meat and on such animal products as eggs, milk, and cheese. These sources of food, however, are more energy and carbon intensive than simply consuming vegetables and plants. One study recently calculated that switching to a plant-based diet would do more to stop global warming than switching from a "sport utility vehicle" to a Toyota Camry,[74] yet this is precisely what most people are *not* doing. They are instead switching to a more meat-centered diet. Global meat consumption has increasing by a factor of 5 in the past 50 years—from 44 million tons in 1950 to 242 million tons in 2002. And per capita meat consumption has more than doubled in the same period, from 17 kilograms (37.5 pounds) per person per year to a worldwide average of 39 kilograms (89 pounds)[75]—more than 100 grams (3.5 ounces) of meat per person per day.[76]

The worldwide proliferation of an animal-centered diet involves not only meat but also milk, eggs, and produce. In 1962, animal produce accounted for only 600 calories per day in industrialized countries, but in 2005 it accounted for almost 800 calories; some regions, such as East Asia, saw an even greater increase (figure 2.11). This shift has four consequences.

First, livestock production and the raising of animals for food require the conversion of land, and are thus connected to massive changes in land use. The livestock sector accounts for 18 percent of global GHG emissions and 80 percent of anthropogenic land use. One study estimated that deforestation and a small amount of desertification are responsible for 35 percent of livestock-associated GHG emissions.[77] Land equivalent in area to Russia and Canada *combined* is currently used exclusively as pasture or cropland to grow animal feed. If this land were converted to growing vegetables for human consumption, or into forests, it would soak up so

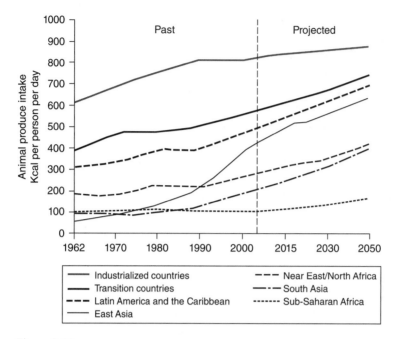

Figure 2.11
Trends in consumption of livestock products per person. Source: Anthony J. McMichael et al., "Food, livestock production, energy, climate change, and health," *Lancet* 370 (9594), 2007: 1253–1263.

much carbon dioxide that it could halve the costs of complying with the Kyoto Protocol, rather than being a source of emissions as it is today.[78]

Second, livestock production, like the growing of corn and rice, is inefficient. Roughly 35 percent of livestock-affiliated emissions are associated with on-farm fuel use, fertilizer, manure, and transport.[79] The estimated energy efficiency of eating animal products—that is, the amount of energy put into raising those animals relative to the energy received from eating them—is shockingly low. One must put in 99 calories of energy to get one calorie from lamb, 19 calories of energy for one calorie from chicken, and 4 calories to get one calorie from beef. (Thus, beef is the most efficient of the three.) These efficiencies are readily comparable to a 2:1 ratio for corn and tomatoes and a *positive* ratio of 1:5 for oats and 1:1.7 for potatoes and oranges (i.e., every unit of energy put into oats gives 5 calories back).[80]

Third, enteric fermentation by ruminants—a nice way to describe the farting of cattle—is responsible for 25 percent of livestock-associated emissions.[81] Because cattle release methane instead of carbon dioxide into the atmosphere, they are a potent contributor to climate change. Animal agriculture produces more than 100 million tons of methane a year. A single cow produces about 80–110 kilograms

(176–243 pounds) of methane in the same period, not including methane released from manure.[82] Significant emissions are also released from the lagoons used to store untreated farm and animal waste. Together, the carbon equivalent emissions from livestock are *greater* than the emissions from all the passenger vehicles in the world.[83]

Fourth, the shift to meat-centered diets is unhealthy, both to the animals involved and to us. Producing a kilogram (2.2 pounds) of meat can require 3–10 kilograms (22 pounds) of grain, and raising animals in dense feedlots and pens contributes to the rise of new antibiotic-resistant strains of foot and mouth disease, salmonella, bovine spongiform encephalitis ("mad cow disease"), and *Escherichia coli* outbreaks, which have resulted in millions of deaths of livestock and poultry and are pathogenic to humans.[84] The manure lagoons at many animal-processing facilities release toxic gases and contaminate water. There are also notable health benefits to eating less meat. Medical studies have shown that a substantial reduction in meat consumption in high-income countries has benefited public health by lowering the risks of heart disease, obesity, and colorectal cancer.[85]

2.3.3 Deforestation and Changes in Land Use

Much of the world's farming, livestock production, and changes in land use has taken place in former forests and tropical forests. (The transformation of such land is sometimes called "land use and land-use change and forestry," abbreviated LULUCF.) The hunter-gatherer lifestyle of the days of old supported about 4 million people, whereas modern agriculture feeds billions. About half of the world's usable land is now committed to pastoral or industrial agriculture, which come with a suite of negative consequences.[86] Conversion of natural landscapes into food-production and agricultural systems has created dangerous monocultures, which are easily wiped out by a single pest or disease. Between 1960 and 1995, global use of nitrogen fertilizer increased by a factor of 7 and phosphorous by a factor of 3.5, yet reliance on these inputs can erode the fertility of the land and degrade soil. The use of both is expected to triple again by 2050.

Insidiously, reliance on fertilizer can create dangerous positive feedback loops. The nitrogen oxides emitted from agricultural soils and through combustion of fossil fuels (correlated with electricity needed for irrigation and mechanized agriculture) increase the concentration of ozone in the local atmosphere, damaging as much as 35 percent of cereal crops, which then need more fertilizer to grow, causing more damage, and so on. Owing to these practices, about 17 percent of the world's vegetated land has undergone human-caused soil degradation.[87]

This transformation of forestland into other types of land is also problematic because it directly releases carbon into the atmosphere. Forests cover about 30 percent of the world's land area and store 683 billion tons of carbon, more than the total amount of carbon in the atmosphere. Through the carbon cycle, forests remove an additional 3 billion tons of carbon dioxide each year through growth; they also absorb about 30 percent of carbon dioxide emissions from combustion of fossil fuels.[88] Forests store about 45 percent of terrestrial carbon and can sequester additional carbon dioxide emissions directly out of the air. As forests grow, they store carbon dioxide from the atmosphere in their roots, branches, and leaves (their "biomass"), in essence making CO_2 a part of the natural landscape.[89] Yet when forests are cleared, are harvested, or catch fire, their stored carbon is emitted back into the atmosphere. About 36 percent of the carbon added to the atmosphere from 1850 to 2000 came from elimination and conversion of forests.[90]

Thus, forests can be a sink of emissions but also a source of emissions, depending on how they are managed. It is helpful to view forests in terms of stocks and flows. The total stock of carbon in all *tropical* forests equals about 300 billion tons. Through deforestation, about 1.5 billion tons of carbon is converted into 6 billion tons of CO_2, which is emitted into the atmosphere [91] In other words, tropical forests alone contribute to about 20 percent of overall human-caused CO_2 emissions per year,[92] which makes them the largest emitter of carbon in the world after the energy sector. This amount is equivalent to the total emissions of China or the US, and it exceeds the emissions produced by all the cars, trucks, planes, ships, and trains on Earth.

Forestry is thus unique in its ability to fight climate change, in that its benefits are reversible. A ton of carbon sequestered in a forest is not permanent, and is of benefit to the atmosphere only if it remains stored. If a tree is felled or a forest is cleared, carbon is released and the temporary benefit is reversed. Partly because of this aspect of forestry, tropical deforestation was excluded from the Kyoto Protocol as an eligible project class. Acknowledging that forests are decreasing at an alarming rate, the Copenhagen Accord—produced but not adopted at the fifteenth Conference of the Parties to the 2009 meeting of the UN Framework Convention on Climate Change—"recognize[s] the crucial role of reducing emissions from deforestation and forest degradation."[93] Yet the rate of deforestation worldwide averaged 13 million hectares a year between 1990 and 2005 (out of total forest coverage of about 4 billion hectares).[94] Indonesia and Brazil account for about half of the emissions from deforestation, which also explains why they are (respectively) the third- and fourth-largest emitters of GHGs, behind China and the United States. As table 2.4 shows, only nine countries account for more than 80 percent of all GHG emissions from deforestation.[95]

The impacts of deforestation extend far beyond climate change. Tropical forests are home to many species of plants and animals that may become extinct as their habitats are destroyed. Forests are essential sources of food, medicine, and clean drinking water, and they provide critical ecosystem services such as flood and

Table 2.4
World leaders in carbon dioxide equivalent emissions from deforestation. Source: D. Boucher, *Out of the Woods* (Union of Concerned Scientists, 2008).

	Share of emissions from deforestation
Indonesia	33.7%
Brazil	18.0%
Malaysia	9.2%
Myanmar	5.6%
Democratic Republic of the Congo	4.2%
Zambia	3.1%
Nigeria	2.6%
Peru	2.5%
Papua New Guinea	1.9%
Total	80.8%

drought control and the regulation of rainfall.[96] More than 1.2 billion people also depend on forests for their livelihood.[97] By 2050, at current rates of deforestation, more than one-third of the Amazon Rainforest will be lost, releasing about 3.5 billion tons of carbon dioxide.[98] If roughly quantified not for damage to the climate but rather for impact on local livelihood and economic services, deforestation causes global net damages of $2–5 trillion (€1.5–3.9 trillion) per year.[99]

Among the forces driving deforestation are economic development, the aggressive timbering and harvesting practices of multinational companies, the push to grow crops for biofuels and food, and the need to expand roads, suburbs, and cities as populations grow. These interests almost always take precedence over the preservation of forestland. For example, in June of 1992 Brunei, Indonesia, Malaysia, the Philippines, Singapore, and Thailand faced a choice between protecting forests and bolstering trade. That month, Austria had passed legislation mandating labeling of tropical timber imports, requiring them to be certified as having been produced under sustainable forestry conditions. Because Southeast Asian countries exported a large amount of timber to Europe but did not practice sustainable foresting, the Association of Southeast Asian Nations asserted that the fundamental "right to development" superseded any type of forestry regulation, and threatened to ban all Austrian exports.[100] Malaysia lodged a formal complaint against Austria with the governing body of the General Agreement on Tariffs and Trade. Austria repealed its legislation in March of 1993, and the Southeast Asian countries endorsed economic growth over the proper management of their forests. The consequences of this decision have been disastrous. In Southeast Asia, deforestation has been 5 times the world average and 10 times the average for the rest of Asia.[101] Indonesia alone is being deforested at a rate of 1.4 million hectares (3.5 million acres) a year and has only 53 million hectares (131 million acres) of total forest area left.[102] The deforestation there has contributed to the forest fires and the degradation of peat land that have made the country a large emitter of GHGs.[103] Roughly 98 percent of the forest cover of Borneo and Sumatra will be "severely degraded" by 2012 and "completely gone" by 2022.[104] Illegal logging is difficult to control there; as a result, three quarters of timber is extracted illegally and milling capacity exceeds legal limits by as much as a factor of 5.[105] And deforestation creates a dual penalty with, for example, the logging carried out to clear land for palm oil plantations that then produce a moderately carbon-intensive fuel for transport in addition to the emissions caused from land clearing. Uncontrolled logging throughout Southeast Asia has also contributed to soil erosion, to a collapse in agricultural productivity, to the emergence of new tropical diseases, to rising food prices, to higher rates of rural poverty, and to species extinction. Scientists have recently predicted that Southeast Asian forests could lose 13–42 percent of all their species by the end of the century, half of which are likely to represent worldwide species extinctions.[106] And the outlook may worsen considerably. Malaysia, the world's largest palm oil exporter, plans to increase production by 45 percent in the next few years and intends to convert 4 million hectares of pristine rainforest into palm plantations. Indonesia's plans are even more ambitious, with a 43-fold expansion in palm oil production planned, requiring an additional 20 million hectares (49.4 million acres) of plantations by 2025.[107]

Many other countries may want to stop deforestation, but lack the resources or capacity to do it. One recent survey of forest-management schemes around the world found that processes were highly uneven and inconsistent, and that implementing agencies were poorly coordinated.[108] Another comparative study of 164 schemes for sustainable forestry management around the world discovered extensive differences in standards, monitoring, and capacity.[109]

2.4   Waste and Water

Water supply is perhaps more closely linked to natural systems than electricity and energy production. Whereas fossil fuels and uranium take millions of years to create, the amount of water on the planet stays the same, changing only in form and quality. Figure 2.12 shows that more than 97 percent of the world's water is in the form of salt water; 2 percent is in the form of glaciers, ice caps, and snow, and less than 1 percent (200,000 cubic kilometers, or 47,983 cubic miles) is easily accessible for human use.[110] Humanity's primary source of fresh water is runoff from precipitation, naturally replenished through the hydrological cycle (the repeated process of the evaporation and redistribution of water in various forms around the world). Water is truly a renewable resource, since the water that was here a million years ago

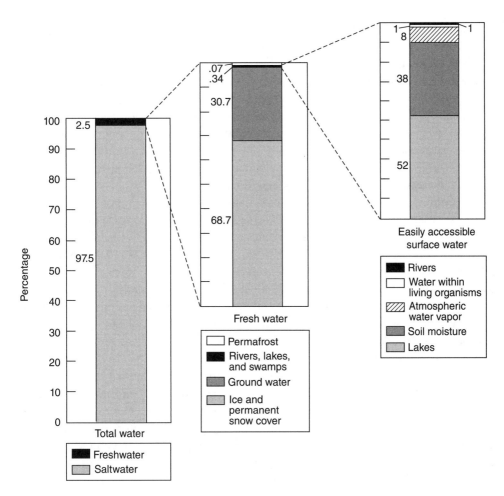

Figure 2.12
Earth's supply of water. Source: Jill Boberg, *Liquid Assets* (Rand Corporation, 2005): 16.

is still here today, continuously moving back and forth between the Earth's surface and atmosphere. Yet even though the total amount of water on Earth remains constant, its physical state is continually changing between three phases (ice, liquid, and water vapor) and circulating through different pathways (ocean, atmosphere, glacier, river, lake, moisture, and groundwater). About 40,000–45,000 cubic kilometers (9,597–10,800 cubic miles) of water are transported each year from the ocean into the atmosphere to replenish fresh-water resources, and the small transference of this water is the part of the hydrological cycle that can be utilized for human purposes each year.[111]

Before humans interfered, the hydrologic system was in long-term equilibrium. Discharges equaled recharges, and the volume of water remained constant. Groundwater levels fluctuated, but only within a small natural range.[112] More recently, however, three trends have perhaps irreparably altered this natural balance: increasing amounts of trash and waste, which can contaminate water and land; population growth, which has created shortages of drinkable water where it is needed most; and greater reliance on thermoelectric power plants, which use and degrade water in a variety of ways.

2.4.1   Increasing Amounts of Trash and Waste

One pronounced threat to the supply of water and preservation of land is the amount of waste that we generate each year. The twentieth century saw significant changes in how humans consume and use goods and services. Today, people commonly replace items not when they wear out, but when styles, images, and fashions change. Many manufacturers build obsolescence into their products so that they can sell more goods, and both producers and users "throw away" many components and items as waste. Many consumers have entered a post-industrial age of "consumption communities"—that is, they make themselves known and enhance their identity through the products they purchase. Today, Americans dispose of some 220 million tons of waste per year, and it is estimated that less than 1 percent of all materials mobilized to serve the market are made into products still in use 6 months after sale.[113] A billion people live in developed countries, and they consume and produce waste at rates 32 times above the rates of consumption and waste production of the 5 billion people in the developing countries.[114] Figure 2.13 shows some of this waste accumulating in the Maldives, a little-developed country in the Indian Ocean.

Almost 13 billion tons of new municipal solid waste were created in 2000. The amount of waste generated in OECD countries increased by a factor of 1.7 from 1980 to 2005 and is expected to double again from 2005 to 2050 (figure 2.14).[115] The US is home to more than 1,600 active landfills (in addition to 40,000 abandoned or full landfills) that must store 657 kilograms (1,450 pounds) of waste per person per year, a number that excludes industrial waste and commercial trash.

The amount of waste per person may not sound like much, but in the course of a year the US produces enough to bury 990,000 football fields under 1.8 meters (5.9 feet) of waste, or form a line of filled-up garbage trucks long enough to reach the moon.[116] Americans throw away enough aluminum (mostly from soft drink cans) to duplicate the entire commercial air fleet of the United States. One of the highest points in the state of Ohio, Mount Rumpke, is a landfill composed entirely of 1,000 feet (305 meters) of garbage.

Other countries are not far behind. The average person living in one of the OECD countries produces about 500 kilograms (1,100 pounds) of garbage per year (about

Figure 2.13
Residents of Malé, the capital of the Maldives and the most densely populated city in the world, at a pile of trash at their local landfill. The pile is more than seven stories tall. Source: B. K. Sovacool.

3 pounds or 1.4 kilogram per day), an amount that has increased by 30 percent in the past few decades.[117] Six European countries also have 55,000 abandoned landfills.[118]

The amount of waste that ends up in landfills varies by country, but China and the United States lead the world. Municipalities in the US send 55 percent of their waste (148 million tons) to domestic dumps and landfills, and in China 43 percent of waste ends up in landfills (the rest is incinerated or recycled). In China, 65 percent of cities are on the brink of running out of landfill space. In parts of Latin America, 40 percent of landfills do not meet minimum standards for environmental safety. In Africa, 80 percent of solid waste is left out in the open, untreated. In the US, only 28 percent of waste is recycled, whereas the best-performing countries in Europe recycle 60 percent.[119]

A worrying proportion of trash never makes it to a landfill or a recycling center, and can end up dumped into the ocean or washed up on coasts and beaches. About 46,000 pieces of plastic litter can be found in every square mile of ocean, and a swirling mass of trash twice the size of the state of Texas currently circulates in the North Pacific. More commonly called the "Pacific Trash Vortex," because it swirls like a whirlpool in an area of heavy currents and slack wind, the garbage patch is made up of abandoned fishing gear (buoys, nets, lines, and so on) as well as common refuse

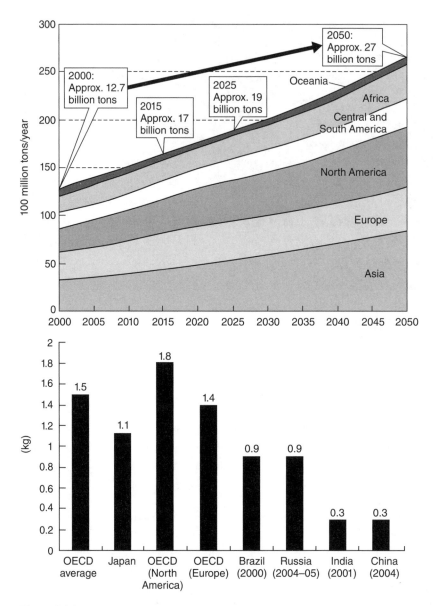

Figure 2.14
Above: World waste generation, 2000–2050. Below: Per capita daily municipal solid waste generation in OECD countries. Source: *The World in Transition and Japan's Efforts to Establish a Sound Material-Cycle Society* (Government of Japan, 2008): 6.

(e.g., plastic bags, light bulbs, toothbrushes, and bottle caps). Similar gyres appear to be forming off the coast of Japan and in a region of the Atlantic Ocean between the latitudes of Cuba and Philadelphia.[120] Every year, 1.6 billion barrels of oil are transformed into 500 billion plastic bags that then take 1,000 years to decompose, and about 11.4 million pieces of trash (weighing almost 7 million pounds) are collected from beaches and coastlines.[121]

The presence of all this trash poses a grave threat to the environment. Landfills often leak pollutants into groundwater, drinking water, and soil. One contaminant from landfills, vinyl chloride, is a known carcinogen. Some medical studies have suggested that people living near landfills suffer higher rates of liver and lung cancer, leukemia, and non-Hodgkin's lymphoma. Landfills also emit "landfill gas," usually containing high amounts of methane and carbon dioxide, directly into the environment. Globally, landfills accounted for more than 730 million metric tons (730 billion kilograms; 1.61 trillion pounds) of carbon dioxide equivalent emissions in 2000, representing 12 percent of the world's methane emissions; that is expected to increase to 21 percent by 2020.[122]

From 20 million to 50 million tons of electronic waste ("e-waste") accumulates per year, which makes it the largest amount of hazardous waste by total weight worldwide.[123] Discarded computers, televisions, telephones, and other electronic devices contain amounts in aggregate of pollutants that rival sources such as factories and power plants, but are not prone to the same regulatory scrutiny. The 500 million computers discarded in the United States in a single decade (1997–2007) contain more than 632,000 pounds (286,700 kilograms) of mercury, whereas power plants emit only 96,000 pounds (43,500 kilograms) of mercury per year.[124] A substantial portion of the electronics waste stream in industrialized countries is exported to Asia, where it is dismantled under horrific environmental and labor conditions, including open burning, backyard acid baths to extract metals from circuit boards, and dismantling without proper ventilation and safety standards.[125]

2.4.2 Shortages of Drinking Water and Fresh Water

Deficiencies in water supply and water quality now cause about 4,500 deaths throughout the world per day, or 1.7 million deaths a year, 90 percent of these in young children. More than a billion people lack access to clean water, and 2.6 billion lack access to sanitation facilities.[127] Some rivers, aquifers, lakes, and other water sources are so polluted that it is more profitable for residents to remove plastic bottles and trash from them for recycling than to fish. The US Central Intelligence Agency believes that more than 3 billion people will be living in water-stressed regions around the world by 2015, a majority of them in North Africa and in China. The water tables of major grain-producing areas in northern China are shrinking at

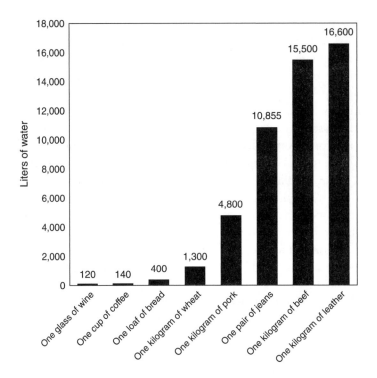

Figure 2.15
Water consumption associated with selected goods and services. Source: *Global Water Conservation and Use Survey* (WHO and UN Environment Program, 2010).

a rate of 5 feet (1.5 meter) per year, and per capita water availability in India is expected to decrease 50–75 percent in the next 10 years.[128]

China's urban population grew from 300 million in 1990 to 550 million in 2005, creating a consequent increase in the demand for water. Surface and groundwater resources surrounding cities have diminished as a result, and more than 100 million urban Chinese consume water contaminated with human and animal waste. The tension between providing water for drinking, manufacturing, and irrigation is especially acute in the Yellow River basin in Northern China, home to 13 percent of China's food production but only 3 percent of its water.[129]

India, heavily dependent on groundwater for irrigation and drinking, has lost 15 percent of its aquifers as a result of contamination and overuse, and is projected to lose 60 percent by 2030. Inefficient diesel water pumps, used for most of the irrigation and the well systems in India, tend to leak (wasting water) and inject pollutants into water sources.[130]

In the United States, sedimentation is reducing reservoir capacity by about 1.5 million acre feet (1.8 billion cubic meters) per year,[131] and water managers in 36 of the 50 states anticipate shortages of fresh water under normal conditions in the near future.[132] Scientists looking at soil erosion and increased droughts in the past few years have warned about the likelihood that the western US will experience another "dust bowl" within the next 150 years.[133]

To show how interconnected the issue of water availability and quality is with other challenges discussed in this chapter, such as waste, deforestation, and agriculture, consider the situation in the Cameron Highlands of Malaysia, a major source of vegetables for all of Southeast Asia. Forty years after the pesticide dichlorodiphenyltrichloroethane (DDT) was banned in the United States and in the European Union, farmers in Malaysia still use it heavily. As a result, DDT at 1,000 times the level allowed by health standards is detected in drinking water, along with unsafe levels of 40 other pesticides. Farmers discharge fertilizer and agricultural waste, and hotels and restaurants discharge human waste and sewage, directly into streams and rivers. Landfills have also begun to seep and leech pollutants into drinking water supplies, and deforestation has contributed to erosion and landslides that have increased the amount of suspended particles and foreign materials, including chlorine, aluminum, and iron, in rivers. The total upper water catchment area for the Camerons has plummeted from 95 percent to 45 percent in the past 30 years. Of the 121 rivers in the region, only 12 now are safe enough to use for drinking water. The sedimentation rate for these rivers has increased by a factor of 9 in the past 25 years, and the tap water consistently tests positive for *Escherichia coli*.[134]

Changes in hydrology and shortages of water can do more than cause health problems; they also spawn regional conflicts in the Middle East and Africa. The genocide in Darfur was partially triggered by an extended drought that destroyed farming and grazing land. Groundwater consumption in Yemen exceeds the natural recharge rate by more than 70 percent. Ninety percent of water use in Egypt, Libya, and Sudan supports irrigation and agricultural systems, which means that droughts can cause widespread shortages of food.[135]

Food processing and the light manufacturing of goods and services can be extremely water intensive. It takes 10 liters (2.6 gallons) of water to make a single sheet of paper, 70 liters (18.5 gallons) to produce one apple, and 91 liters (24 gallons) to make a pound (0.45 kilogram) of plastic.[126] As figure 2.15 shows, a number of other activities are even more water intensive. When compared to the water we drink, the result is astounding: humans must consume 2–4 liters (0.5–1.1 gallon) of water per day to remain healthy but must use 2,000–4,000 liters (528–1,057 gallons) of water to produce their daily food requirements. For every liter required to stay alive, humans consume 1,000 liters of water to meet their daily food requirements.[136] Figure 2.15 shows that a number of other activities are even more water intensive.

Complicating this picture is climate change, which is slowly but steadily altering precipitation and water patterns. For instance, if global warming induces the rise in sea level that many climatologists and scientists expect, the intrusion of salt water could contaminate fresh-water aquifers on both coasts of the United States, possibly reducing supplies of potable water by 45 percent.[137] Warmer temperatures resulting from global climate change will also increase energy demands in urban areas and require more air conditioning, in turn increasing the water needs of power plants. Hotter weather increases the evaporation rates for lakes, rivers, and streams, and thus accelerates the depletion of reservoirs, and causes more intense and longer-lasting droughts as well as more wildfires, which in turn requires vast quantities of water for firefighting.[138] The National Academy of Sciences has projected that con-tinued climate change will result in winter decreases of available precipitation by 15–30 percent, with reductions of snowpack concentrated in the Central Valley of California and along the northern Pacific Coast of the United States.[139] Continued rising temperatures are likely to cause substantial reductions in snowpack in the Sierra Nevada Mountains, with significant reductions in Californian stream-flow, water storage, and expected supply.

2.4.3   The Electricity-Water Nexus

Thermoelectric power plants—those relying on coal, oil, natural gas, biomass, waste, or the use of uranium in nuclear reactors—withdraw water from rivers, lakes, and streams to cool equipment before returning it to its source, and they consume water through evaporative loss. The average power plant uses about 25 gallons (95 liters) of water per kilowatt-hour generated (table 2.5).[140] Since the world consumed about 17,000 terawatt-hours of electricity in 2007, power plants ostensibly used 425 tril-lion gallons (1.61 quadrillion liters) of water that year. This means that, on average, thermoelectric generators use more water than the agricultural and horticultural industries combined.[141] The water use of individual power plants is even more strik-ing. A conventional 500-megawatt coal plant consumes about 7,000 gallons (26,498 liters) of water per minute, or the equivalent of 17 Olympic-size swimming pools per day.

The intimate relationship between electricity generation and water use could cause severe water shortages and crises if population and electricity-consumption trends continue, as well as a variety of environmental impacts (including impinge-ment, entrainment, eutrophication, and radioactive contamination of water sup-plies). At the point of intake, thermoelectric plants bring water into their cooling cycles through specially designed structures. To minimize the entry of debris, water is often drawn through screens. Seals, sea lions, manatees, crocodiles, sea turtles, fish, larvae, shellfish, and other riparian or marine organisms are often killed as they are trapped against the screens in a process known as impingement. Organisms small

Table 2.5
Water use (consumption and withdrawals) of selected power plants (gallons/liters per kilowatt-hour). Source: Benjamin K. Sovacool and Kelly E. Sovacool, "Identifying future electricity water tradeoffs in the United States," *Energy Policy* 37, 2009, no. 7: 2763–2773.

	Withdrawals	Consumption	Withdrawals	Consumption	
	(Combustion/downstream)		(Production/upstream)		Total
Nuclear	43/163	0.4/1.5	0	0.11/0.4	43.5/165
Coal (mining)	35/133	0.3/1.1	0.17/0.6	0.045/0.2	35.5/134
Coal (slurry)	35/133	0.3/1.1	0	0.05/0.2	35.3/134
Biomass/ waste	35/133	0.3/1.1	0.03/0.1	0.03/0.1	35.3/134
Natural gas	13.75/52	0.1/0.4	0	0.01/0.04	13.9/53
Solar thermal	4.5/17	4.6/17	0	0	9.1/34
Hydroelectric	0	0	0	4.5/17	4.5/17
Geothermal (steam)	2/7.6	1.4/5.3	0	0	3.4/13
Solar photovoltaics	0	0	0	0.3/1.1	0.3/1.1
Wind	0	0	0	0.2/0.8	0.2/0.8
Energy efficiency	0	0	0	0	0

enough to pass through the screens can be swept up in the water flow, after which they are subject to mechanical, thermal, and toxic stress in a process known as entrainment. Smaller fish, fish larvae, spawn, and a tremendous volume of other marine organisms are often pulverized by power plants' cooling systems. One study estimated that more than 90 percent are scalded and discharged back into the water as lifeless sediment that clouds the water around the discharge area, blocking light from the ocean or river floor, which further kills plant and animal life by curtailing light and oxygen.[142]

Impingement and entrainment account for substantial losses of fish and have severe consequences. Environmental studies of entrainment during the 1980s at five power plants on the Hudson River in New York—Indian Point, Bowline, Roseton, Lovett, and Danskammer—estimated grave year-class reductions in fish populations (the percent of fish killed within an age class).[143] Authorities noted that power plants were responsible for age reductions as high as 79 percent for some species, and an updated analysis of entrainment at three of these plants estimated year-class reductions of 20 percent for striped bass, 25 percent for bay anchovy, and 43 percent for Atlantic tom cod. Other researchers have evaluated entrainment and impingement impacts at nine facilities along a 500-mile (805-kilometer) stretch of the Ohio River. The researchers estimated that approximately 11.6 million fish were killed per year through impingement and 24.5 million through entrainment. The study calculated recreational related losses at about $8.1 million (€6.3 million) per year.[144]

Thermoelectric power plants also alter the temperatures of lakes, rivers, and streams. The data on temperature intake and discharge points for 150 thermoelectric power plants with "once through" cooling systems in the United States, for example, revealed that they had summer or winter discharges with water temperature deltas (differences between intake and discharge water temperatures) greater than 25°F (3.9°C). In some cases, the thermal pollution from centralized power plants can induce eutrophication—a process whereby the warmer temperature alters the chemical composition of the water, resulting in a rapid increase of nutrients such as nitrogen and phosphorous and less oxygen. In riparian environments, the enhanced growth of algae can choke vegetation and collapse entire ecosystems, and this form of thermal pollution has been known to decrease the aesthetic and recreational value of rivers, lakes, and estuaries and to complicate the treatment of drinking water.[145]

Nuclear generation of electricity can create wastewater contaminated with radioactive tritium and other toxic substances that can leak into nearby groundwater sources. Twenty-seven of the 104 nuclear reactors in the United States were recently found to have tainted groundwater with dangerous levels of tritium.[146] In December 2005, for example, Exelon Corporation reported to authorities that its Braidwood reactor in Illinois had released millions of gallons of tritium-contaminated waste water into the local watershed since 1996, prompting the company to distribute bottled water to surrounding communities while local drinking water wells were tested for the pollutant. Similarly, in New York, a faulty drain system at Entergy's Indian Point plant caused thousands of gallons of radioactive waste to leak into underground lakes. The Nuclear Regulatory Commission accused Entergy of not properly maintaining two spent-fuel pools that leaked tritium and strontium-90 (cancer-causing radioactive isotopes) into underground watersheds at rates as high as 50 gallons a day.[147]

Collectively, the water requirements of new power plants could precipitate a series of full-blown water crises. In the United States, thermoelectric power generation accounts for approximately 39 percent of total fresh-water withdrawals. The evaporative loss from thermoelectric power generation sector is about 3.3 billion gallons a day. The average (weighted) evaporative consumption of water for power generation over all sectors is approximately 2.0 gallons per kilowatt-hour. Consumptive water use is particularly high in hydro, nuclear, coal, and oil power plants; even where used cooling water is returned to the environment, water withdrawals can have a deleterious effect on fresh-water ecosystems.[148]

One assessment of population growth, electricity use, and shortages of water during the summer in the United States found that 22 counties and 20 large metropolitan areas could experience severe water shortages by 2025. The study noted that power plants could deplete the water available from Lake Lanier in Georgia and exacerbate interstate litigation between Tennessee, Alabama, and Florida. Fish and other marine species could perish along the Catawba-Wateree River Basin in North Carolina. Chicago could find itself embroiled in domestic and international legal disputes over the consumption and withdrawal of water from Lake Michigan. Households and businesses could run out of water from the South Platte River in Colorado. Rivers could stop recharging the groundwater needed for drinking and irrigation in Texas. Lake Mead and the Colorado River could continue to suffer drought, drastically affecting the state of Nevada and inducing an agricultural crisis in California and Mexico. Fisheries along the Hudson River in New York could collapse. The Delta Smelt could become extinct in the San Joaquin River Basin in California. These impending but avertable risks serve an important reminder that climate change is not the only serious environmental issue that the electricity industry must confront.[149]

2.5   Climate Change

Addressing climate change promises to be one of the most significant socio-technological challenges of the twenty-first century. The atmosphere has been the world's principal waste repository because the simplest and least costly approach to waste management has been discharging exhaust up through smokestacks, tailpipes, and chimneys. Changing these practices will require scientific and engineering genius to produce entirely new energy systems that avoid emitting GHGs and simultaneously power global economic growth. Success will also necessitate institutional, economic, social, and policy innovations to foster the widespread and rapid deployment of transformational systems.

2.5.1   The Science behind Climate Change

At the end of 2009, Earth's atmosphere had a carbon dioxide concentration of 388 parts per million (ppm),[150] higher than it has been in at least 800,000 years. The atmospheric concentration of carbon dioxide is increasing at 2.5 ppm per year; if untempered, concentrations could surpass 750 ppm—an extremely dangerous situation—by the end of the century. The expected warming from a doubling of carbon dioxide is between 3.6 and 8.1°F (2.0 and 4.5°C). Recent modeling research indicates that industrialized countries will have to reduce their carbon emissions by 50–80 percent by 2050 to prevent temperatures from rising more than 2–3°C.[151] If, instead of throttling back on carbon dioxide emissions, everyone achieved lifestyles equivalent to those of today's average US citizen, Earth would need an atmosphere 9 times the size of its actual atmosphere to handle the resulting pollution and carbon dioxide.[152]

Many chemical compounds found in the atmosphere act as greenhouse gases. These gases allow sunlight to pass through the atmosphere, but when the sunlight is re-radiated back toward space as infrared radiation they absorb the infrared radiation and trap the heat in the atmosphere and at the surface. GHGs allow Earth to be habitable—without them, the average temperature of the planet would be $-18°C$ ($-64°F$) instead of $15°C$ ($59°F$). However, human actions are increasing the atmospheric concentrations of GHGs, resulting in an "enhanced greenhouse effect" that is warming the climate and increasing the frequency and magnitude of extreme weather events (figure 2.16).

The most abundant GHGs are naturally occurring. They include water vapor, carbon dioxide (CO_2), methane (CH_4), nitrous oxide (N_2O), and ozone (O_3). In addition, there are numerous highly potent, anthropogenic GHGs that are entirely human made. Three classes of these—chlorofluorocarbons (CFCs commonly called freons), hydrochlorofluorocarbons (HCFCs), and bromofluorocarbons (BFC, commonly called halons)—also deplete the stratospheric ozone layer. These ozone-depleting substances are controlled under the 1987 Montreal Protocol on Substances

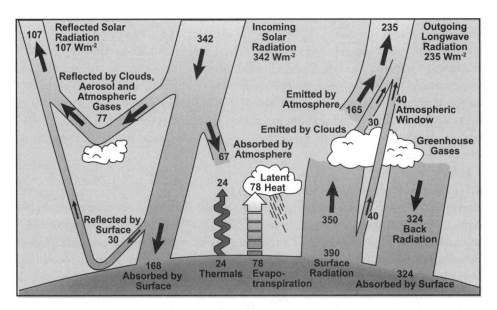

Figure 2.16
Global mean energy balance, now in disequilibrium. Source: IPCC, *Climate Change 2007* (Cambridge University Press, 2007), FAQ 1.1: 96, figure 1.

That Deplete the Ozone Layer, and as a result their impact on both ozone and the greenhouse effect has been greatly reduced. Other anthropogenic fluorine-containing halogenated substances do not deplete stratospheric ozone but are potent GHGs. The most important of these, collectively called the "F-gases," include hydro-fluorocarbons (HFCs), perfluorocarbons (PFCs), and sulfur hexafluoride (SF_6). Numerous other minor trace gases complete the inventory of GHGs.[153]

The contribution of a gas to the greenhouse effect is a result of both its characteristics and its atmospheric concentration. For example, molecule-by-molecule, CH_4 is a much stronger GHG than CO_2; however, it is present in much lower concentrations, so its total contribution is smaller. When these gases are ranked by their contributions to the greenhouse effect, the most important ones (excluding ozone-depleting substances) are water vapor, CO_2, CH_4, and O_3, followed by N_2O and the three F-gases.

Radiative forcing is used to quantify the effect of increased concentrations of GHGs on the climate. It is a measure of the global energy balance relative to that in pre-industrial times. "Forcings" are influences that "push" the climate toward overall warming (positive forcing) or cooling (negative forcing). This phenomenon is measured in watts per square meter alteration in global-average atmospheric radiation flow. The "sensitivity" of the climate to forcings is defined as the change in average surface temperature produced by a forcing of one watt per square meter (W/m^2).

Through the study of ice cores from Antarctica and Greenland, it is possible to compare concentrations of CO_2 in the atmosphere with temperature variations in the past 800,000 years. A visual comparison indicates a very tight connection between concentrations of GHGs and temperature. These ice-core records support a sensitivity of about 0.75°C per W/m^2.[154] Sensitivity is also sometimes expressed as the change in average surface temperature that would be produced by a doubling of CO_2 from its pre-industrial concentration of 278 parts per million by volume (ppmv), which corresponds to a forcing of 3.7 W/m^2. On the basis of this estimate, a forcing of 3.7 W/m^2 would result in a global temperature rise of approximately 2.7°C (4.9°F).

According to the Fourth Assessment Report of the IPCC and Nicholas Stern's report *The Economics of Climate Change*, about 300 billion metric tons (gigatons) of carbon have been released into the atmosphere since the onset of the Industrial Revolution. Global anthropogenic GHG emissions amounted to 49 billion metric tons of carbon dioxide equivalent in 2004, meaning that the world's emissions have increased 24 percent since 1990 rather than stabilizing or decreasing as stipulated by the Kyoto Protocol.[155]

We now know that human influences have dominated the climate forcings that brought about global climate change in the period 1750–2005. The warming influence of anthropogenic GHGs and absorbing particles is about 10 times the warming influence of the estimated change in input from the sun.[156] With the Industrial Revolution, CO_2 concentrations began to rise from 280 parts per million by volume (ppmv) in the early 1800s to the current level of 388 ppmv. Studies of air bubbles trapped in ancient ice cores show that CO_2 has not been as high as it is today in 400,000 years. The concentration of carbon dioxide in the atmosphere has increased more rapidly in the past 10 years than in any other 10-year period since continuous atmospheric monitoring began in the 1950s. The overall concentration of CO_2 is now almost 40 percent above pre-industrial levels. The concentration of methane is roughly 2.5 times the pre-industrial levels.

In 2005, countries belonging to the OECD were responsible for 48 percent of the world's CO_2 emissions, and the United States alone was responsible for 21 percent. According to the U. S. Energy Information Administration, world CO_2 emissions will increase from 28 billion metric tons in 2005 to 42 billion metric tons in 2030. A majority of this growth will take place in non-OECD countries—especially China, India, and the Middle East. Almost all (97 percent) of the projected increase in global CO_2 emissions between now and 2030 is likely to come from non-OECD countries—three-fourths from China, India, and the Middle East alone. By 2030, only 37 percent of CO_2 emissions will be coming from OECD countries.

An assessment of per capita emissions helps put the above figures into context. The United States consumes much more energy than many other developed countries—about 20 metric tons of carbon dioxide per capita, whereas the global average is about 4 metric tons per capita (figure 2.17).[157] Despite recent improvements, the carbon and energy intensity of the US is comparably high relative to the world average and relative to many other developed countries (such as Japan, which produces a dollar of GDP with less than half of the energy than is required in the US).[158] Although China overtook the US in 2006 to become the world's largest carbon emitter, the US probably will remain one of the most carbon-intensive and energy-intensive countries in the world well into the future.

2.5.2 The Consequences of Climate Change

The consequences of climate change are severe and increasingly certain. According to the IPCC, its impacts will include the following:

- global warming (already 0.6°C higher than before the Industrial Revolution)
- rising sea levels (a minimum of 0.3–0.6 meter, or 1–2 feet, by 2100)
- intensification of tropical cyclones
- decreases in meridional overturning of the Atlantic Ocean
- declining ocean pH (by 0.14–0.35—already down 0.1)
- decreasing snow cover, permafrost, and sea ice

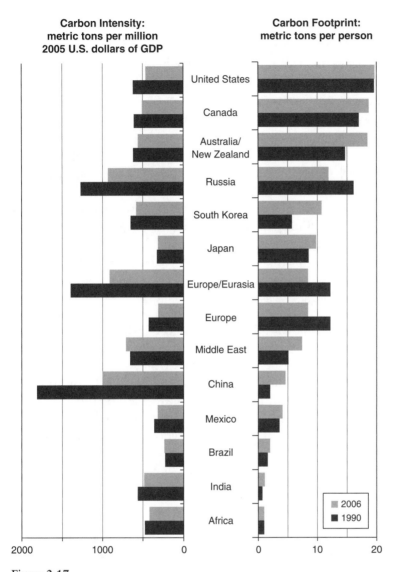

Figure 2.17
Carbon intensities and carbon footprints of selected countries. Source: World Bank/United Nations, 2005.

- more frequent and more intense extreme weather events
- increasing rainfall in high latitudes and decreasing rainfall in the subtropics
- changing micro-climates that will affect food production.

Consider the likely impacts on the acidity of the oceans and on food production. The acidity of the oceans has increased about 30 percent since the time of the Industrial Revolution, the greatest rate of increase in the past 55 million years. At the base of the marine food chain, acidification is rapidly depleting algae and plankton. Increased acid is bleaching and damaging coral reefs. For example, clown fish will lose their ability to "smell," and the reproductive processes of brittle stars will be damaged, which in turn will shrink stocks of herring. Ocean acidification is also causing a decline in the levels of aragonite and calcium carbonate, constituents of almost all marine skeletons and shells. Rather than being limited in scope, the threats from acidification are global. The greatest degree of acidification is occurring in the Atlantic, in the North Pacific, and in the Arctic Ocean, all crucial summer feeding grounds for billions of organisms.[159]

Climate change also has grave implications for the production, processing, and distribution of food. According to one study published in the medical journal *The Lancet*, as many as 40 of the least-developed countries, with a total population of 3 billion people, could lose 20 percent of their cereal production by 2080. Alterations to the ranges of agricultural pests and diseases with warmer winters could cause infestations of locusts, whiteflies, and aphids, resulting in extensive declines in crop yields. In the past 30 years, precipitation across the Sahel in Africa has declined by 25 percent, contributing to hunger and malnutrition in the Niger Delta, in Somalia, and in Sudan. Most experts anticipate severe climate-induced shortages of food in Angola, Burkina Faso, Chad, Ethiopia, Mali, Mozambique, Senegal, Sierra Leone, and Zimbabwe that will cause nearly 90 million people to starve.[160] Another study warned that as many as 250 million people in Africa could be exposed to increased water stress by 2020 as yields from rain-fed farms fall by 50 percent.[161]

Taken collectively, the climate-change consequences of "business as usual" could be nothing short of catastrophic for the planet. A consensus of studies from the IPCC,[162] the United Nations,[163] and various climatologists[164] warn that continued emissions of GHGs will contribute directly to the following:

- major changes in wind patterns, rainfall, and ocean currents, resulting in severe alterations to the distribution, availability, and precipitation of water and drinking water shortages for millions of people
- destruction of ecosystems, species, and habitats (especially coral reefs, beaches, and intertidal zones) and widespread deaths of numerous migratory species

• a significant loss of agricultural and fishery productivity, along with a shift in the growing seasons for crops and increased drought in areas with marginal soils that have low buffering potential

• increased damage from floods and severe storms, especially in coastal areas

• deaths arising from changes in disease vectors, particularly diseases regulated by temperature and precipitation

• the risk of abrupt and catastrophic changes such as the sudden release of pockets of deep sea methane or collapse of the North Atlantic's thermohaline circulation and Greenland's ice sheet.

A research team headed by the NASA Goddard Institute for Space Studies and the Columbia Center for Climate Systems Research recently confirmed the likelihood of these impacts.[165] After analyzing 30,000 sets of data relating to biological and physical changes affecting the planet over a span of 30 years, the researchers identified how many of the adverse consequences from climate change are already occurring. Plants are flowering earlier, birds are breeding prematurely, polar bear populations are declining, precipitation patterns are changing in South America and the Alps, fisheries and forests are collapsing in Europe and Africa, and migration patterns are shifting in Asia, Australia, and Antarctica. Another research team at the US Geologic Survey found that glaciers are melting much faster than anyone could have imagined a few years ago. Their assessment of the South Cascade glacier in Washington and the Wolverine and Gulcana glaciers in Alaska, long considered benchmarks for the condition of glaciers worldwide because they are located in different climate zones and at various elevations, found that the rate of surface loss has greatly accelerated for each of them in the past 15 years, even doubling for some glaciers.[166]

The impacts of global warming on industrialized economies are likely to be historic in magnitude, challenging such previous global threats as AIDS, the Great Depression, the proliferation of weapons of mass destruction, and terrorism. The Pew Center on Global Climate Change estimates that in the Southeast and the Southern Great Plains of the United States the financial costs of climate change could reach $138 billion (€106.8 billion) by 2100. Pew researchers warn that "waiting until the future" to address global climate change might bankrupt the US economy.[167] Stern projected that the overall costs and risks of climate change will be equivalent to losing at least 5 percent of the world's gross domestic product, or $3.2 trillion (€2.5 trillion), every year, now and forever, and that these damages could exceed 20 percent of the world's GDP ($13 trillion; €10 trillion) if more severe scenarios unfold.[168] Stern's figures have since been criticized for being too *conservative* and for relying on an improperly low discount rate.[169] Global economic damages from natural catastrophes, many of them climate related, have also doubled every 10 years and have reached about $1 trillion (€773 billion) in total damages in the past 15 years. Annual weather-related disasters have increased by a factor of 4 from 40 years ago, and insurance payouts have increased by a factor of 11 in the same period, rising by $10 billion (€7.7 billion) per *year* since 2000.[170]

The climate-change risks faced by developing countries are staggering in magnitude. Today, developing countries are home to 80 percent of the world's population; by 2060, that percentage could be 90. Developing economies rely heavily on agriculture, tourism, and forestry, and thus are affected directly and significantly by changes in temperature and precipitation and by extreme weather. In addition, developing countries tend to lack advanced health-care and transportation systems, leaving them at greater risk to the adverse impacts of a changing climate. They are also, for a variety of geographic and economic reasons, located in regions at the greatest risk of rising sea level, deteriorating ecosystem services, social tensions, and the creation of environmental refugees. Less affluent countries have fewer ways to recover the economic losses induced by climate change, as their assets are less likely to be fully insured, and the poorest countries rarely have access to systems of recourse other than humanitarian aid donations.

Indeed, one recent study explored the climate-change-related risks of a sample of eight countries—China, Guyana, India, Mali, Samoa, Samoa, Tanzania, the United Kingdom, and the United States—with a range of geographic locations, economies, and lifestyles.[171] It found that the negative impacts from climate change could cost some countries as much as 19 percent of their GDP by 2030, the developing countries affected most severely. Some locations, such as Maharashtra, India, will be prone to droughts that could wipe out 30 percent of food production, inducing $7 billion (€5.4 billion) in damages among 15 million small and marginal farmers. The Asian Development Bank also warned in a separate study that, because of their unique geography, Indonesia, the Philippines, Thailand, and Vietnam will suffer more from climate change than the global average. These four countries alone are expected to lose 6.7 percent of combined GDP by 2100 ($86 billion, €67 billion), more than twice the rate of global average losses if emissions of greenhouse gas continue as expected.[172] Another assessment concluded that climate change will disproportionately affect developing countries in Asia. Using detailed projections of tropical cyclones, drought, floods, landslides, and rising sea level confirmed by satellite imagery and historical records, that study concluded that all of the Philippines, the Mekong River Delta, all of Cambodia, northern and eastern Laos, and the Indonesian islands of Sumatra and Java will be "especially vulnerable" to climate change because of their lack of adaptive capacity.[173] Similarly, Tuvalu, the Marshall Islands, Papua New Guinea, and low-lying parts of the Caribbean and Bangladesh could be submerged within 60 years if sea level continues to rise. The Republic of Kiribati, a small island country in the Pacific, has already had to relocate 94,000 people living in shoreline communities and coral atolls to higher ground.[174] The Republic of Maldives could lose 80 percent of its land as a result of rising sea level and has already begun purchasing land in Sri Lanka for its "climate refugees."[175] Melting glaciers will flood river valleys in Kashmir and Nepal, and reduced rainfall will worsen water and food insecurity so that 182 million people could die of disease epidemics and starvation attributable to climate change.[176] Developing countries are likely to suffer most from the effects of climate change but are least able to invest in mitigation and adaptation efforts. They are also least responsible for the emissions that caused climate change to occur in the first place. The United States and Western Europe account for two-thirds of the primary buildup of carbon in the atmosphere, yet the entire continent of Africa is responsible for just 3 percent of the world's emissions since 1900.[177] Developed countries, however, spent only $40 million (€31 million) in 2007 on adaptation measures in the world's poorest regions, yet expended billions of dollars in technology and infrastructure to prepare their own countries for climate change.[178] To put it succinctly, the costs of climate change will befall the weakest and least-developed countries, while the benefits, if there are any, probably will accrue to the rich and powerful.[179]

Skeptics continue to argue that the threats from climate change are exaggerated, that humans are not to blame, and that Earth is not warming. Doubts were fueled by the 2009 "Climategate" scandal over emails from the Climatic Research Unit at the University of East Anglia, followed by a particularly cold winter in North America and Northern Europe and a controversy over whether Nepal's glaciers will melt in the next 35 years. Such skepticism underscores the complexity of deciphering the greenhouse effect, much like trying to fit together constantly changing pieces of a jigsaw puzzle. Among the complexities is El Niño, a quasi-periodic climate cycle that causes the tropical Pacific Ocean to sometimes release heat and sometimes store heat, affecting global temperatures. The sun's brightness further fluctuates typically on an 11-year cycle (and was at the bottom of its cycle in 2009); moreover, climatologists have since calculated that sunspots can at most only cause a 0.1°C change in global temperature.[180] The past 10 years have seen volcanic activity that has helped reflect heat back into space, again creating a pocket of lower temperatures. Aerosols from smokestacks, tailpipes, and volcanoes were once presumed to have only a cooling influence, but are now known to have warming and cooling effects depending on their color and composition.[181] And despite the skepticism, the year 2009 tied as the second warmest year in the 130-year record of global instrumental temperature records, according to a surface-temperature analysis conducted by the NASA Goddard Institute for Space Studies. For the Southern Hemisphere, 2009 was the warmest year on record. Overall, the period from January 2000 to December 2009 was the warmest decade on record.[182] Surface records, tree rings, ice-core samples, and coral reefs also confirm worrying changes in temperature and climate.[183] As Paul Edwards has eloquently written, though climate science can certainly get better, it is already as good as it has to be for us to know conclusively that we are seriously damaging the planet.[184]

Projections published in late 2009 suggest that the impacts from climate change could be *twice* as severe as predicted by the IPCC and other groups. These new calculations, based on improved economic modeling that also accounted for volcanic activity, ocean temperature, the rapidly thawing ice sheets of the Arctic, the Antarctic, and Greenland, and the interaction of soot emissions and aerosols in the atmosphere, predict a median probability that surface temperature will increase by 5.2°C (9.1°F) by 2100. They also anticipate a 90 percent probability of an increase ranging from 3.5°C to 7.4°C (from 6.3°F to 13.3°F).[185] As one of the co-authors of the study bluntly noted, "there's no way the world can or should take these risks."[186]

2.5.3   Toward Carbon Lock-In

The obstacles that thwart climate-friendly policies, however, are many, and they involve almost all aspects of modern society.[187] The climate issue is technically complicated and exceedingly difficult to comprehend, peppered as it is with complex terms such as "albedo" and "sinks." Almost 300 terms had to be defined for the glossary of the IPCC report alone. Climate scientists communicate about risks, scenarios, and ranges, which means they speak about probability rather than certainty. People generally believe that scientists lack consensus, an idea promoted by fictional works such as those of Michael Crichton, by the ambiguous language used in official climate documents and reports, and by the perceived uncertainty arising from computer models and measurements from field research. Climate change unfolds gradually and has yet to induce specific and identifiable points of crisis, instead producing unevenly distributed and discrete environmental and social consequences. Combating it involves altering our energy and industrial sectors, which have immense political power. Consumers remain uninformed about energy production but are very concerned about any attempts to raise energy prices.

The news media, seeking to balance their presentation of two sides to every issue, have contributed to the confusion by giving voice to the minority of climate skeptics. They also present misleading and inconsistent information about climate change. (See table 2.6.) Moreover, a strong "environmental-disinformation industry" has been funded and promoted by industries that are heavily reliant on carbon-intensive forms of manufacturing.

The complicity of consumers in this global climate crisis is perhaps the most insidious, and perhaps surreptitious, influence. Many citizens adhere to values that are inherently materialistic (valuing commodities above people and nature), anthropocentric (concerned primarily with humans), and contempocentric (preoccupied with

Table 2.6
Contradictory discourses of climate change found in UK newspapers. Source: H. Doulton and K. Brown, "Ten years to prevent catastrophe?" *Global Environmental Change* 19, 2009: 191–202.

	Basic theme	Primary actors	Solution proposed
Optimism	Climate-change science is uncertain, and can be ignored.	Some skeptical think tanks and lobbying firms in developing countries	We need not worry about climate change and should do nothing about it.
Rationalism	Climate change is a problem, but severe impacts are unlikely.	Economists from developing countries	Focus on other current pressing problems, not climate change.
Mitigation	Climate change is manageable as long as we begin mitigating (lowering and stopping) greenhouse gas emissions.	Governments, scientists, and stakeholders of most developed countries	Developed countries should take the lead on mitigation and researching new low-carbon technologies.
Adaptation	Climate change is already here, models show it will worsen despite mitigation.	Governments, scientists, and stakeholders of most developing countries	Developed countries should take a leadership role on adaptation and pay for adaptation in the developing countries.
Crisis	Dangerous climate change is already here and imminent environmental collapse is possible.	Ecologists and some climate scientists	Develop a new ethic of global stewardship.

the short term instead of the long term), failing to properly appreciate the intrinsic worth of ecosystems and future generations.

2.6   Conflicts and Complementarities

Although we have enumerated various crises as though they were independent and distinct, their interactions are quite strong, resulting in both conflicts and complementarities. Without a coordinated strategy that treats the challenges relating to electricity, transportation, agriculture and forestry, waste and water, and climate change as connected, unanticipated consequences can result. For example, protecting the shipping lanes used by oil tankers with military force can safeguard supplies of oil, but diverts attention and resources from pursuing alternatives to petroleum such as energy efficiency or biodiesel. Increasing the production of corn-based ethanol can reduce petroleum consumption, but can damage and degrade the land through the widespread use of fertilizers and destructive farming practices. Stockpiling petroleum and natural gas in strategic reserves can serve as a buffer against price shocks, but also offers just the kind of centralized targets that terrorists and saboteurs find attractive.

Other types of actions have double penalties. Building highways can incentivize people to use cars instead of public transport, increasing emissions, and can also contribute to deforestation, further increasing emissions. One project paving a highway in the Amazon Rainforest from Manaus to Porto-Velho in Brazil, for example, would not only have GHG emissions associated with vehicle use but would also enable the logging and harvesting of the forest, releasing an additional 950 million–4.9 billion tons of CO_2 into the atmosphere.[188]

Unfortunately, many countries continue to adhere to particularly narrow energy policies, pursuing a solution to one type of challenge that often occurs at the expense of solutions to other challenges. A study of large-scale energy projects in Thailand, Myanmar, and Laos found that, although the construction of regional interstate natural-gas pipelines and hydroelectric projects enhanced the availability of energy supply, such projects have exacerbated social tension, widened the gap between rich and poor, hastened environmental degradation, and intensified various manifestations of human insecurity, ultimately making electricity and energy more expensive.[189] International funding by the European Union on the production of liquid fuels from coal ("coal-to-liquids") has helped some countries reduce dependence on foreign sources of oil, but this strategy conflicts with efforts to fight climate change.[190] Similarly, the United States has begun to shift from the use of coal to natural gas in the power sector to reduce GHG emissions, enabled partly by the growing promise and use of shale gas, but this unconventional source of natural gas faces water management challenges, particularly the effective disposal of fracture fluids.[191]

Before you put the book down and give up in despair, however, you should note that these challenges are not nearly as bleak or daunting as they seem. First, only a few sources are responsible for a majority of the world's emissions, and these come from relatively few countries. Brazil, Canada, China, the European Union, India, Indonesia, Japan, Mexico, the Russian Federation, and the United States emit 81 percent of the world's GHGs, including those from changes in land use.[192] Second, although climate change cannot be addressed independently from development and electricity, agriculture, waste, water, and transport problems, these issues can be broken down into a series of more manageable challenges. As the chapters to come document, each challenge has readily available technology and policy solutions.

3

Technologies for Mitigating Climate Change

In his novel *Slaughterhouse-Five*, Kurt Vonnegut asks readers to imagine what the firebombing of Dresden during World War II would have looked like in reverse.[1] American bombers, full of holes and wounded men, would have taken off backward from airfields in England. German fighter planes would have sucked bullets and shells out of the aircraft over France. The bombers would have opened their doors and "exerted a miraculous magnetism" to shrink flames, restore buildings, and bring the dead to life. What amazing machines, Vonnegut is musing, the tools of war would be if they instead did the opposite.

The same could be said about the existing technologies that emit greenhouse gases into our atmosphere. What a wonderful sight it would be to witness factories, power plants, refineries, and vehicle exhaust systems sucking emissions and pollutants right out of the air and transforming them into easily stored and transported tons of coal, cubic feet of natural gas, and barrels of oil. Standing by would be electric utilities, energy traders and distributors, and attendants at gasoline stations ready to return these fuels to their sources. Once they arrived at their destinations, coal miners and petroleum geologists would rush to place them back deep into Earth, where they would no longer harm our atmosphere and environment.

Instead of being regarded as fantastic fiction, a bold system such as this is precisely what we will need if humanity is to avert the disastrous impacts of impending climate change and the consequences of deteriorating energy security. We need to implement the proper tools, shape the proper values, and strengthen the proper institutions that can undo climate change, restore ecosystems, and eradicate energy poverty. Although we have nothing yet that resembles the Vonnegut-style system of climate restoration, this chapter describes the closest equivalent available today, and an array of technologies that are well along the R&D pipeline. Such technologies range from off-the-shelf best practices to technically feasible next-generation technologies.

In this chapter we steer clear of opining about possible new technological breakthroughs that might offer remedies in the distant future. Though the "frontiers of science and technology" are critical to designing the next generation of solutions that will allow deep cuts in GHG emissions, it is difficult to speculate about which of them currently in laboratory research will emerge victorious.[2] (The economist John Kenneth Galbraith once joked that the primary function of predicting future economic and technological developments was to make astrology look respectable). At the other end of the spectrum, we also do not dwell on characteristics of "typical current use." These include technology choices made years ago under different price and policy environments with limited options and less concern for energy costs and global warming. Though this legacy constrains future decisions, it is not our principal focus in this chapter. Instead, we examine two stages of technologies that are either currently deployable or are nearly ready (figure 3.1): *best practices* and *technically feasible technologies*. "Best practices" means the most advanced climate-change-mitigation technologies that are cost effective and available at this time. "Technically feasible technologies" means the best-performing technologies being prototyped and demonstrated that are technically feasible but have not yet been proved and indeed may not yet be cost competitive. These stages of technology development tend to move toward higher levels of performance as the result of improvements from research and "learning by doing."[3] A continuing issue in most countries is how much emphasis to put on moving the technology frontier farther out, on advancing the current best practices, and on moving typical technologies closer to current best practices.

When evaluating the potential for additional climate-change mitigation in the future, three measures are relevant: *technical potential*, *economic potential*, and *achievable potential*. Technical potential refers to improvements that could be achieved as a result of the complete penetration of all applications that are technically feasible. Since many of these options (such as cellulosic ethanol, solid oxide fuel cells, and carbon dioxide refrigerants) are not cost competitive, realizing the technical potential could be quite costly. Economic potential is defined as that portion of the technical potential that is judged to be cost effective. What is cost effective in one location may be unaffordable elsewhere, depending on the prices of labor, materials, and especially energy. For example, with residential electricity prices in 2010 ranging from 8 cents per kWh in Idaho (comparable to Malaysia) to 28 cent per kWh in Hawaii (comparable to Sweden), the economic potential of energy options varies widely. Since consumers have many barriers to adopting technologies, even when they are judged to provide an attractive return on investment, realizing the full economic potential is unlikely. Achievable potential is the amount of economic potential that could occur with aggressive policies and programs.[4] Translating economic potential into achievable mitigation depends on many conditions, including an informed and motivated public and the absence of regulatory barriers, infrastructure constraints, and other over-riding obstacles (more on that in later chapters).

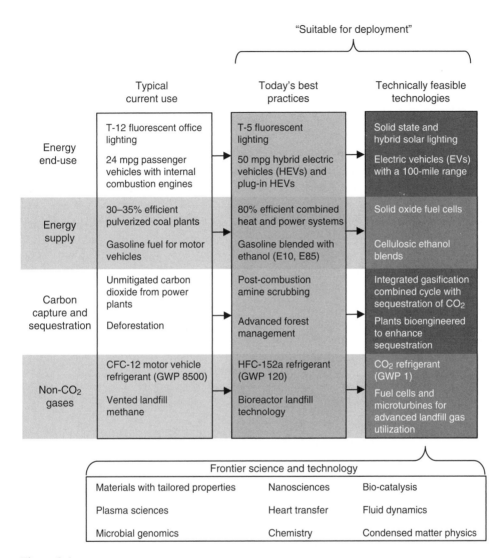

Figure 3.1
State of technology development, with illustrative technologies.

A particularly notable constraint on energy systems is the longevity of existing technologies. Long-lasting equipment can deter investments in new and improved technologies, whereas rapid stock turnover enables quicker technology transformation. It turns out that the lifetime of energy-related capital stock is highly variable.[5] Generally, energy end-use measures such as refrigerators, motors, and cars are more short-lived than the long-lived energy infrastructure systems of buildings, power plants, highways, and transmission lines (figure 3.2). When incumbent technologies (e.g., coal power plants) are more long-lived than alternative technologies such as insulation and efficient appliances, transformation from one to the other can be sluggish. Infrastructure and product longevity contributes to the "lock-in" of incumbent technologies.[6]

Using an appealing typology described in the *US Climate Change Technology Program Strategic Plan*,[7] we divide technologies into four categories: reducing CO_2 emissions from energy end use, reducing CO_2 emissions from energy supply, capturing and sequestering CO_2, and reducing emissions of non-CO_2 greenhouse gases.[8]

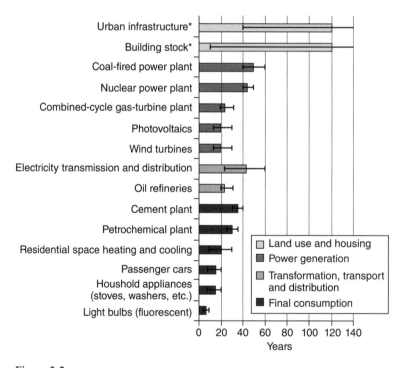

Figure 3.2
Typical lifetimes of energy-related capital stock, with bars indicating average lifetimes, lines indicating typical ranges, and an asterisk indicating that the upper range exceeds 140 years. Adapted from figure 1.6 of International Energy Agency, *World Energy Outlook 2008*.

That strategic plan identified about 400 technologies with the technical potential to reduce GHG emissions. Of these, the biggest technical potential to fight climate change was assigned to the first two categories (energy use and supply); smaller but still meaningful savings were assigned to carbon capture and non-CO_2 gases. Recognizing the value of merging portfolios of technologies into holistic approaches, we add to this typology the option of systems integration. The result is a table with five technology goal areas and 16 types of technologies (table 3.1). This chapter explores each of these five technology goal areas, focusing on today's best practices and also on technically feasible technologies, taking special care to highlight the technologies and practices with the greatest achievable potential to reduce GHGs and simultaneously improve energy security.[9]

3.1   Energy End Use

End-use energy efficiency offers some of the greatest near-term opportunities for large-scale GHG mitigation. Using energy wisely has been shown to be a valuable "front-line" strategy against global climate change because it offers a "no regrets" approach. Investments in energy efficiency can save consumers money while

Table 3.1
Sixteen types of GHG-reducing technologies.

Energy-efficient systems		
Energy-efficient transportation		
High-performance buildings		
Transformational industries		
Electric grid		
Low-GHG production		
Low-carbon energy supply	Carbon capture and sequestration	Non-CO_2 greenhouse gases
High-efficiency fossil power	Carbon capture	Methane from energy and waste
Hydrogen power and fuels	Geologic storage	Methane and nitrous oxide from agriculture
Renewable electricity and fuels	Terrestrial sequestration	High-GWP industrial gases
Nuclear fission		N_2O from combustion and industrial sources
Integrated systems		
Sustainable communities		
Renewable hybrid systems		

reducing pollution and GHG emissions and stretching global energy resources. Nevertheless, though many concede the virtues and necessity of energy efficiency, its priority among leaders and industries is often more rhetorical than real.

Energy efficiency does not necessarily mean "doing less" or "suffering without," but instead what the physicist Amory Lovins calls "doing more with less through smarter technologies."[10] It does not mean cold showers and warm beer. Rather, energy efficiency means getting more energy services per unit of energy consumed—with light bulbs that need less power, weatherstripping around doors and windows, hybrid electric vehicles instead of gas guzzlers, properly inflated automobile tires, more efficient industrial motors, and the recovery of waste heat in industrial plants. Here are some relevant definitions: *Energy efficiency* refers to permanent changes in equipment that result in increased energy services per unit of energy consumed. *Energy conservation* refers to behavioral changes that reduce energy use. *Energy productivity* is the ratio of gross domestic product (or some other measure of output such as the value of a firm's shipments or the square footage of an office building) per unit of energy consumed. (The inverse of energy productivity is energy intensity.)

In addition to reducing carbon emissions, energy-efficiency improvements tend to be more labor intensive than traditional energy-production options. For instance, utilities providing natural gas and electricity employ four or five people per million dollars (774,000 euros) of spending. However, sectors vital to energy-efficiency improvements support 8–13 jobs per million dollars of spending—more than twice as many.[11] Increasing energy efficiency requires new jobs for home energy auditors and inspectors, electricians, civil and mechanical engineers, and hybrid electric autoworkers and apprentices. Politicians from across the political spectrum have highlighted the potential of these "green jobs" for stimulating economic recovery and expanding employment.

In characterizing energy resources, both the International Energy Agency and the Energy Information Administration use categories of fuels such as coal, oil, natural gas, nuclear, and renewable resources. The omission of energy efficiency from this mix reinforces the perception that a megawatt saved (a "negawatt") is not as valuable as a megawatt generated. Yet numerous studies indicate just the opposite:

• Enkvist, Naucier, and Rosander of McKinsey & Company produced global supply curves for CO_2 abatement that highlighted low-cost and no-cost efficiency options.[12] A follow-up assessment of energy-efficiency potential in the US economy concluded that an investment of $520 billion (€402.4 billion) in efficiency measures through 2020 had the "economic potential" to save $1.2 trillion (€929 billion) in energy costs and bring energy consumption in 2020 back to current levels.[13]

• A study titled Scenarios for a Clean Energy Future analyzed energy policy and technology alternatives for addressing multiple energy-related challenges facing the United States.[14] Conducted by five national laboratories, it concluded that advanced policies implemented in 2000 had the "achievable potential" to cut US electricity consumption in 2020 by 24 percent with no net cost to the economy. The peer-reviewed study was funded by the US Department of Energy and by the US Environmental Protection Agency.

• A Meta-Analysis of Recent Studies conducted by Steven Nadel, Anna Shipley, and R. Neal Elliott under the auspices of the American Council for an Energy-Efficient Economy compared the technical, economic, and achievable energy potential identified in more than a dozen studies in the past several years.[15] The meta-evaluation concluded that a reduction of 10–33 percent in electric efficiency is "achievable," depending on the time frame and the programs.

Other studies, one by the National Action Plan for Energy Efficiency (NAPEE) "Leadership Group" and one by the National Academies, have estimated similar levels of achievable potential.[16]

Comparing the energy productivity of different countries is also illuminating. Measured in gross domestic product per energy consumption, Japan has one of the world's most energy-efficient economies—more than 5 times as energy efficient as Russia and 4 times as efficient as China.[17] Many European countries also rate high on this efficiency indicator, outperforming the United States. Although the energy productivity of the US has improved in recent years, it is approximately half that of Japan. And although China has overtaken the United States and Europe to become the world's largest carbon emitter, the US is forecast to remain the biggest emitter of CO_2 per capita.[18]

Specific energy-efficient technologies can be loosely divided into four subcategories: transportation, buildings, industry, and the electric grid.

3.1.1 High-Efficiency Transportation

Transportation of people and goods accounts for approximately 25 percent of the world's energy consumption[19] and 28 percent of its energy-related CO_2 emissions.[20] In the next few decades, the transportation sector is expected to be one of the fastest-growing sources of GHG emissions. Much of the projected increase is attributed to the rapidly growing demand for petroleum-based transportation fuels in non-OECD economies, which is forecast to increase more than 2 percent per year whereas the demand in OECD countries is forecast to increase less than 1 percent per year.[21]

Many technically feasible technologies and practices can reduce the consumption of petroleum as a fuel for transportation. These include lightweight materials, cylinder deactivation, electric-fuel engine hybrids ("hybrid-electric" vehicles and "plug-in hybrids"), clean diesel engines, and the use of hydrogenated low-sulfur gasoline. In aviation, the next-largest consumer after highway use, GHG emissions could be lowered through improved engine designs, fuel blends, and air-traffic-management systems. Further reductions could result from modal shifts (e.g., from highway modes to rail, facilitated by improved intermodal connections), higher load factors, more intelligent transportation systems, more efficient freight hauling, reduced idling by heavy vehicles, and more sustainable land-use configurations with mixed uses and denser development. Of course, eliminating trips through remote banking, on-line shopping, and telecommuting is perhaps the most energy-frugal and climate-friendly transportation option, but such lifestyle changes face strong behavioral impediments, as will be discussed in chapter 5.

Several vehicle designs have either already begun to sell into mass markets or have near-term potential to penetrate the market. (See figure 3.3.) Sales of gasoline-electric hybrids, for example, have grown significantly in recent years. Plug-in

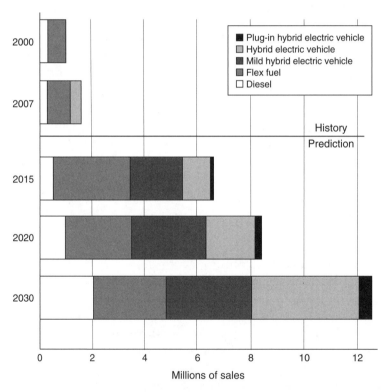

Figure 3.3
Actual and predicted sales of various types of light-duty vehicles in United States. Drawn from data in EIA Annual Energy Outlook 2009 Reference Case Presentation, December 17, 2008.

hybrid-electric vehicles (PHEVs) are capable of even greater fuel economy and gasoline displacement. Lithium-ion batteries provide electric power for existing and next-generation hybrids. Though the potential petroleum savings are substantial, analysis has shown that PHEVs in large number could cause local or regional constraints on the electric grid. Additional electricity capacity or significantly increased efficiency might be required, and the type of generation used to recharge the vehicles will determine the total impact on fossil-fuel consumption and carbon emissions.[22]

Flex-fuel vehicles, which can utilize ethanol or gasoline, are increasingly available today. (See chapter 8 for a case study of their use.) The technology for vehicles using other alternative fuels, such as biodiesel, natural gas, and propane, currently exists for commercial use, as well. In addition, prototype hydrogen fuel cell and hydrogen internal-combustion-engine vehicles, and hydrogen-refueling infrastructure, have been developed and demonstrated.[23]

Whatever the power train or the vehicle, lightweight technologies can profoundly improve fuel efficiency. Reducing an automobile's weight by 50 percent can increase its energy efficiency by more than 30 percent. Aluminum, carbon fiber, polymer composites, and other lightweight materials are in use today in both light-duty and heavy-duty vehicles, in aviation, and in marine transportation.

More intelligent transportation systems, such as high-speed and automated toll collection, adaptive signal controls, incident-management systems, and travel information systems, can reduce traffic and increase fuel efficiency. (See chapter 8 for a case study of Singapore's advanced transportation system.) Advanced screening technologies can make inspection of commercial vehicles more efficient and save fuel. Transit-oriented development and mixed-land-use urban designs also hold significant potential for reducing the miles traveled by vehicles.

Despite the availability of numerous GHG-mitigation technologies and practices, the accelerating pace of global mobilization in the form of ever-expanding auto-centric systems is a worrisome trend. Future development paths must diverge from the past, embracing ultra-high-efficiency transportation in combination with low-carbon fuels and taking a fresh look at multi-modal options and land use.

3.1.2 High-Performance Buildings

The built environment—consisting of residential, commercial, and industrial structures—accounts for about one-third of primary global energy demand[24] and is the source of 35 percent of the world's energy-related CO_2 emissions.[25] Over the long term, buildings are expected to continue to be a significant component of energy use and emissions, owing in large part to the continuing trends of urbanization, population and economic growth, and the longevity of building stocks. A growing body of evidence suggests that improving the energy efficiency of the existing building stock and that of new buildings is a low-cost approach to mitigating GHG emissions.[26]

Fortunately, lighting, office equipment, appliances, and some structural components of buildings (windows, doors, insulation) are typically replaced and upgraded often. Therefore, the short-term potential for improving the energy efficiency of the existing building stock is high. Since residential and commercial building energy use differ considerably among nations and climates, the magnitude of specific opportunities also varies. In both China and the United States, the largest user of energy in residential buildings is space heating (29 and 32 percent, respectively), followed in China by water heating and appliances and in the United States by other uses (primarily electric appliances) and water heating. In contrast, the share of household energy used for space heating is as high as 80 percent in Germany, where few homes are air conditioned.

The principal uses of energy in commercial buildings also differ. In China, heating is by far the largest energy use, followed by water heating and lighting. In the US, the largest energy use is plug loads such as office equipment and small appliances, followed by lighting and space heating (figure 3.4).[27] Many of these uses are increasingly met by electricity in both countries.

Although the built environment is a complex mix of heterogeneous building types and functional uses, most buildings have common features, each of which may benefit from energy-saving technological advances. On the appliance and equipment side, there are numerous GHG-saving opportunities:

• Best practices for heating and cooling include air-source and ground-source electric heat pumps, gas-fired absorption heat pumps, centrifugal chillers, and desiccant air pre-conditioners. Integrated heat-pump systems, which merge cooling, heating, hot water, and dehumidification into a single system to lower cost and increase efficiency, are technically feasible but are not yet cost competitive.

• Instantaneous and heat-pump water heaters are commercially available but not widely used. In addition to being highly efficient, heat-pump water heaters can be particularly attractive in hot and humid environments, because they also dehumidify the surrounding air.

• Solar systems for heating air and water have been commercially available for decades. Whereas solar space heating has seen only limited use to date, solar water heating has taken hold in many countries where solar radiation is strong, including Spain (which mandates solar water heating) and China (where one of the most populous provinces, Jiangsu, requires solar water heating in all new low-rise residential buildings).[28] One of the most popular uses in the United States is the heating of swimming pools: a majority of the heated pools sold in recent years are solar heated.[29]

• Solid-state lighting (including light-emitting diodes) is an important emerging technology with significant energy-saving potential. Producing light by passing

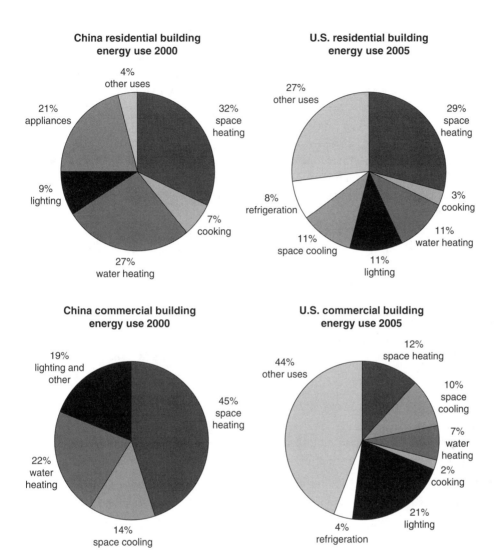

Figure 3.4
Energy use in buildings in China (2000) and the United States (2005). Redrawn from Intergovernmental Panel on Climate Change, *Climate Change 2007: Mitigation of Climate Change* (Cambridge University Press, 2007): 393, figure 6.3.

electricity through a semiconductor material, it generates very little heat and therefore is highly efficient. (Incandescent light bulbs produce light by passing electricity through a filament that glows, and compact fluorescent bulbs produce it by passing electricity through a gas.) Compact fluorescent lamps are a major improvement over incandescent lamps with respect to efficacy (about 60 lumens per watt vs. 15 lumens). If every home in the US were to replace just one incandescent light with a compact fluorescent, the country would save enough energy to light 3 million homes for a year, accrue about $600 million in annual energy savings, and avoid 4 million metric tons of GHG emissions each year.[30] LEDs offer even greater advantages, but they are not yet cost competitive in most applications. The best white LEDs are now more efficient than fluorescent lamps, and they are expected to reach 150 lumens per watt and to become cheaper as their production increases.[31]

• Various electronic lighting controls, including dimmers, motion sensors, occupancy sensors, photo sensors, and timers, are commercially available but have not yet cornered a dominant share of the marketplace.

• Combined heat and power (CHP) systems (sometimes referred to as "cogeneration") hold great promise. They require much less fuel to achieve the same energy output as separate heat and power systems. A traditional system of separately producing heat and power typically operates at 45 percent efficiency, while a CHP system can raise that efficiency to 80 percent or even higher.[32] Any industrial facility that consumes heat and electricity in sufficient quantities can benefit from integrated recovery and use of heat. CHP is a popular practice in Europe and Japan—several European countries produce more than 20 percent of their power from CHP systems. Market penetration is much lower in the United States, where project entrepreneurs find it difficult to overcome regulatory and other hurdles.[33]

Numerous building-envelope technologies and integrated designs are also ready for use:

• Improved shell designs and insulating systems can reduce energy use significantly, and improved training of construction professionals can enable builders to achieve the promise of these gains.

• Low-emissivity ("low-e") windows are economically attractive in many regions, but they have not yet fully penetrated the market, and electrochromic glazed windows (featuring the dynamic control of infrared energy) and other advanced windows have only recently become commercially available. Current research focuses on the complex vapor deposition manufacturing process to optimize for high-volume and low-cost production. Efforts to gain a more thorough understanding of the interactions of glazing, shading, lighting, controls, and occupants should contribute to wider deployment of dynamic windows and skylights for daylighting and energy-management purposes.[34]

• Other technologies available today to improve the energy efficiency of building envelopes include radiant barriers and reflective roofing materials. "Smart roofs" promise to deliver new advances by deploying nano- and micro-technologies that will change the reflectance and infrared emissivity of roof materials as a function of temperature so as to retain heat in winter and reflect heat in summer.[35] "Green roofs" not only reduce energy use in buildings; they can also provide space for recreation and mitigate the urban heat-island effect (caused by asphalt and other materials that absorb light and heat).

• Integrated building-design tools and control technologies have proved effective at optimizing building equipment and envelope systems, but they remain under-utilized. With effective policies, the achievable potential for such technologies appears to be large.

Through technological advances such as these, by 2020 it may be possible to achieve a 60–70 percent reduction from the average energy use of an equivalent home today. Incremental cost estimates for these advanced building systems could be partly offset by cost savings from the downsized HVAC system.[36] A few large prototype buildings incorporating such technology have been constructed with incremental costs of only 5–7 percent, which is generally recovered from reduced energy bills in less than 5 years.[37] Solar photovoltaic panels offer the possibility of "net-zero-energy" buildings when combined with 60–70 percent whole-building energy reductions. The estimated cost premium for such a system today is approximately 25 percent, which is a strong economic disincentive.[38]

A net-zero-energy building in 2020 probably would have a site plan designed to optimize the use of solar energy and take advantage of any sheltering terrain. It would also likely be a super-insulated and airtight structure with high-performance building components with low embodied energy but high thermal storage. Its equipment might include heat-recovery air exchangers and exhaust systems; integrated low-GHG-emitting energy systems that produce electricity on site while productively using waste heat; high-efficiency appliances and heating, ventilating, and air conditioning (HVAC) equipment; and smart sensors and controls. (See figure 3.5.)

Over the longer term, developments in thermoelectric materials may open up new possibilities, such as transforming low-grade heat into electrical energy that would enable self-powered sensors for control systems and the recovery of waste heat from appliances. With adequate breakthroughs in material science research, thermoelectrics could also serve as localized heating and cooling systems, thereby reducing the need to heat and cool large volumes of air.[39]

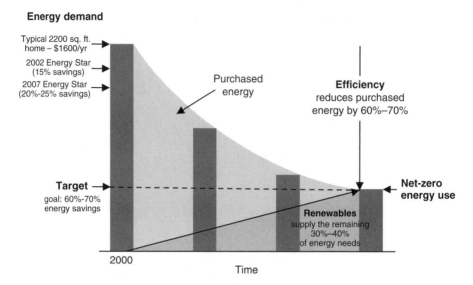

Figure 3.5
Pathway to net-zero-energy homes. Based on figure 3 of *Net-Zero Energy, High-Performance Green Buildings* (National Science and Technology Council Committee on Technology, 2008).

One way to identify "best practices" in the buildings sector is to consult the ENERGY STAR product list. ENERGY STAR is an international standard for energy-efficient products that was first created by the US government in 1992. Australia, Canada, Japan, New Zealand, Taiwan, and the European Union have now adopted the program. The level of upgrade varies by appliance. Devices carrying the ENERGY STAR logo save 20–30 percent on average. ENERGY STAR-qualified room air conditioners are at least 10 percent more energy efficient than the minimum US federal government standards, and accounted for 50 percent of US sales in 2007. ENERGY STAR-qualified fluorescent lighting uses 75 percent less energy, lasts up to 10 times as long as normal incandescent lights, and accounted for 20 percent of sales in 2007. ENERGY STAR-qualified refrigerators use 20 percent less energy than the federal standard, and accounted for 30 percent of sales in 2007. ENERGY STAR-qualified dishwashers deliver energy savings of at least 41 percent, and accounted for 77 percent of sales in 2007. However, ENERGY STAR-rated appliances have not yet penetrated more than half of the market for most types of appliances in the US. The limited market penetration of these cost-effective products suggests that behaviors, social, institutional, and other such influences play a strong role in consumer decision-making, as will be documented in chapter 5.

Japan's Top Runner Program shows how adept and nimble government interventions can be at promoting the widespread use of best practices by closely monitoring technological advances.[40] Japan established the Top Runner program in 1999 as a means of tying energy standards, by category, to the currently most efficient commercially available appliance. Then, within a time period mandated by the government, the average of the weighted sales of each manufacturer and distributor must meet that new energy standard. This approach appears to be much more effective than the more sluggish conventional approach, which involves an arduous process of public review and produces minimum standards of performance.

Currently, the Top Runner program applies to a number of major appliances, including air conditioners, space heaters, and refrigerators, with more products falling under its umbrella each year. The program's goals are also strengthened by the Japanese concern for corporate image: "naming and shaming" companies for failing to meet standards is a highly effective tool used by the Japanese government. Few programs are grounded in such intimate market intelligence about the progression from current use to best practices and technically feasible technologies. The close oversight of marketplace trends in the Top Runner program differentiates this effort from those in other countries.

3.1.3   Transformational Industries

The industrial sector is the largest consumer of energy worldwide, accounting for an estimated 36 percent of global primary energy in 2006[41] and producing a slightly larger share of CO_2 emissions, partly because of the use of fossil fuels as feedstocks in the production of chemicals and other industrial products (such as cement) that release CO_2 during their production.[42] Global energy consumption and CO_2 emissions from this sector are projected to increase rapidly through 2030,[43] driven by expansion in the economies of China, India, and other Asian countries.

A handful of large industries are highly energy intensive in most countries of the world in which they operate. These industries include petroleum refining and the production of chemicals and fertilizers, the metals industries (including iron, steel, and aluminum), the pulp and paper industries, the mineral products industries (including cement, lime, limestone, and soda ash), and the glass industry. Light manufacturing (which includes the manufacture and assembly of automobiles, appliances, electronics, textiles, and food and beverages) generally requires less energy per dollar of shipped product. As a result, managers of plants in light manufacturing industries pay much less attention to their energy requirements, even though light manufacturing remains a large fraction of economic output and contributes significantly to global emissions.

Many commodities that once were produced in the United States and in other OECD countries are now manufactured elsewhere. According to a recent analysis, the products imported into the United States in 2002 had an embodied energy content of about 14 quads, whereas the embodied energy of exports amounted to about 9 quads.[44] This imbalance highlights some of the accounting challenges of assigning responsibility to energy and carbon emissions. In addition to combustion-related emissions, industry is responsible for several process-related GHG emissions from sources such as aluminum production, cement, ammonia, and lime manufacturing.[45]

Advanced industrial technologies could therefore make significant contributions to reducing CO_2 emissions. In the long term, fundamental changes in energy infrastructure could dramatically shrink industry's carbon footprint:

• Many boilers in use today are more than 40 years old.[46] Revolutionary changes in *energy conversion and utilization* may include novel heat and power sources and systems, such as microwave processing of materials and nano-ceramic coatings, which show great potential for boosting the efficiency of industrial processes.[47] A combination of enhanced design features could raise the fuel-to-steam efficiency of industrial gas-fired boilers by 10 percent, from 85 percent to 95 percent. For improved heat transfer, "super boilers" use advanced firetubes with extended surfaces that help achieve a compact, lightweight design. The advanced heat-recovery system combines compact economizers, a humidifying air heater, and a patented transport membrane condenser. High-efficiency, low-nitrogen-oxide-emission burners such as radiation stabilized burners and forced internal recirculation burners also are more efficient than conventional equipment.

• There are *improvements in the efficiency of industrial processes* that can be deployed more widely today, such as the introduction of heat exchangers within distillation columns in the chemical and petroleum refining industries.[48] Cokeless ironmaking using Mesabi nugget technology and a cokeless oven/blast furnace could result in an energy saving of 30 percent in steel production in the United States.[49] In glass manufacturing, "oxy fuel" firing, which employs oxygen instead of air, can reduce fuel use by 15–45 percent.

• Innovative *enabling technologies* for energy-efficient and low-CO_2-emission products and processes may take advantage of developments in sensors and controls, in catalysis, in nanotechnology, and in micro-manufacturing. For example, adoption of advanced sensors and process controls offers industry-wide energy-saving potential by providing intelligent feedback control through continuous monitoring and diagnosis. In the papermaking industry, for example, fiber-optic and laser sensors can monitor water content, sheer strength, and the bending stiffness of paper, saving energy and improving paper quality.[50]

• Advances in *resource recovery and utilization* can cut GHG emissions while also reducing waste streams. Many of these approaches provide multiple ancillary benefits, such as improved productivity, product enhancements, and lower production costs. For example, in the cement industry, blending recycled materials such as fly ash and steel slag with cement could cut energy consumption by 20 percent. Avoiding the production of clinker yields both energy savings and reductions in CO_2 emissions.[51] Industry can also make greater use of waste heat through distributed energy generation, combined heat and power, and cascaded heat.[52] (See figure 3.6.)

Industries could cut their energy consumption by using cost-effective technologies that are available in the marketplace today. (See table 3.2.) Industry can also be a source of new technologies. For example, developing a new generation of fuel cells may lead to greater savings in motor vehicles. Deploying ink-jet printing systems to manufacture complex three-dimensional building panels may minimize thermal losses in buildings because materials are joined in the production process and not

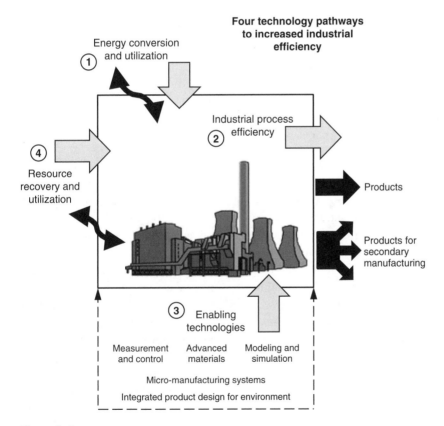

Figure 3.6
Pathways for reducing industrial greenhouse gases. Adapted from figure 4-6 of US Climate Change Technology Program, *Strategic Plan*, US Department of Energy Report DOE/PI-0005, 2006.

Table 3.2
Economic potential for energy-efficiency improvements in industry in 2020 (quads). Source: National Academies, *Real Prospect for Energy Efficiency in the United States* (National Academies Press, 2009). Global estimates from International Energy Agency, *Tracking Industrial Energy Efficiency and CO_2 Emissions* (2007).

	Energy-saving potential in 2020			
	Quadrillion Btus			Percent
	CEF study scaled to AEO 2008	McKinsey & Company (2008)	Other US studies	Global estimates from IEA (2007)
Petroleum refining	N/A	0.3	0.73–1.46 to 1.68–3.94	13–16%
Pulp and paper	0.14	0.6	0.53 to 0.85	15–18%
Iron and steel	0.21	0.3	0.76	9–18%
Cement	0.08	0.1	0.04 to 0.65	28–33%
Chemical manufacturing	N/A	0.3	0.19 to 1.1	13–16%
Combined heat and power	2.0	0.7	4.4	
Total industrial sector	7.7 (22.4%)	4.9 (14.3%)	—	18–26%

when they are assembled "on the job."[53] Still other ways in which industries could cut their energy consumption include using diodes that emit light for 20 years without bulbs, using deprintable and reprintable paper, using piezoelectric polymers that can generate electricity from the heel of a shoe or the force of a wave, and having roads double as collectors of solar energy.[54] As corporate sustainability has become better understood, industries around the world have taken a much broader view of their energy and environmental responsibilities, extending their concerns to the sustainability of the products and services they offer and to the sustainability of their suppliers. The dual goals of advancing energy efficiency at industrial plants and advancing product innovation for broader use are both critical to mitigating climate change.

Some industries integrate efficiency efforts with redesigned manufacturing processes to reduce waste drastically. Gessner AG, a subsidiary of Lantal Textiles, operates a production facility in Wädenswil, Switzerland, that minimizes waste, improves efficiency, and helps the environment through a focus on "cradle to cradle" design. The facility produces biodegradable upholstery fabric "safe enough to eat." To produce the new fabric known as Climatex, company managers redesigned almost every aspect of the production process, including the dyes and chemicals used, the factory workspace, the sourcing of raw materials, and the procedures for disposal and recycling. Managers first consulted with environmental experts and with the chemical dye manufacturer Huntsman, the fiber manufacturer Lenzing, the chemical producer Clariant, and other major suppliers, sharing information about chemicals and fabrics and identifying possible substitutes. They then designed a new production process to eliminate 8,000 chemicals and the need for additives and corrected processes, taking special care to select ingredients that would not damage the environment. After testing 1,600 dyes, company managers decided that only 16 met their standards. They also incorporated a number of natural fibers, including ramie (a tall fibrous plant grown in Asia) and wool from free-range sheep, into their product. When the fabric went into production, regulators tested the factory and thought their instruments were broken, as the effluent coming out of the factory was cleaner than that going in, with no registered pollutants. The water coming out of the factory was safer than the town's primary water supply, and workers no longer had to wear protective gloves and masks within the factory. The elimination of regulatory paperwork has reduced production overhead by 20 percent. The fabric decomposes naturally in the soil or in a compost heap after it is used, and trimmings are turned into a biodegradable mulch that is used to insulate plants at nearby strawberry farms.[55]

3.1.4   The Electric Grid

As the world's population grows and standards of living rise, the global demand for electricity is projected to continue its rapid expansion in both developing and industrialized economies. Nearly 2 billion people lack access to the electric grid, but the demand for electricity is expected to increase rapidly. As a result, the electric grid will need an infusion of investment in its transmission and distribution (T&D) infrastructure. The International Energy Agency forecasts worldwide investments of $6.8 trillion (€5.3 trillion) in T&D upgrades between 2007 and 2030.[56] These investments may create a supply crunch and a credit squeeze once the global economy recovers. As these grid upgrades and expansions are being made, advanced materials and technologies can be deployed to reduce carbon emissions by lowering line losses and by enabling access to low-carbon sources of electricity such as wind turbines and solar panels.

To accommodate these trends, the electricity-transmission infrastructure will have to extend its capacity and evolve into an intelligent and flexible system with a wide and varied set of baseload, peaking, and intermittent generation technologies. Enhancements for grid reliability probably will go hand in hand with improved efficiency.[57] Advanced storage concepts and high-temperature superconducting (HTS) wires and equipment are technically feasible technologies with great promise. HTS materials can transmit electricity with half the energy loss of conventional cables and are envisioned to carry perhaps 150 times as much electricity as conventional copper wires of the same gauge.[58]

Digital sensors, smart meters, and autonomous controls may eventually enable real-time responses to system loads. In the future, power electronics will be able to provide significant advantages in processing power from distributed energy sources using fast response and autonomous control systems designed to perform well under significant uncertainties. By facilitating the diagnosis of local faults and coordination with grid-protection schemes, these technologies enable the identification and isolation of faults before they cascade through the system. As the grid incorporates more distributed generators, such controls will be increasingly important.

To accommodate the increasing demand for and the greater reliance on regionally concentrated renewable sources, the electricity-transmission infrastructure will have to evolve into a more intelligent and flexible system. Modernization of the current grid combined with the continued electrification of the world's population will require vast sums of investment capital and materials, but may substantially improve the quality of life for billions of people.

3.2   Energy Supply

Making a transition from increasingly scarce, high-emissions fossil fuels to low-carbon energy supplies is one of the most critical challenges facing modern society. Prominent among the energy-supply options with inherently low life-cycle CO_2 emissions is a suite of renewable energy technologies. To the extent that these technologies emit greenhouse gases, the emissions generally occur during manufacturing and deployment and not during the combustion of fuels.[59] The inherently low-carbon nature of these technologies is attributable to the fact that most renewable energy technologies are powered by the sun: Plants and algae require sunlight for photosynthesis before they can be converted to biofuels or biopower. Hydropower capitalizes on rain and snowfall from evaporation and transpiration of water. Wind generates electricity directly by turning a turbine or indirectly in the form of ocean waves, but the wind itself is driven by the sun. Ocean thermal energy conversion uses the temperature differential between surface water warmed by the sun and cold deep water to drive a turbine and make electricity.

Tidal and geothermal energy are the only renewable energy resources that do not result directly from solar energy. Tides go up and down because of the moon's gravitational effect on the oceans. Some of the heat trapped in Earth is leftover heat from formation of the planet; some of it is due to radioactive decay of uranium, thorium, and other elements within the crust.

Energy-supply options with relatively small carbon footprints also include hydrogen for power production and fuels (if the hydrogen is produced sustainably), nuclear power (depending on how the uranium is produced and how the spent fuel is managed),[60] and high-efficiency fossil-fuel power. Several advanced technologies could significantly increase the efficiency of fossil-fuel power plants, thereby decreasing carbon emissions. When accompanied by carbon dioxide capture and sequestration (CCS) (discussed in the next section), this is a potentially attractive alternative solution.

We do not include in our list of viable options the production of liquid fuels from coal. A recent report issued by the National Academies persuasively argues that "unlike liquid fuels from biomass, liquid fuels from coal cannot, even with the use of carbon capture and storage, offer any greenhouse-gas benefit relative to gasoline."[61] We also do not include fusion energy, since it is still decades away from being deployed.

3.2.1 High-Efficiency Fossil Fuels and Electricity

Fossil fuels are expected to maintain a large share of the energy market in the next several decades because of their relatively high energy density, their ease of conversion to usable mechanical energy, and existing infrastructure such as pipelines, tankers, and rail transport. The International Energy Agency projects that fossil-based sources will continue to constitute more than 80 percent of the primary energy market through 2030.[62] Low-emission technologies—especially high-efficiency coal power plants, such as integrated gasification combined-cycle (IGCC) and supercritical pulverized coal (SCPC) plants—can dampen the associated growth in GHG emissions as this market grows.

Similarly, it is unlikely that fossil fuels will be replaced in the United States by non-emitting sources in the near future. Despite the rapid growth projected for biofuels and other non-hydroelectric sources of renewable energy and the expectation that orders will be placed for new nuclear power plants for the first time in more than 25 years, oil, coal, and natural gas still are projected to provide roughly the same share of the total US primary energy supply (86 percent) in 2030 as in 2005.[63] These opportunities to improve fossil-fuel use remain important; they include advanced power systems (such as coal gasification, ultra-supercritical power plants, oxygen-enhanced combustion, and co-production of hydrogen) as well as distributed generation. If successful, these innovations will allow continued global reliance on fossil resources for transportation fuels and electric power, particularly if CCS proves workable.

Advanced power systems that allow for greater efficiencies in power generation from fossil resources are expected to be significant emissions reducers for both GHG and criteria pollutants. Several successful full-scale demonstration projects suggest that IGCC systems could capture the environmental benefits of gas-fired generation with the thermal performance of a combined-cycle plant, yet with the low fuel cost associated with coal. Relative to pulverized coal power plants, not only has IGCC demonstrated a 20 percent reduction in CO_2; it also has made capturing and sequestering carbon easier.[64]

By operating at higher temperatures, supercritical coal plants offer efficiency improvements and emission reductions. Recent reports suggest that compared with subcritical coal plants that operate at about 34.3 percent energy efficiency, supercritical coal plants operate at 39.3 percent, and ultra supercritical coal plants can provide a boost to 43.4 percent energy efficiency.[65]

Oxygen-enhanced combustion can reduce emissions of nitrogen oxides (NO_x) and facilitate carbon sequestration.[66] Improvements in the efficiency of fossil-fuel power plants by a few percentage points can have a significant impact on aggregate GHG emissions. However, for these advanced coal systems to produce the scale of carbon emissions reductions needed globally, it will require supplemental post-combustion capture of carbon dioxide. The US Department of Energy is developing an oxy-combustion project that involves re-powering a coal plant in Meredosia, Illinois, using technology that could be deployed to retrofit as many as 590 units nationwide. Oxy-combustion increases flue gas CO_2 concentration from 13 percent to perhaps 70 percent, providing the potential for a 99 percent CO_2 capture and saving many aging plants from being shut down because of environmental requirements.[67]

Co-production of hydrogen with electricity offers additional flexibility in use of coal resources and improves overall plant efficiency to around 50 percent. Though not commercially viable now, traditional fossil-fuel plants have been configured to produce hydrogen and other marketable products such as chemicals and fuels.

Distributed generation of electricity using fossil fuels includes cogeneration and combined heat and power. This family of technologies utilizes natural-gas engines and turbines, as well as other power systems that can be installed directly on the consumer's premises or located nearby in district energy systems, power parks, and mini-grids. Outfitting these systems with digital sensors, information technologies, and controls could further increase their overall efficiency. Today's customers for distributed generation include hospitals, industrial plants, Internet server hubs, and other businesses that have high costs associated with power outages.

Markets are likely to grow as wealth increases and more consumers are willing to pay more to avoid the inconvenience of blackouts. Distributed energy resources are used at remote sites and as a cost-reduction measure for dealing with on-peak electricity charges and price spikes. The demand for distributed generation is likely to increase as the worldwide digital economy continues to expand. It is possible that in the next 50 years the demand for ultra-reliable power service will increase far more rapidly than the demand for electricity itself. This demand could be met by distributed energy resources.

Carbon emissions can also be reduced through distributed generation, using fuel cells and microturbines, if the waste heat generated is usefully employed on site to improve overall system efficiency. With such cogeneration systems, the US industrial and commercial sectors could cut their energy consumption by one quad with this technology alone in the next 10 years.[68]

Though not all distributed energy systems are "climate friendly" (and "back-up" diesel-generator sets are particularly polluting), numerous technologies for distributed generation offer significant potential for reducing emissions of CO_2 and local air pollutants, as well as fuel flexibility, reduced T&D line losses, enhanced power quality and reliability, and more control by end users. Many experts believe that these potential advantages will bring about a "paradigm shift" from central power generation to distributed generation.[69]

3.2.2 Hydrogen

Hydrogen, used in conjunction with fuel cells, has the potential to be an attractive non-carbon energy carrier for both the transportation sector and stationary

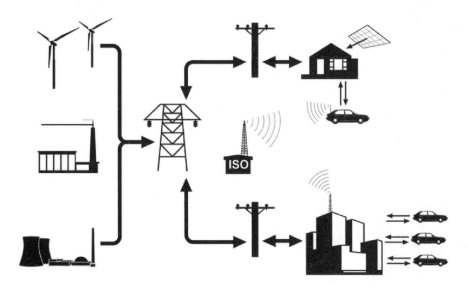

Figure 3.7
A distributed grid of the future. Source: Joseph Romm, "Energy myth four—The hydrogen economy is a panacea to the nation's energy problems," in *Energy and American Society—Thirteen Myths*, ed. B. Sovacool and M. Brown (Springer, 2007).

electricity applications. The use of hydrogen fuel cell vehicles can achieve large reductions in oil consumption and CO_2 emissions, although several decades will be needed to convert this technical potential into achievable benefits.[70] As such, there is widespread international interest in advancing hydrogen technology to address the growing emissions, supply, and energy-security concerns associated with conventional fuels. Governments and industries around the world are engaged in research and development programs whose goal is to make hydrogen technology cost effective. In the United States, the strategy for deploying hydrogen-based technologies focuses primarily on the transportation sector, while Japan and other countries are focusing more on the use of hydrogen in stationary fuel cells for residential heating and cooling.

There are three areas of technology development required for the hydrogen economy to succeed: hydrogen production, transportation of hydrogen, and use of hydrogen in fuel cells.

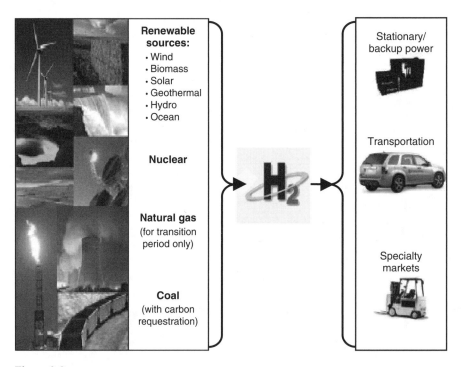

Figure 3.8
Options for hydrogen production. Source: US Department of Energy, Committee on Climate Change Science and Technology Integration, *Strategies for the Commercialization and Deployment of Greenhouse Gas-Intensity Reducing Technologies and Practices*, 2009: 49.

Since hydrogen is not a natural resource, it must be manufactured from other primary energy sources, such as natural gas or coal, or from electrolysis of water, which requires electricity. Currently, more energy is needed to produce hydrogen than can be recovered from the hydrogen.[71] More than 90 percent of the hydrogen produced in the United States for industrial purposes is derived from steam reforming of natural gas.[72] With reformer technology commercially available, and with greater economies of scale, it is likely that this technology will gain market share. As figure 3.9 depicts, well-to-wheels analysis shows that fuel cell vehicles operating on hydrogen from natural gas emit about 20 percent less CO_2 than gasoline-electric hybrid vehicles and 45 percent less than vehicles with only gasoline-fueled internal-combustion engines.[73]

Distributed hydrogen production from natural gas employing small-scale steam methane reforming technology offers several advantages over centralized hydrogen production technologies to enable a near-term transition to hydrogen. In particular, it allows hydrogen production to occur at consumer-refueling sites, thereby reducing the immediate need for a large and expensive hydrogen infrastructure.[74] Success of steam methane reforming for hydrogen production depends critically on the price of natural gas.

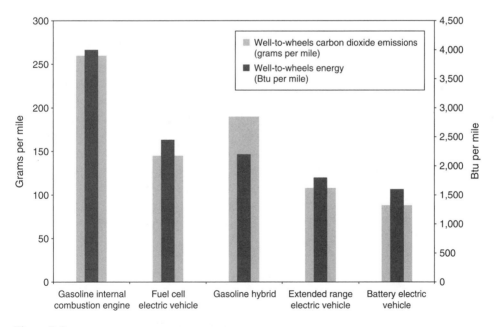

Figure 3.9
"Well-to-wheels" energy and carbon emissions for selected vehicles. Source: William J. Mitchell et al., *Reinventing the Automobile* (MIT Press, 2010): 88.

In the long run, it will be necessary to produce hydrogen in a way that emits little or no CO_2 if hydrogen technology is to be competitive in the market for clean energy options. Cogeneration of low-carbon hydrogen and electricity in energy plants using gasification processes may be an option by 2025.[75] Other options include electrolysis using wind, solar, geothermal and hydroelectric power; various biological and chemical processes such as partial oxidation and auto thermal reforming; water shift reactions with coal and natural gas, accompanied by CO_2 capture and storage; thermal and electrolytic processes using nuclear energy; and direct photo conversion of biomass such as algae or bacteria.

Unlike electricity, hydrogen can be stored for long periods of time without significant losses. Today, hydrogen is stored as a cryogenic liquid or a compressed gas and is transported in refrigerated tanks, in high-pressure trucks, and, to a limited extent, in pipelines. Other storage technologies are emerging, such as metal hydrides. High-pressure composite tanks that can store hydrogen at pressures up to 10,000 pounds per square inch have been certified but have seen only limited commercial application. Although pipelines are in moderate use today to transport hydrogen to areas where refineries and chemical plants are concentrated, advances in pipeline materials could offer greater capacity and require less maintenance for hydrogen transport.

Fuel cells are an important enabling technology for a hydrogen economy and offer cleaner, more efficient alternatives to the combustion of gasoline and other fossil fuels. Fuel cells have the potential to replace the internal-combustion engine in vehicles and provide power in stationary and portable power applications because they are energy efficient, clean, and fuel-flexible.[76] In vehicular applications, proton exchange membrane (PEM) fuel cells are being demonstrated in bus and taxi fleets around the world, and in indoor-operating forklifts they are replacing trucks powered by lead-acid batteries, with notable advantages of reduced charging time, longer operable times, and greater stability in performance.[77] A significant number of companies are manufacturing PEM fuel cells, and numerous other firms are conducting R&D.[78] It is estimated that automotive fuel cells could be produced—in high volumes (300,000 units per year or greater)—for about 100 US dollars (77 euros) per kilowatt.[79] Using platinum to catalyze chemical reactions in fuel cells appears to constitute a major cost-reduction challenge; this precious metal currently represents a majority of fuel cell stack costs.[80]

Fuel cells are being supplied as commercial products for stationary applications in specialty markets. Hospitals, credit-card-processing facilities, telecommunication firms, and investment banks are using hydrogen fuel cells for backup power.[81]

Though hydrogen is a commodity fuel already serving various niche markets, widespread deployment of fuel cells is inhibited by numerous technical and market barriers. One of these barriers is concern about hydrogen's safety. The explosion of the dirigible *Hindenburg* in 1937 left a lasting fear of hydrogen-based transport that will require some effort to overcome. The industry also needs the promulgation of harmonized international standards. With success, hydrogen has the potential to supplant hydrocarbon fuels for transportation and for electricity generation.

3.2.3   Renewable Energy

Production of renewable energy is increasing at double-digit rates around the world.[82] Although they are starting from a small base, renewable energy technologies are the world's fastest-growing source of energy.[83] Much of the growth is in hydropower, solar photovoltaics, wind power, and biomass.

Hydropower, onshore wind turbines, and electricity from landfill gas have generally achieved "grid parity"—that is, they have reached a point at which they can generate electricity at or below the cost of traditional grid power. Other renewable energy technologies are not cost competitive under current pricing regimes. As a result, government policies and incentives typically are the primary drivers behind the growth of renewable energy technologies.[84] Many OECD countries have government policies to incentivize the use of renewable energy, including feed-in tariffs, tax incentives, and renewable portfolio standards (called market-share quotas in Europe) that encourage the construction of renewable electricity facilities. (Feed-in tariffs in Germany have transformed that country's market for renewable energy. See chapter 8.) The extension of production tax credits in the 2005 US Energy Policy Act, and the standards for renewable fuels in both the 2005 Energy Policy Act and the 2007 Energy Independence and Security Act are expected to result in moderate acceleration in the use of renewable energy technologies in the United States.

Renewable electricity technologies are generally modular and can be used to help meet the energy needs of a stand-alone application (such as a building), an industrial plant, or a community, or the larger needs of a regional or national electrical grid network. Because of this flexibility, interconnection standards (for connecting individual renewable technologies with individual loads or buildings and with the electric grid) are critical to the safe, reliable, and efficient performance of renewable energy systems.

The flexibility of hybrid fossil/renewable systems, the diversity of sources of renewable energy offers a broad array of technology choices that can reduce CO_2 emissions. The generation of electricity from solar, wind, geothermal, or hydropower sources contributes essentially no CO_2 or other greenhouse gases directly to the atmosphere; and renewable resources generally emit fewer GHGs than conventional fuels. Increasing the contribution of renewable energy to any country's energy portfolio will directly lower GHG emissions in proportion to the amount of carbon-emitting energy sources displaced.

Renewable energy technologies are in various stages of market readiness. Within solar, wind, geothermal, ocean, biomass, and hydropower, each resource includes mature technologies and emerging systems still in various stages of development, ranging from laboratory testing to prototype demonstrations. A few examples of technologies that have made successful strides toward commercialization are described below.

Biomass power and heat differ in their sources as well as in their energy-conversion processes. Sources tended to be divided into agricultural wastes, residues, and wood wastes; energy crops; and trash and garbage. Processes tend to be thermo-chemical (i.e., combustion, which burns biomass in some way to produce heat or steam to turn a turbine) or biological (i.e., digestion, which lets waste decompose to produce methane, which is then captured and converted into electricity). Direct-firing involves burning the biomass material to create steam and drive a turbine, whereas co-firing involves mixing biomass with coal in a coal-fired power plant; both of these technologies are already quite mature. Gasification and pyrolysis entail using high temperatures in an environment with little or no oxygen to produce a gas or a liquid for use. Anaerobic digestion is generally used to produce methane in a controlled environment. After pre-sorting to remove plastic, steel, and other nonbio-degradable substances, the digestion process mimics the human one: waste is digested by bacteria, which excrete both gaseous waste (methane) and solid waste.

Wind energy has demonstrated robust market growth in recent years: from 2004 to 2008, worldwide wind turbine capacity grew by 250 percent. In 2009 the United States led the world in added and total wind power capacity, surpassing Germany (the longtime leader in wind). Net installed capacity of wind power in the US increased by 39 percent in 2009, equal to nearly 10 gigawatts.[85] However, wind remains a significantly under-tapped resource. Advances in materials, moorings, turbines, and blade design will increase wind energy's potential for commercialization and allow for commercial development offshore and places where wind speed is low. Improved integration of wind power into the electric grid is another need because of wind's intermittency. Deploying more than 20 percent renewable generation in a single system requires greater transmission capacity, advanced control technologies, and the installation of fast-responding generation to offset the intermittency of wind and solar energy. Denmark has illustrated how this can be done. As described in chapter 8, Denmark has the largest portfolio of wind projects integrated into its power grid of any country in the world (more than 20 percent), and some parts of the country often supplies more than 40 percent of their electricity from wind turbines. Even with a record-breaking quantity of cogeneration being deployed to meet Denmark's electricity needs, the country's electricity system has operated reliably for many years.

Most wind development to date has been onshore—offshore wind capacity reached only 1.5 gigawatts in 2008, virtually all of it in Europe. Nevertheless, it is experiencing strong growth, with 200 megawatts added in 2007 and 360 megawatts in 2008. The United Kingdom became the offshore wind power leader in 2008, sur-passing the historic leadership of Denmark. Building on this European experience base, the Netherlands announced €160 million ($200 million) per year for 15 years to support offshore wind power. Experts and advocates have argued that offshore wind possesses important advantages: offshore wind turbines can be placed out of sight, with minimal noise obstruction, where winds blow faster, and near to urban markets. At the same time, offshore development faces the challenge of inadequate and costly service environments and deep-water substructures that are challenged by severe ocean conditions, as well as expensive, high-voltage underwater transmission cables. In the United States, offshore wind also faces numerous regulatory issues dealing with siting and imbalance penalties.[86] Though deep-water costs may remain noncompetitive in the next 10–20 years, shallow water wind farms have been fore-cast to reach grid parity in 2020.[87] Recognizing this potential, US Secretary of the Interior Ken Salazar and the governors of ten East Coast states recently signed a Memorandum of Understanding establishing an Atlantic Offshore Wind Energy Consortium in order to promote the development of wind resources on the Outer Continental Shelf.[88]

Solar photovoltaic (PV) systems currently in the market are one of the most rapidly growing forms of renewable power. From 2004 to 2008, solar PV capacity increased by a factor of 6 to more than 16 gigawatts, and by 2009, global grid-connected solar PV capacity reached 21 GW (a 52 percent increase since 2008). Germany is the world leader in the use of grid connected solar photovoltaics with 47 percent of existing global capacity in 2009. Recent years have experienced signifi-cant expansion of building-integrated PV, which is a small but fast-growing mar-ket—particularly in Europe.[89] Because of its aggressive policies (reviewed in a case study in chapter 8), Germany's solar industries have grown exponentially despite having solar resources similar to those in Alaska. The case study of Bangladesh's Solar Home Program, also described in chapter 8, shows how a combination of financing, maintenance, training, and other capacity building assistance can support the rapid uptake of solar panels in a developing country that suffers from electricity poverty. Strong policies and subsidies have also helped China's solar manufacturing, enabling it to muscle its way into a young but sure to be significant generation resource in coming decades. In 2010, China produced half of the world's solar panels with the assistance of massive government subsidies.[90]

In the future, many anticipate that thin-film semiconductors will provide increased solar PV production volumes at reduced costs and greater efficiency. For example, cadmium telluride thin-film technology is actively commercialized with cell efficien-cies of more than 16 percent in the laboratory and commercial module efficiencies projected to be about 9 percent, and thin film PV has grown to nearly 7 percent of worldwide PV shipments in only 4 years of commercial production.

Spurred on by various incentives, concentrating solar power (CSP) has resurged in recent years, with five 50+-megawatt plants coming on line in the United States and Spain in 2008 and more under development. According to the *Renewables Global Status Report 2009 Update*, future plants are also being planned in Abu Dhabi, Algeria, Egypt, Israel, Italy, Portugal, and Morocco. This technology has the advantage of thermal storage (often using molten salt), which allows the power to be dispatched during off-peak hours.

Geothermal energy encompasses three types of systems: (1) geothermal direct heating by surface or near-surface geothermal energy; (2) hydrothermal resources that exist down to a depth of about 3 kilometers and are accessed by geothermal heat-pump systems; and (3) enhanced geothermal systems (EGS) associated with heated rocks at depths down to 10 kilometers.

Worldwide, geothermal direct heating provides approximately 28 gigawatts of energy. Its use is concentrated in tectonically active regions such as Iceland, where five geothermal power plants generate nearly one-fourth of the country's electricity, and geothermal technology meets the heating and hot water needs of more than 80 percent of Iceland's buildings.[91]

On the other hand, the potential for major expansion of geothermal direct heating is seen to be small because of the limited number of suitable sites.[92] Hydrothermal systems generate around 10 gigawatts, and its potential for expansion is large. In the United States, for instance, geothermal usage in 2006 was about 7.5 gigawatts, and its economic potential is estimated to be greater than 66 gigawatts by 2025.[93] EGS represents an even larger resource base;[94] however, EGS technology is not currently in large-scale operation because it is expensive and technically complicated. Research and demonstration projects are needed to reduce the technical and financial risks of EGS if the market is to achieve commercial viability.[95]

Hydropower is the largest renewable source of electricity worldwide; it represented 83 percent of world renewable electric power capacity in 2008, with large hydro accounting for a majority (860 of the 1,140 gigawatts). However, small hydropower (defined as sites smaller than 30 megawatts and often reaching to KW scales) is growing rapidly; at 85 gigawatts, it currently represents about 7 percent of all world renewable power capacity, especially in developing countries. "Pico" (meaning under 5 kilowatts) hydropower, in particular, is gaining popularity in many developing countries in Asia and Africa. In Laos, for example, small-scale "pico-hydro" systems are deployed in rivers and streams to generate electricity for households and small communities. These clean facilities lack the environmental destruction associated with mega-dam projects, offer cheap electricity solutions for the rural poor that lack grid connectivity, potentially showing them a way out of poverty.[96]

The United States has approximately 77 gigawatts of hydropower,[97] but proportionately less of it is small hydropower—totaling 3 percent (3 gigawatts) in 2008.[98] Assessments have shown that most of the best locations for large hydropower are already developed; however, small conventional hydropower sites offer the potential for expansion, particularly at non-powered dams where generation can be added. Smaller, low-impact hydro that does not require dams or water reservoirs is also becoming more accepted; electricity is generated by a turbine at the end of a penstock (water pipe) in parallel with the stream. In both of these cases (small hydropower added to an existing dam and low-impact hydro) minimal environmental impacts result from the added power production.[99]

Hydrokinetics can further expand a community's hydro resources by harnessing the kinetic energy of moving bodies of water. For example, tidal energy can be collected from tidal streams, which entails harvesting energy from the underwater current in entrances to bays or other narrow passages with turbines similar to those used for wind energy. In addition, wave energy can be converted to electricity by moving river water through motors connected to generators. There are several installations in the world, including a commercial wave energy plant located in Portugal[100] and the grid-connected Roosevelt Island Tidal Energy project in New York City's East River.[101]

Pumped hydro storage generates electricity by reversing the flow of water between two water sources, typically involving an elevated reservoir or water tower. Such storage technologies can deliver more than a gigawatt of capacity and can respond quickly with relatively low operating costs during periods of peak demand when purchasing power at spot market prices can be expensive. The Helms Pump Storage Facility near Fresno, California, for example, has three units totaling 1,200 megawatts of generation capacity.[102] Worldwide more than 90 gigawatts of pumped hydro storage facilities operated in 2007, with 22 gigawatts located in the United States.[103]

Liquid biofuels are dominated by ethanol, the most widely used renewable transportation fuel today. Worldwide, more than 17 billion gallons of ethanol were produced in 2007.[104] Ethanol is made by converting the carbohydrate from biomass into sugar, which is then transformed into ethanol in a fermentation process similar to brewing beer. A variety of other fuels can be produced from biomass resources including liquid fuels, such as biodiesel, methanol, Fischer-Tropsch diesel, and gaseous fuels, such as hydrogen and methane. Biofuels displace approximately 1.1 million barrels of oil per day, just over 1 percent of the world's total production of liquid fuels.[105] Almost 500 million gallons (1.9 billion liters) of biodiesel are produced each year.[106]

Although the United States is one of the world's largest producers of biofuels, with 9 billion gallons produced in 2007, biofuels account for a small fraction of the transportation fuels consumed in the US. In addition, reliance on corn-grain ethanol is not sustainable, since growing corn requires large amounts of fertilizer and using corn to produce transportation fuels competes with its use for food and feed.

Brazil's program shows how to produce ethanol at scale. In 2007, Brazil produced 6.5 billion gallons of ethanol, meeting approximately half of its transportation fuel requirements. Brazil's success is attributable to many factors, including the choice of a feedstock—sugarcane—that enables a much higher energy balance (i.e., renewable output per unit of fossil energy used in the process) than US corn-grain ethanol. Brazilian sugarcane ethanol is also less land intensive and is produced at lower cost than US corn-grain ethanol (table 3.3). Brazilian mills are self-sufficient in steam and electrical energy production, and in many cases surplus energy is exported to the electric grid and sold as a by-product. (See the case study in chapter 8, which describes how public policies created a strong renewable fuel alternative for Brazil, a country that once relied almost entirely on imported petroleum to fuel its transportation sector.)

Cellulosic ethanol is defined as fuel derived from cellulose or hemicelluloses from biomass, and its lifecycle GHG emissions are 60 percent less than the baseline of blended gasoline. Other "advanced biofuels" are anticipated to be produced mostly from cellulosic biomass such as woody or fibrous feedstocks rather than from starch feedstocks such as corn. Though cellulosic and advanced ethanols are not produced commercially today, they are being explored worldwide. Among the advances being studied are fast pyrolysis, chemical refining, algae farms, and the use of alternative enzymes and microorganisms to break down plant fibers.

Biodiesel can be made from vegetable oils, animal fats, or microalgae oils. It is produced by a process in which organically derived oils are combined with alcohol (ethanol or methanol) in the presence of a sodium or potassium hydroxide catalyst to form ethyl or methyl ester. This process has a glycerol by-product that can be used

Table 3.3
Comparison of US and Brazilian ethanol production. Source: Elena Berger Harari, "Biofuels," presented at Georgia Institute of Technology, March 4, 2009.

	US	Brazil
Feedstock	Corn grain	Sugarcane
Energy balance (renewable output/fossil input)	1.4	10.2
Approximate yields (gallons/acre)	370–430	590
Cost of production (2006) ($/gallon)	1.05	0.81
Retail price of gasoline in November 2006 (cents/gallon)	2.30	4.60

in soaps and in many other commercial applications. These esters can then be blended with conventional diesel fuel or used as a neat fuel (100 percent biodiesel).[107] Biodiesel emits 64 percent fewer GHGs than diesel fuel in terms of pounds/gallon produced from the production and use of fuel.[108]

Table 3.4 compares the pros and cons of corn-grain ethanol with cellulosic ethanol and biodiesel, along with plug-in hybrid electric vehicles and hydrogen. As summarized in a recent National Academy of Sciences panel report,[109] corn-grain ethanol can only be regarded as a transition to cellulosic biofuels or other biomass-based liquid hydrocarbon fuels. Cellulosic biomass could be sustainably produced today at about 400 million dry tons per year with minimal adverse impacts on food and fiber production. The food-fuel-fiber competition is managed in large part by the assumption that this biomass would be grown as dedicated energy crops on some of the 27 million acres (10.9 million hectares) of land that is currently idle as part of the US Department of Agriculture's Conservation Reserve Program. This dedicated biomass would be supplemented by the use of forestry and agriculture residues and urban wastes.

By 2020, the National Academy panel forecasts that the country could sustainably produce as much as 550 million dry tons per year. This 2020 estimate could theoretically result in up to 30 billion gallons (113.6 billion liters) of gasoline-equivalent fuels, or almost 2 million barrels per day. Under the Energy Independence and Security Act of 2007, the United States is committed to a fivefold increase in its use of transportation biofuels from 9 billion gallons (34 billion liters) of biofuels in 2008 to 36 billion gallons (136 billion liters) in 2022 (figure 3.10). Pipeline transport is generally seen as the preferred option for transporting large volumes of conventional liquid fuels over long distances. However, transporting biofuels such as ethanol by pipeline may pose a daunting challenge.[110]

3.2.4   Nuclear Fission

There are signs of renewed interest in nuclear power spurred by a coalescence of motivating trends. Much like renewable energy and energy efficiency, nuclear power can help meet the rapid growth of demand for electricity, avoid rate increases caused by escalating fossil-fuel prices, reduce dependence on imported oil by supporting plug-in electric cars, and reduce GHG emissions. Nuclear power is a technology option with the potential for reducing GHG emissions at a large scale. Each new plant could avoid as much as 6 million metric tons per year of GHG emissions, in view of the current mix of energy sources used in US electricity generation.[111] Over an expected operating lifetime of 50 years, this would amount to about one-third of a gigaton (a billion metric tons) of CO_2.

Nuclear power, more precisely defined as nuclear fission, is today a significant source of low GHG-emitting "baseload" electricity production, depending on the

Table 3.4
Summary of renewable fuel options. Adapted from table 4-1 in *Ending the Energy Stalemate* (NCEP, 2004).

	Hydrogen	Corn-grain ethanol	Cellulosic ethanol	Biodiesel	Plug-in hybrid electric
Energy security (ample domestic resource)	YES Can be produced from fossil fuels or water via electrolysis	NO Land limitations (used 18% of corn crop in 2008)	YES Greater diversity of feedstocks: agriculture waste, idle land	YES Crop animal waste, vegetable oils, used grease	YES Domestic supply more available
Low carbon	DEPENDS More carbon intensive if produced with electricity from today's grid; less with sequestration, nuclear or renewable energy	YES Roughly 20% lower greenhouse gases than gasoline	YES Has potential to achieve near-zero net carbon emissions	DEPENDS on energy inputs and feedstock; could have very low carbon emissions	DEPENDS (like hydrogen) on how the electricity is generated
Infrastructure compatible	NO Would require a new national distribution infrastructure	DEPENDS Pipelines for 10–15% blends; 85% needs barges or trucks	DEPENDS Same as corn ethanol	YES Compatible with diesel engines and infrastructure	YES PIHVs can be recharged using existing electric grid
Competitive with gasoline by 2020	NO Depends on technology breakthroughs	NO Depends on tax credits, import duties, and portfolio standards	YES Depends on technology breakthroughs, carbon price	DEPENDS on feedstock costs	DEPENDS on battery technology

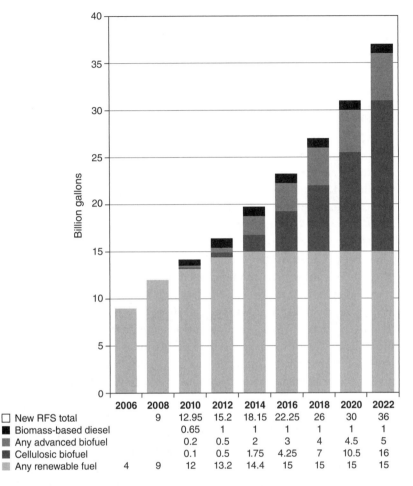

Figure 3.10
Renewable fuel standard mandated volumes from the Energy Independence and Security Act. Drawn from data published on website of Renewable Fuels Association.

fuel cycles used. Nuclear power plants are particularly well suited for generating "baseload" power, because they are most economical when operated at or near full capacity and with little load-following variability.

The expansion of nuclear power worldwide is advancing. In 2010, there were 438 nuclear power reactors operating worldwide, producing about 15 percent of the world's electricity.[112] Approximately 61 nuclear units are currently under construction, and most of these are being built in non-OECD countries—particularly China, Russia, and India, where ambitious nuclear power programs are being pursued.[113] The Energy Information Administration forecasts that by 2039 China will have added 47 gigawatts of nuclear capacity, Russia will have added 21 gigawatts, and India will have added 18 gigawatts. A few OECD countries, most notably South Korea and Japan, are also expected to increase their nuclear capacity.[114]

In contrast, no new nuclear power plants have been built in the United States since the rapid buildup of nuclear power during the 1970s and the 1980s. Progress in recent years has been limited to the refurbishment of one shut-down reactor and the completion of one previously ordered but not completed reactor (Watts Bar Unit 2 by the Tennessee Valley Authority).[115] The Nuclear Regulatory Commission (NRC) is actively reviewing 22 reactors for licensing, and three design certifications are in progress. "First movers" have begun site preparations and have ordered long-lead-time components. The Nuclear Energy Institute expects four to eight new reactors to be in commercial operation in the US in the years 2016–2018. Four major challenges to increased use of nuclear power remain, including cost, safety, waste management, and proliferation risk.[116]

The cost of electricity from new nuclear plants is expected to be higher in most countries than the cost of new gas-fired and coal-fired power along with lower cost electricity from renewable resources such as wind and biomass. Nuclear power's position will improve if a price is placed on GHG emissions. It is also less costly in South Korea, which constructed its last six nuclear units in only 53–58 months.[117]

Despite the improved safety record of today's nuclear reactors, if the number of reactors in the world is to double, defenses against terrorist attack as well as precautions against malfunction and human error will have to be improved.[118]

No countries other than Finland and Sweden currently plan for managing high-level wastes; rather, dry cask spent fuel storage has been implemented on a large scale at reactor sites. Despite the fact that both open and closed fuel cycles require the geologic disposal of some radioactive waste, little progress in high-level waste disposal has occurred.[119]

The risk of the proliferation of weapons-grade radioactive materials remains an issue worldwide, especially with commercial nuclear energy programs in Iran and North Korea.

In the United States, 104 operating nuclear plants produce about 19 percent of the electricity,[120] a market share that has been maintained for several decades by capacity upgrades and plant life extensions. During the past 50 years, US nuclear power plants have operated relatively safely and reliably with steadily improving economic performance. Significant advances have been achieved in reducing maintenance and operation costs and improving plant availability. Due primarily to its low fuel cost, historical nuclear power production costs are lower than those of other forms of baseload electricity generation. Likewise, the total kilowatt-hours of electricity generated by these reactors has steadily increased, and their operating "capacity factors" of more than 90 percent is the highest in the power industry.[121] It is expected that the carbon-mitigation benefits of these facilities will be extended as their licenses are renewed for a total of 60 years of operation.

In addition to electricity, nuclear plants produce large quantities of thermal energy[122] that could be usefully employed for controlled heating in industrial processes, to desalinate seawater, or as an energy alternative to electrolytic hydrogen production. In the future, some high-temperature advanced nuclear technology designs may offer significantly improved efficiencies in both power generation and hydrogen production. Research and development into autonomous, long-lived small nuclear power plants that do not require refueling could provide safe and reliable electric power to remote locations.

From a deployment perspective, nuclear fission technology benefits from a well-functioning commercial industry that is global in reach. The expansion of nuclear energy, however, appears to be contingent upon satisfactory resolution of several factors. Public opposition to nuclear energy became strongly rooted in many American and European cultures in the 1970s after the 1979 partial meltdown of a reactor core at Three Mile Island Unit 2 (the worst accident at a US commercial nuclear plant) and the 1986 core destruction and resulting contamination at Chernobyl Unit 4 in Ukraine (the worst accident in the history of nuclear power).[123] The former resulted in the creation of the Institute of Nuclear Power Operations to promote excellence in utility operations of nuclear power plants; the latter resulted in the formation of the World Association of Nuclear Operators to maximize the safety and reliability of nuclear power plants worldwide.

On March 11, 2011, the fourth-largest earthquake in recorded history hit the east coast of northern Japan. Fifteen minutes later, a 14-m (46-ft) tsunami wave struck the Fukushima Daiichi nuclear power plant. The wave destroyed backup power systems, pumping equipment, and the electrical and venting systems for four of the plant's six units. Reactors 1, 2, and 3 appear to have experienced partial core meltdowns. Hydrogen explosions destroyed the upper cladding of the buildings housing reactors 1, 3, and 4 and damaged the containment of reactor 2. Spent fuel rods began to overheat as water levels in their storage pools dropped, resulting in hydrogen explosions, failures of fuel rod cladding, and releases of radioactivity to the environment. The event was given the maximum rating of 7 on the International Nuclear and Radiological Event Scale, the same as Chernobyl, although Japanese authorities estimate that radiation released at Fukushima is only 10 percent of the amount released from the Ukrainian plant and although the public health consequences of the Fukushima event are much less severe. As this book went to press, clean-up and remediation were beginning and were expected to take many years, and operators of nuclear reactors all over the world were considering how to maintain the safety of their plants with additional safeguards.[124]

Since the late 1980s, the image of Nuclear Safety Inspector Homer Simpson and the dangerously ignorant operators of the Springfield Nuclear Power Plant under the ownership of the greedy Mr. Burns on the television program *The Simpsons* has given comic expression to some real concerns the North American public has with nuclear energy. As a result of public sentiment, new reactor construction has been barred in 13 US states, although several of these are reconsidering their bans. In sum, it is worth noting the 2009 MIT report's conclusion: "The sober warning is that if more is not done, nuclear power will diminish as a practical and timely option for deployment at a scale that would constitute a material contribution to climate change risk mitigation."[125]

3.3 Capturing and Sequestering Carbon

Coal is Earth's most abundant fossil fuel, and it is distributed more evenly across all major regions than uranium, natural gas, and oil. Because of its relative abundance and low delivered price, coal is likely to be a mainstay of both developing and industrialized economies well into the future. Its high carbon content, however, imposes considerable technical challenges: the International Energy Agency estimates that carbon releases from the use of coal in the next 25 years will nearly double that of the previous 25 years, releasing a staggering amount of emissions into the atmosphere.[126] As a result, technologies are being developed to capture and store the carbon being released from the use of coal.

Standard power plants around the world burn coal in a boiler at atmospheric pressure. The heat generated by the coal combustion transforms water into high-pressure steam, which turns a steam turbine, whose mechanical energy is converted to electricity by a generator. In this "workhorse of the coal industry," a mixture of exhaust (or flue) gases exits a tall stack after having its sulfur removed. Only about 15 percent of the flue gas is carbon dioxide; most of the remainder is nitrogen and water vapor (the result of burning coal in air, which is almost 80 percent NO_2). Separating CO_2 from such a dilute gas stream is difficult. Storing or sequestering the CO_2

once it has been captured is also a challenge, since the storage system must be able to prevent the CO_2 from escaping and re-assimilating into the atmosphere—hence the emergence of the combined terminology "carbon capture and sequestration" (CCS).

Globally, a transformation of fossil-fuel-based combustion systems into low-carbon energy processes would enable the continued use of plentiful coal and other fossil energy resources. Such a transformation requires further development and application of technologies to capture CO_2 and store it for the long term using safe and acceptable means that prevent its emission into the atmosphere. Deployment of technologies to capture, store, or sequester GHGs is intricately linked with the goal of reducing carbon emissions. This is especially true in the near to medium term as we transition from emitting to non-emitting energy-supply technologies and fuels. Carbon capture and geologic storage technologies are expected to work together to reduce GHG emissions from large concentrated sources. The Intergovernmental Panel on Climate Change's special report on CCS postulates that CCS applied to conventional modern power plants could reduce CO_2 emissions to the atmosphere by 80–90 percent.[127] By turning coal into a combustible gas that could be cleansed of virtually all of its pollutant-forming impurities and burned in a gas turbine, coal could rival natural gas in terms of environmental performance. The IPCC report further concludes that CCS has the technical potential to account for a majority of the carbon-mitigation effort worldwide in the next 100 years.[128]

Though there is limited experience with the full-scale operation of CCS—especially in the power sector—four large-scale CCS projects are currently in operation around the world, each separating about a million tons of CO_2 per year[129]: Sleipner (Norway), Snøhvit (Norway), In Salah (Algeria), and the Great Plains Synfuels Plant (North Dakota). The first three of these projects involves the separation of CO_2 from produced natural gas, which often has higher CO_2 content than is allowed to enter natural-gas distribution networks. The Great Plains Synfuels Plant captures CO_2 from the manufacture of synthetic natural gas from coal. At approximately a million tons of CO_2 per year, each of these projects is capturing the equivalent of the annual output of a 150-megawatt coal-fired power plant.[130]

3.3.1 Carbon Capture

It is easiest for CO_2 to be sequestered from power plants and other industries that have major CO_2 emissions—if it can first be captured as a relatively pure gas. CO_2 is routinely captured as a by-product of ammonia and hydrogen production and from limestone calcination. In the future, capture and storage systems for power plants in different regions of the country with different coal ranks and geologic storage opportunities will be needed. In September 2009, the first US project that both captures and sequesters carbon dioxide emissions from a coal-fired power plant began operations at American Electric Power's Mountaineer Power Station in West Virginia.[131] This project scales up Alstom Power's post combustion chilled ammonia capture process to a 20-MW pre-commercial scale, with storage of the CO_2 to involve injection into two on-site wells. A second project that is also being conducted in collaboration with the Electric Power Research Institute will deploy an advanced amine CO_2 capture technology at the Southern Company's Plant Barry in Alabama. It will evaluate the integration of CCS at about 25 megawatts. The carbon storage portion of the project is under way with support of the US Department of Energy. These two projects are part of a power industry collaborative program of research on critical-path CO_2-reducing technologies.[132]

There are three basic processes for capturing CO_2: pre-combustion capture, post-combustion capture, and oxyfuel combustion.[133]

Pre-combustion capture involves processing the fossil fuel to separate CO_2 and hydrogen, such as in gasification reactions. The CO_2 is then removed before combustion occurs. Pre-combustion capture is already commercial on a limited basis and is widely applied in fertilizer and hydrogen production. In these industries, the fossil fuel is partially oxidized, often in a gasifier, and the resulting syngas (composed of CO and H_2) is converted to CO_2 and H_2 through the addition of steam to produce the "water gas shift reaction." The CO_2 is then removed before combustion occurs using a carbon-free fuel (hydrogen). In the power sector, about 28 IGCC plants are in operation worldwide; seven of these use coal as the feedstock; the others use petroleum coke.[134]

A coal gasification facility in North Dakota that was built to produce synfuels more than 30 years ago is using pre-combustion carbon capture. Specifically, the Great Plains Coal Gasification Plant captures more than 200 million standard cubic feet of carbon dioxide per day in a 96 percent pure stream. Part of this stream of carbon dioxide is sent via a 320-kilometer (199-mile) pipeline to Weyburn, Canada, and sold for use in an international research project focusing on CO_2 storage and enhanced oil recovery.[135] An assessment of the first phase of this enhanced oil recovery (EOR) project suggests that the CO_2 is likely to be stored securely for thousands of years.[136]

Post-combustion capture is a separation process that extracts CO_2 from the flue gases produced in the conventional air-fired combustion process—after combustion of coal or natural gas. This is the typical approach that has been used for several decades in other industrial applications (although at somewhat smaller scales) and would be used in conventional power plants. The most common technology employed today is the use of an amine-based chemical absorbent. Specifically, the power plant's smokestack is replaced by an absorption tower, in which flue gases come in contact with droplets of amines that selectively absorb CO_2. In a second reaction column, known as a "stripper tower," the amine liquid is heated to release concentrated CO_2 and to regenerate the chemical absorber. This amine-based chemical absorbent process is currently used in the natural-gas production facilities at Sleipner in Norway and In Salah in Algeria to remove CO_2 impurities from the natural-gas production stream and then inject the CO_2 into geologic formations. The cost of this type of CO_2 capture is high because flue gases are only 10–12 percent CO_2 by volume for coal power plants, and even less (3–6 percent) for gas-fired power plants.[137] Project developers are beginning to promote new conventional coal plants as "capture ready," but this simply means that space is being made available at plant sites to accommodate future stripper towers to capture relatively pure streams of CO_2.

Oxyfuel combustion utilizes pure oxygen instead of air in the combustion process. The use of oxygen creates a flue gas of mainly CO_2 (typically over 90 percent concentration), which is much greater than with traditional air-fired combustion and facilitates post-combustion capture of the CO_2. Oxyfuel combustion with CO_2 capture and sequestration has not yet been tested in a large-scale facility; however, the US Department of Energy does plan to capture and sequester CO_2 at the repowered plant in Meredosia, Illinois, perhaps with a regional CO_2 repository.[138] Considerable research on combusting coal in oxygen and capturing CO_2 from the flue gas is also underway; the energy company Alstom is working on this in Germany with the Swedish utility company Vattenfall. Rather than generating electricity, the steam is being piped to a nearby industrial estate. After cleaning the flue gas of small particles and sulfur dioxide, the almost pure CO_2 gas stream is compressed, stored, and ultimately will be shipped by tanker to a geologic storage site.[139]

Carbon capture holds considerable promise as part of a series of technologies that will allow countries with large coal reserves (such as the United States, Russia, China, India, and Australia) to continue to take full advantage of their fossil-fuel resources in pursuit of continued economic development and prosperity. However, carbon capture technologies are only just beginning to demonstrate this potential. The public and private sectors are expending a great deal of effort, often in public-private partnerships to research, develop, and demonstrate carbon capture technologies in conjunction with geologic storage.

The lack of experience with CCS in the power sector makes it difficult to estimate its incremental costs. Recent estimates of CO_2 capture costs range from \$100 (€77) to \$150 (€116) per ton of CO_2 for first-of-a-kind coal plants and \$30 (€23) to \$50 (€39) per ton of CO_2 for ninth-of-a-kind coal plants. These cost premiums are with reference to a conventional supercritical pulverized coal plant.[140] They exclude the additional costs of carbon transport and storage, but also do not account for possible revenue generated by enhanced oil recovery or methane. They are higher than the range of future CO_2 prices estimated for the next few decades and for many alternative carbon-mitigation options; they are also higher than many previous estimates, perhaps reflecting the inflation in capital costs experienced in 2007–08. Estimates today would probably be lower as a result of the global economic downturn that began in late 2008.

Still, there is general agreement that carbon capture carries a significant energy penalty. Power plants' energy efficiencies decline significantly because of the energy requirements needed for the capture technologies: from 34.3 percent to 25.1 percent for subcritical pulverized coal plants, from 39.3 percent to 29.9 percent for super-critical pulverized coal plants, and from 43.3 percent to 34.1 percent for ultra critical pulverized plants.[141] As a result, environmental problems associated with coal mining and the added consumption of limestone (for flue gas desulfurization systems for SO_2 control) and ammonia (for selective catalytic reduction systems for NO_x control) all increase proportionately.

The possible industrial uses of captured CO_2 in various energy sectors and in the food and beverage industry are likely to absorb only a small portion of the amounts currently emitted. Therefore, the deployment of carbon capture requires the concurrent development of storage technologies. In addition, successful CCS will necessitate considerable investment in infrastructure to transport the captured gases from their source to their storage site. Typically, CO_2 has been transported by pipeline, which is generally the cheapest form of transportation.[142] Only 5,800 kilometers (3,600 miles) of CO_2 pipeline operate in the US today, less than 1 percent of the natural-gas and hazardous liquid transmission pipelines that crisscross the country.[143] Although CO_2 pipeline networks already operate for EOR, "developing a more expansive national CO_2 pipeline network for CCS could pose numerous new regulatory and economic challenges," according to the Congressional Research Service.[144]

3.3.2 Geologic Storage

Scientists around the world are looking for ways to store, rather than emit, CO_2 produced by fossil-fuel combustion and industrial processes. Long-term storage of GHGs in geologic formations is one possible way to avoid emissions, even with continued production of GHGs. Such geologic formations, located deep underground, could store injected CO_2 much as natural gas and oil have been stored naturally for millennia. Moreover, a great deal of experience exists worldwide for dealing with geologic formations like those currently being considered for potential CO_2 storage. The Sleipner project is an example of a major carbon capture and storage effort that has been operating since 2000. This project captures the CO_2 from the natural gas stream produced from a North Sea gas reservoir and injects it into a saline aquifer below the reservoir, avoiding the release of more than a million tons of CO_2 per year. CO_2 could also be injected into deep ocean masses (which could exacerbate ocean acidification through the accelerated production of carbonic acid, H_2CO_3), or it could be solidified in mineral carbonates (which may be the ultimate approach to preventing leakage back into the atmosphere).

In the long term, geologic storage of carbon dioxide in the form of a supercritical fluid could allow continued energy conversion through combustion of coal resources with very low GHG emissions. Saline formations, found under much of the United States and around the world, may offer extensive storage capacity. Additionally, storage can be combined with efforts to improve recovery of other valuable energy commodities, such as oil and natural gas.

Three primary types of geologic opportunities exist for trapping and storing CO_2: oil and gas reservoirs, saline formations, and unmineable coal seams.

The natural-gas and petroleum industries have a long history of injecting CO_2 into depleted or underperforming wells to boost production. For instance, in the United States, CO_2 is injected into oil wells for EOR programs. Gas injection has been used since 1972 and currently accounts for 50 percent of EOR projects in the US.[145] In addition, these industries have vast experience with site identification, transportation of gases, and subsurface gas injection that can benefit the advancement of geological storage of CO_2.

Since CO_2 is soluble in saline water and will dissolve in such a solution upon contact, it can be injected into saline formations, where it becomes suspended. Current research suggests that geochemical trapping mechanisms, in combination with impermeable caprocks, could significantly reduce the risk of CO_2 leakage. (Large leakage of sequestered CO_2 can be deadly, although it probably would be quite localized.)[146] In contrast with storage in oil fields or coal beds, no by-product will provide revenues to offset the storage cost.

Substituting CO_2 for nitrogen for injection into a deep seam coal bed displaces methane from the surface of the coal. This methane can then be captured and used for both energy and industrial processes, while the CO_2 is left stored in the coal seam. Pilot projects have also shown that CO_2 injection can improve the recovery of natural gas from deep unmineable coal seams.[147]

As part of the Regional Carbon Sequestration Partnerships Phase II activities, the US Department of Energy is conducting 25 pilot-scale geologic storage projects, including nine EOR tests, ten saline formation tests, five enhanced coal bed methane tests, and one test of enhanced gas recovery. There are also plans to conduct several large volume tests in North America with injection rates of up to a million tons per year for several years. Scaling these up to near-commercial levels will provide insights into important operational and technical issues within different formations.

Many injection and storage technologies required for storing CO_2 are borrowed from the petroleum industry, which uses CO_2 injection for enhanced oil recovery. Technologies for CO_2 processing, transport, compression, and subsurface reservoir engineering and characterization can also be leveraged from the petroleum industry.[148]

In addition, the US Geologic Survey is developing a method for quantifying the amount of CO_2 that can be stored in underground geologic formations. The *2008*

Carbon Sequestration Atlas estimates the CO_2 storage resource of the United States and Canada. The total CO_2 resource is the sum of the three geologic reservoirs (oil and gas reservoirs, deep saline formations, and unmineable coal seams) noted earlier. It is estimated that the region has between 3,600 billion and 12,900 billion metric tons of storage capacity.[149] At the current rate of US CO_2 production (6.0 billion metric tons of CO_2 emitted from energy-related resources in 2007), even the lower estimate of 3.6 trillion metric tons of storage capacity would meet the storage needs of the US for 600 years.

To limit GHG emissions, geologic storage will work in conjunction with carbon capture technologies, which are just beginning to show their potential. Existing capture technologies are currently very expensive, however, and a considerable effort to lower capture cost through the development and demonstration of advanced technology will be important if CCS is to become a major GHG-mitigation option. In addition, widespread deployment will require sufficient investment in infrastructure to transport the CO_2 from the source to the storage site.

3.3.3 Terrestrial Sequestration

Terrestrial carbon sequestration is the conversion of atmospheric CO_2 to carbon stored in vegetation and soils through photosynthesis and the prevention of net CO_2 emissions from terrestrial ecosystems into the atmosphere (figure 3.11). Enhancing the natural processes that remove CO_2 from the atmosphere is characterized as one of the most cost effective carbon-mitigation strategies.[150] Relative to geologic storage, which is typically envisioned at point sources of CO_2, terrestrial sequestration is less limited to specific locations and is widely available as a carbon-mitigation option. Terrestrial sequestration currently offsets about 13.4 percent of all GHG emissions in the United States.[151] Most of this is terrestrial sequestration from above- and below-ground tree biomass in US forests. Only a small fraction of this sequestration results from activities undertaken specifically to sequester carbon.

Globally, the CO_2 released by deforestation overwhelms the CO_2 stored in forests; forestry accounts for about 17 percent of global GHG emissions.[152] Russia holds 22 percent of the world's forestland, more than any other country. Its forestland provides a reservoir for storing 40 percent of the carbon produced in the Northern Hemisphere. However, it is estimated that Russian forestland is disappearing at a rate of approximately 2 million hectares (4.9 million acres) per year.[153] Brazil has already lost 40 million hectares (the largest forest loss of any country worldwide), and Indonesia 30 million hectares.[154] In Argentina, the burning of forests now generates more greenhouse gases than motor vehicles.[155]

The logging industry, the expansion of agriculture, and natural disasters (fires and floods) are major causes of deforestation. In many countries, poor forestry management harms forest ecosystems. In Russia, for instance, it is estimated that 90

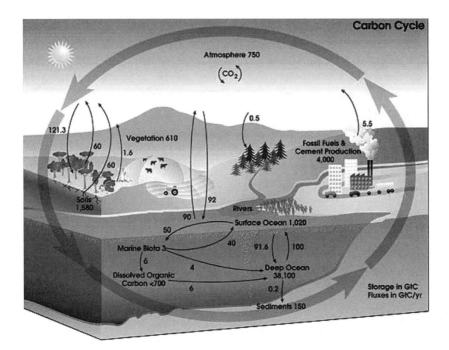

Figure 3.11
The terrestrial carbon cycle.

percent of logging is done by clear cutting, which causes soil erosion and exposure of permafrost to light.[156] As soil erodes and permafrost melts, methane and carbon are emitted into the atmosphere and the soil is depleted of nutrients. The building of roads to reach uncut forests further contributes to soil erosion and permafrost melting.[157]

Terrestrial sequestration has distinct economic and environmental advantages as a GHG-mitigation strategy. It can provide a positive force for improving land management and generating significant additional benefits to society, such as improvements in wildlife and fisheries habitat, enhanced soil productivity, reduction in soil erosion, and improved water quality.[158,159]

Many technologies and practices that sequester carbon have already been widely adopted for other reasons. These include measures to improve soil conservation, reduce soil erosion, and increase crop yields. At the same time, soil carbon data have been used to estimate the carbon sequestration potential of these technologies and management practices. The best practices deployable today to enhance terrestrial sequestration include a range of crop and forest-management technologies and practices.

Cropland management using precision agricultural techniques can increase productivity and reduce the rate at which CO_2 is released into the atmosphere. "No-tillage," nutrient, and water management can mitigate CO_2 release into the atmosphere. Cover crops, buffer strips, and biomass energy crops are also valuable approaches to climate-change mitigation.

Forest management, including afforestation, reforestation, and the prevention of deforestation, mitigates atmospheric CO_2 levels by increasing or maintaining carbon stocks in forests. Appropriate forest management and harvest techniques can maintain higher stand-level forest carbon stocks than traditional practices, and can minimize carbon loss by reducing erosion, collateral tree damage, and slash burning.[160]

In the future, biotechnology may be able to increase vegetation sequestering capacity by modifying the chemical makeup of plants.[161] Other valuable mitigation approaches include active forest management, active wildlife habitat management, low-impact harvesting, precision use of advanced information technologies, genetically improved stock, wood products life-cycle management, and advanced bio-products.

3.4   Reducing Other Greenhouse Gases

Reducing emissions of greenhouse gases other than CO_2 affords significant near-term opportunities for addressing the underlying causes of climate change. Many of these GHGs have global warming potentials far higher than that of CO_2. The families of synthetic chemicals known as hydrofluorocarbons (HFCs) and perfluorocarbons (PFCs) are particularly potent, as is sulfur hexafluoride (SF_6)—the most potent GHG evaluated by the IPCC. Their long atmospheric lifetimes result in their essentially irreversible accumulation in the atmosphere.[162] A diverse array of primary technologies can be deployed today to mitigate emissions of CH_4, nitrous oxide (N_2O), and other harmful synthetic gases.

Just as the global contribution of CO_2 emissions is dominated by developed countries, the global contribution of non-CO_2 GHGs is highest in developed countries (figure 3.12). Though reductions in CH_4 emissions from coal mining and landfills in industrialized countries have resulted from environmental controls, the emissions of developed countries remain much higher than those of developing countries. The dispersed geography of these emissions highlights the need to deploy mitigation measures and strategies worldwide.

3.4.1   Methane from the Energy and Waste Sectors

Methane emissions from the energy and waste sectors accounted for 31 percent of global non-CO_2 GHG emissions, and nearly 50 percent of global methane emissions, in 2000. Major emission sources include coal mining, oil and natural-gas systems,

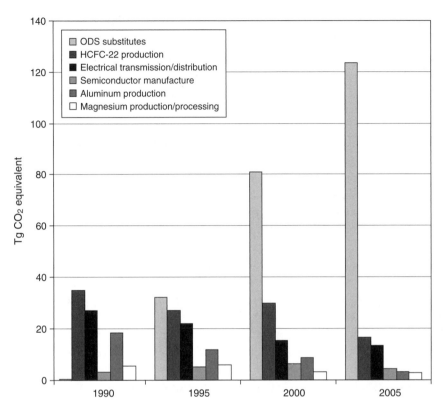

Figure 3.12
High-GWP gas emissions in the United States, by source. Source: *Inventory of US Greenhouse Gas Emissions and Sinks: 1990–2005*, US Environmental Protection Agency Report 430-R-07-002, 2007.

landfills, and wastewater treatment. Among the energy-related and waste-related methane sources, landfills and oil and gas systems are the largest, accounting for 11 and 15 percent, respectively, of global emissions.

Most landfill methane emissions come from developed countries, where sanitary landfills facilitate the anaerobic decomposition of organic waste. In the United States, methane emissions from the energy and waste sectors have declined by about 5 percent since 1990,[163] largely as the result of voluntary deployment of available, cost-effective technologies. Landfill methane emissions, however, are expected to increase in developing countries and in countries with economies in transition as solid waste is increasingly diverted to managed landfills.[164] Three dominant sources of methane exist: emissions from oil and gas systems, landfill emissions, and emissions from coal mining.[165] In many cases, reducing emissions of methane (the primary component in natural gas) can be cost effective because of the market value of the recovered gas.

In Brazil, the Bandeirantes Landfill Gas to Energy Project captures methane that would otherwise be vented into the atmosphere and converts it to electricity. The city of São Paulo produces nearly 15,000 tons of waste per day, and half of it goes to the Bandeirantes landfill—one of the world's largest, with a current capacity of about 30 million tons (equal to 175 football fields filled with 8 meters, or 26 feet, of trash). The Bandeirantes landfill can hold about 20 years' worth of city rubbish, but it was also responsible for emitting 808,450 tons of carbon dioxide equivalent per year. Working with the city, Biogás Energia Ambiental SA built a system of degassers, pipes, heat exchangers, and 24 Caterpillar engines to capture the methane and use it to generate about 20 megawatts of electricity, enough to run the homes of about 400,000 people. From its inception in 2006, the facility has worked with a flare efficiency of 99.997 percent (meaning that it captured almost 100 percent of the methane) and has reduced the metropolitan region's carbon emissions by 11 percent.

In the oil and gas sector, cost-effective methane emission reduction technologies and practices already exist, but there is still opportunity for their broader deployment. There is also opportunity for the increased development and use of leak detection and measurement systems. Advanced methane measurement and detection technologies measure actual emission rates, hand held optical imaging cameras can visualize methane leaks, and remote sensing technologies can detect fugitive methane emissions (such as from gas pipelines).

Long-term reductions in landfill gases will result from innovations such as bioreactor landfills. The first commercial full-scale anaerobic and aerobic bioreactor technology was operational in 2002. The National Energy Technology Lab funded a study of the Yolo County Pilot Bioreactor Landfill Demonstration to look for new ways to capture GHGs from a bioreactor landfill that accelerates the decomposition of organic matter in the waste stream via enhanced microbiological processes. The results showed a tenfold increase in methane recovery and an associated reduction in time required for waste stabilization and composting of the landfill.[166]

For coal mine emissions, advances in coal mine ventilation air methane (VAM) and new coalbed methane drilling techniques could help to reduce emissions. Advanced coal mine VAM technologies include flow reversal reactors, concentrators to increase the methane concentration to levels that will support oxidation, use as combustion air in small-scale reciprocating engines or mine-mouth power plants, and as a co-combustion medium with waste coal. Technologies for recovering coal mine methane include improved mine drainage systems through better directional drilling technologies, in-mine hydraulic fracturing techniques, and development of nitrogen and inert gas injection techniques.

The energy and waste sectors offer some of the most promising and cost-effective near-term opportunities for the reduction of methane emissions. Numerous approaches have already been successfully deployed and many are appropriate to industrialized as well as developing countries. One successful example of methane capture and biomethanization comes from Singapore, where, IUT Global—a company that specializes in waste management, recycling, and landfill remediation—operates the largest organic waste biomethanization plant in all of Asia. Singapore incinerates 6,450 metric tons of waste per day, but a substantial portion (about 1,500 tons) is organic food waste. Operators at the IUT plant collect food waste from restaurants and shopping malls with refuse trucks, which then deliver the waste to a biomethanization facility near Jurong. The plant filters and sorts the waste and then lets it decompose through anaerobic digestion in large tanks. The methane gas that the process creates is used to operate gas turbines that generate 8 megawatts of electricity (enough to meet the plant's energy needs and export a small amount to the Singaporean grid) and 50 tons of compost per day. The compost is sold to local nurseries and horticulturalists. The IUT facility is expected to pay for itself in about 7 years through four income streams: electricity generation, compost sales, tipping fees from trash collection, and the sale of carbon credits. By removing food from the waste cycle, moreover, the plant has extended the lifespan of the Pulau Semakau landfill by 20 years and reduced the need for new trash incineration plants from once every 5–7 years to once every 10–15 years.[167]

3.4.2   Methane and Nitrous Oxide from Agriculture

Globally, agricultural sources of CH_4 and N_2O account for nearly 60 percent of non-CO_2 emissions and 48 percent of US non-CO_2 GHGs, primarily from crop and livestock production.[168] Enteric fermentation is the largest anthropogenic source of methane emissions in the United States, accounting for nearly one-fourth of the total. This source of methane continues to increase with the growth of livestock operations. CH_4 emissions from manure management have also been increasing in the past 10 years, mostly as the result of an increase in the use of liquid systems in swine and dairy cow manure management.[169] Though these emissions cannot be entirely eliminated, they can be reduced.

Promising technologies for reducing N_2O and methane emissions from agriculture include deploying technologies and improving practices that increase overall nitrogen efficiency while maintaining crop yields; increasing the use of anaerobic digestion systems that increase methane collection, which provides additional odor-control and energy benefits (such as producing electricity); and improving livestock production efficiency through breed improvements, diet management, and strategic feed selection.

Two specific best practices suitable for widespread deployment are slow or controlled-release nitrogen products and precision agriculture. Slow or controlled-release nitrogen products contain nitrogen fertilizer in a form that delays its availability for plant uptake and use after application, or extends its availability to the plant significantly longer than rapidly available nitrogen products such as ammonium nitrate or urea (which can degrade to gaseous forms of nitrogen, including nitrous oxide).[170] Precision agriculture provides tools for tailoring production inputs to specific plots within a field, thus potentially reducing input costs, increasing yields, and reducing environmental impacts by better matching inputs to crop needs. Information technologies used in precision agriculture can be classified into three categories: data collection or information input, analysis or processing of the precision information, and recommendations or application of the information.[171] Technologies involving precision imagery, sensing, and control are available to more precisely determine how much fertilizer is needed, minimizing over-fertilization practices that lead to emissions. These technologies can also help farmers apply fertilizers under conditions that would increase nitrogen absorption by plants while decreasing nitrogen transformation.

Advanced technologies and practices offer ways to reduce methane and nitrous oxide emissions from agriculture, and many are already in use today.

3.4.3   Gases with High Global-Warming Potential

Emissions of gases with high global warming potential (GWP) are expected to increase significantly worldwide as a result of the growing demand for refrigeration and air conditioning and the industrialization of developing economies.[172] High-GWP gases, such as HFCs, perfluorocarbons (PFCs), and sulfur hexafluoride (SF_6), are powerful global warming agents, commonly hundreds to thousands of times as potent as CO_2. These synthetic gases represent about 4 percent of global and 14 percent of US non-CO_2 emissions.[173] Emissions of high-GWP gases result from their direct use by a range of industries, each with its own unique technical requirement for performance and safety. Some even occur as unintentional by-products, most notably PFCs and HFC-23 from aluminum and HCFC-22 production, respectively. However, there are numerous options for curbing the growth of high-GWP gases.

Ozone-depleting substances are being phased out as a result of the Montreal Protocol. However, the high-GWP gases used as substitutes for ozone-depleting chemicals (such as chlorofluorocarbons) are a growing emissions source globally. HFCs are a prominent class of chemicals that can be substituted for ozone-depleting substances; though they do not deplete the ozone layer, they are greenhouse gases. Figure 3.13 shows the rapidly increasing emissions from substitutes for ozone-depleting substances. It should be noted, however, that the ozone-depleting substances being replaced are also potent GHGs and that in many applications their replacements are non-GHG substitutes such as hydrocarbons. Despite the rapid growth in emissions from substitution of ozone-depleting substances, HFC-23 emissions from the pro-

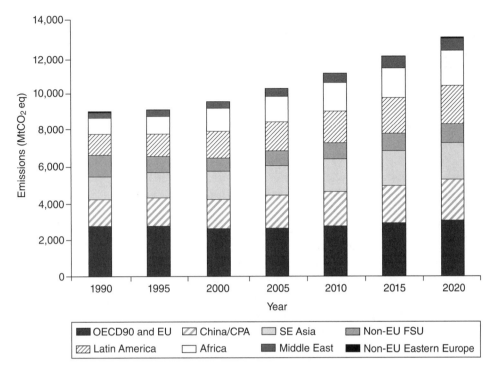

Figure 3.13
Total non-CO_2 emissions by region (million metric tons of CO_2 equivalent). Source: US Climate Change Technology Program, *Strategic Plan*, US Department of Energy Report DOE/PI-005, 2006 (http://www .climatetechnology.gov).

duction of HCFC-22 decreased in the United States by 53 percent from 1990 through 2007.[174]

In the near term, development of safe, high-performing, cost-effective technologies could significantly reduce the emission of high-GWP gases, including the possible elimination of some of them from industrial use. These long-term emission reductions will require improved technologies such as environmentally friendly alternative cover gases to replace SF_6 for magnesium melt protection, improved process controls to reduce PFC emissions from aluminum smelting, the use of molecular fluorine (F_2) to replace SF_6 and PFCs in chemical vapor deposition chamber cleaning and plasma etching processes, and alternatives to the use of SF_6 in high-voltage electric equipment.[175]

Numerous technologies developed in industrialized countries are suitable for immediate deployment. For instance, low-GHG approaches to supermarket refrigeration and motor vehicle air conditioning are available and suitable for more widespread utilization, which could extend to developing countries.

Improvements to supermarket refrigeration include distributed refrigeration (which reduces the need for excessive refrigerant piping, and hence reduces emissions, and secondary-loop refrigeration (which segregates refrigerant-containing equipment to a separate central location while using a benign fluid to transfer heat from food display cases).

Advances in motor-vehicle air conditioning include low-GWP refrigerants, mainly CO_2 (GWP = 1) and HFC-152a (GWP = 120). These have been tested with some success in full-scale vehicles and are being used to replace CFC-12, an ozone-depleting substance with a GWP of 8,500.

3.4.4 Nitrous Oxide from Combustion and Industry

Globally, stationary-source and mobile-source combustion and industrial production of acids accounted for about 4 percent of non-CO_2 emissions.[176] Stationary sources include steam boilers and other systems used for power and heat production; the primary mobile sources include transportation systems (e.g., trucks, cars, buses, trains, ships). Combustion of fossil fuels by mobile and stationary sources is the largest non-agricultural contributor to N_2O emissions; NO_x from combustion is chemically transformed in the atmosphere and deposited in the form of nitrogen compounds, resulting in emissions of N_2O similar to those from fertilizer application. In the production of industrial acids, nitric acid is currently the largest contributor to nitrous oxide emissions.

US N_2O emissions from combustion and industrial acid production accounted for nearly 10 percent of total non-CO_2 GHG emissions, with combustion accounting for over 70 percent. US emissions of N_2O associated with adipic acid production have decreased 61 percent since 1990 as a result of the use of control technologies.[177]

Production of nitric acid for fertilizer and for other applications is the largest industrial source of N_2O emissions. Non-selective catalytic reduction (NSCR) is very effective at controlling nitrous oxide, but only 20 percent of nitric acid plants use it today. Additional research is needed to develop new catalysts that reduce N_2O with greater efficiency, and to make NSCR a more appealing alternative to other control options.

3.5 Integrated Systems

Integrative systems combine technologies, concepts, and disciplines, and sometimes engage multiple sectors of the economy in ways that can offer unprecedented opportunities to mitigate GHG emissions and enhance energy security. They involve bundling concepts into more efficiently functioning systems and merging suites of technologies into more holistic approaches in which the technologies can be jointly optimized. As a result of these complexities, they often require new partnerships and alliances, and they can be difficult to implement.

After an extensive process of scientific input that included a review of its current R&D and technology-development portfolio, the US Climate Change Technology Program identified a number of integrated suites of technologies as high-potential approaches to carbon mitigation. A selection of these concepts (such as including comprehensive approaches to sustainable urban design, integrated waste management, and integrated renewable and fossil energy systems) is described in table 3.5. Two of these integrative systems concepts are described below.

Table 3.5

Illustrative integrated systems. Adapted from material presented in M. A. Brown et al., *Results of a Technical Review of the US Climate Change Technology Program's R&D Portfolio*, Oak Ridge National Laboratory Report ORNL-6976, 2006.

• Sustainable communities use systems engineering and comprehensive urban planning designed to reduce vehicle travel and allow co-location of activities with common energy, water, and other resource requirements.

• Hybrid renewable systems integrate different renewable energy technologies, and can combine renewable energy with energy-efficiency measures, or with fossil-fuel or nuclear energy facilities, to create reliable and jointly optimized systems.

• Integration of energy used to support water distribution and treatment, and more efficient use of water to support energy supply. Technologies to minimize energy requirements for water use include grey-water re-use, efficient desalination technology, and integration of water storage and treatment with intermittent renewable energy supplies. Adaptation of nuclear power plants to "dry" closed-loop systems is an example of a technology to minimize the use of water in power production.

• Integration of plug-in hybrid electric vehicles with zero-energy buildings and utility peak-shaving can reduce GHGs from vehicles using low-carbon power such as nuclear or renewable resources. This technology could also help optimize the use of intermittent energy sources such as solar and wind.

• Biomass gasification to produce electricity merged with carbon sequestration and production of clean diesel (dimethyl ether) as an alternative transportation fuel.

• Systems approach to integrated waste management: Reduce the magnitude of landfill waste through tagging and sorting technologies, distribute waste processing for conversion to power and fuels, and use engineered bacteria to process waste without producing methane.

• Underground energy extraction from hydrocarbons (oil and coal) integrated with the production of heat and power that is brought to the surface. Extract energy underground from hydrocarbons (oil and coal) so that heat and power can be brought to the surface. Approaches might include in-situ refining, which will require materials for high-temperature and pressure or underground in-situ gasification.

3.5.1   Sustainable Urban Systems

Tackling climate change at metropolitan and community scales unleashes sizable opportunities to reduce energy consumption and curb carbon dioxide emissions. About 50 percent of the global population lives in urban areas.[178] If they continue to increase by an estimated 67 million people per year, these urban areas are forecast to become home to 5 billion people by 2030.[179] Although they account for less than 1 percent of Earth's surface,[180] urban areas are responsible for roughly 67 percent of the world's energy demand,[181] three-fourths of all carbon emissions from combustion of fossil fuels, cement manufacturing, and wood use,[182] and about 80 percent of the world's GHG emissions.[183] In developing countries, urban carbon footprints tend to be larger on a per capita basis than their national counterparts because most wealth and consumption is concentrated in urban areas.[184]

In the United States, two-thirds of the people reside and three-fourths of the economic activity takes place within the 100 largest metropolitan areas, yet the per capita carbon footprints of these centers are smaller than that of the United States as a whole; metropolitan areas are more energy efficient and more carbon efficient than rural areas and small towns.[185] The proportion of the US population living in urban areas has grown from 40 percent in 1900 to over 80 percent today, and is expected to reach 87 percent by 2030, which suggest the potential for possible transformation to more environmentally sustainable lifestyles.[186]

Urban areas serve as centers of culture, entertainment, innovation, education, knowledge, and political power worldwide.[187] They form the backbone of economies, and they are the traditional home to technological, entrepreneurial, and policy innovations. Access to capital and a highly trained workforce have enabled metropolitan areas to play a pivotal role in increasing business opportunities and partially addressing environmental challenges. With supportive policies, metropolitan areas could provide the low-carbon, climate-smart leadership that is required to meet targets and timetables necessary to avoid dangerous levels of atmospheric GHGs. In 2008, for the first time, the *World Energy Outlook* included a chapter on "Energy Use in Cities," recognizing the premier role cities are playing in energy systems and carbon emissions. Analysts have also begun to quantify the carbon footprints of large samples of metropolitan areas.[188]

Still, the prevailing method of urban and city planning remains one of "projecting and providing" what city dwellers want, rather than constraining energy consumption, GHG emissions, and environmental damage.[189] Developers focus on meeting zoning regulations rather than on promoting sustainable design. Engineers devise roads to speed the flow of automobiles but ignore the negative social aspects of the spaces created.[190] Real estate agents market houses and apartment complexes that have the lowest cost, not necessarily the most energy-efficient architecture or appliances.[191] A combined effort by urban designers, transport planners, energy providers, and environmentalists is needed to promote less GHG-intensive cities and urban areas and to identify best practices, yet seldom is such an effort made.[192]

"Urban form" (that is, the density and design of the cityscape) links the energy consumed in different building designs, densities, and land-use configurations to the energy required to support daily travel, provide freight pickups and deliveries, and support a rapidly growing number of on-the-job service trips. Transit-oriented development with mixed land uses, sidewalks, and bike paths can reduce automobile use while also improving air quality and reducing congestion. Community-scale planning can also promote passive solar design, combined heat and power, and other features that increase energy performance while creating more livable urban spaces.

The city of Bogotá, Colombia, provides a good example of this potential. In the 1990s, Bogotá's mayor, Enrique Penalosa, undertook a massive urban renewal campaign to rebuild the city for people and not cars. In just 3 years, the city had created or renovated hundreds of miles of bike paths and more than 1,000 parks. An expanded bus system now serves 1.4 million passengers per day, cutting the city's air and noise pollution and even its crime levels perhaps because people are out and about spending more time enjoying the city and less time behind the wheel of a car.[193]

Sprawl has been the norm of urban development in many countries in the past 50 years rather than the exception. However, this spreading out of cities has been far from uniform. A single central business district has given way to a variety suburban and peripheral mixed commercial/retail employment activity centers of various sizes, creating polycentric, or poly-nucleated urban areas, as well as ribbon-like urbanized areas up to 100 miles (161 kilometers) from end to end.

One systematic analysis of the potential for compact development to reduce CO_2 emissions in the United States estimated that shifting 60–90 percent of new growth to compact development would reduce vehicle-miles traveled (VMT) by 30 percent and cut transportation CO_2 emissions by 7–10 percent by 2050, relative to a trajectory of continued urban sprawl.[194] An analysis of the 100 largest metropolitan areas in the US estimated that per capita carbon emissions from highway transportation and residential energy use could be cut in half by increasing urban densities from 0.2 person to 6.7 persons per developable acre.[195]

With shared walls and generally smaller areas, households in buildings with five or more units consume only 38 percent as much energy as single-family homes.[196] Compact urban development enables district energy systems to provide space heating and cooling through highly efficient HVAC equipment in combination with thermal distribution networks. Similarly, compact land use allows distributed energy resources to provide highly efficient power production by enabling the use of waste heat for on-site heating, cooling, and dehumidification. Distributed generation, by requiring less infrastructure investment and by reducing transmission line requirements, could be particularly advantageous in newly settled areas. Since the construction of the first district heating and cooling system in Osaka in 1970, Japan has replicated these systems as a means of reducing pollution, conserving energy, and contributing to improved urban planning and renewal. The centralization of facilities in one location saves valuable urban space and contributes to attractive cityscapes. Japan's 87 utilities provide district heating to 148 districts, constitute the third-largest energy utility in aggregate, following electric and gas companies.[197]

Studies have also shown that household VMT varies with residential density and access to public transit.[198] Higher residential and employment densities and mixed land use with a balance between jobs and housing are associated with shorter trips and lower automobile ownership and use.[199] In a comparison of two households that are similar in all respects except residential density, the household in a neighborhood with 1,000 fewer housing units per square mile drives almost 1,200 miles (1,900 kilometers) more and consumes 65 more gallons (246 liters) of fuel per year than its peer household in a higher-density neighborhood.[200] Generally, knowing a metropolitan area's overall density helps predict its carbon emissions.[201] Dense metropolitan areas such as New York, Los Angeles, and San Francisco stand out for having the smallest transportation and residential footprints. Alternatively, low-density metropolitan areas such as Lexington (Kentucky), Nashville, and Oklahoma City are prominent among the ten largest per capita metropolitan emitters in the United States.[202]

Transit-oriented development (TOD) is also an important tool for shrinking carbon footprints by reducing VMT and associated fuel use.[203] TODs are higher-density, mixed-use building developments designed to maximize access to rail and bus transit stations, while encouraging pedestrian activity within such areas, in lieu of vehicle trips to other locations. TOD households own about half as many vehicles as other households, and are almost twice as likely not to own an automobile at all, resulting in higher transit mode shares and reduced VMT.[204]

Though there are skeptics,[205] pursuing planned, compact growth is viewed by most analysts as a low-cost climate-change-mitigation strategy because it involves shifting investments that have to be made anyway. This is particularly true in the rapidly growing cities in Asia. Each day this urban expansion requires the construction of more than 20,000 new dwellings, 155 miles (250 kilometers) of new roads, and distribution services for 1.6 million gallons (6 million liters) of water.[206]

3.5.2   Hybrid Renewable Energy Systems

Different renewable energy technologies can be integrated together (or with energy-efficiency measures and technologies, or with fossil-fuel or nuclear energy facilities) to create very reliable hybrid systems. Most of the experience to date has been with renewable-renewable hybrids that rely on at least two separate renewable energy technologies in unison. For example, installing wind turbines at geothermal power plants creates effective base-load systems as wind data already exist at plant locations to site cooling towers, and plant designs allow for suitable spare land. These plants can rely on geothermal electricity to backup or offset unexpected shortfalls in wind.[207] Similarly, wind farms can be coupled with biomass plants to eliminate their intermittency using agricultural wastes and residues, methane from landfills, energy crops, and trash as sources of fuel.[208]

A far more extensive hybrid system, called the Kombikraftwerk (meaning "combined power plant"), exists in Germany. Operated by Schmack Biogas AG, Solar-World AG, and Enercon, this plant dispatches power from an integrated network of 36 wind, solar, biomass, and hydropower facilities dispersed across Germany. Wind and solar units generate electricity when those resources are available; a collection of biomass and biogas plants and a pumped hydro facility make up the difference when they are not available. The system can immediately adapt to a shortfall in any one resource by drawing on the others. As of early 2009, the 23.2-megawatt Kombikraftwerk consisted of eleven wind turbines at three wind farms, four combined heat and power biogas units, 23 distributed solar systems, and a pumped hydro storage plant linked via central control (figure 3.14). In 2008, the facility produced 41.1 gigawatt-hours of electricity, enough to supply 12,000 households in the town of Schwäbisch Hall without a single interruption in supply. This project shows, quite clearly, that a combination of different renewable energy technologies could meet the entire electricity demand of Germany. The project size was chosen to represent the German electricity demand on a scale from 1 to 10,000. The Kombikraftwerk also lowered the region's dependence on oil and natural gas, and produced no GHG emissions.

A similar hybrid system exists in the Saxony-Anhalt district of Germany near the Harz district, where 6 megawatts of wind are connected to an 80-megawatt pumped hydro facility used to back up wind output by pumping water up when the wind is available, and then using gravity to power two 40-megawatt turbines to balance the system when the wind is not. The wind-hydro system is in the process of being integrated with distributed solar power plants in the village of Dardesheim, six biogas systems, and a large 5-megawatt cogeneration unit fuelled by recycled vegetable oil. The resulting wind-hydro-solar-biogas-vegetable oil facility, integrated via a digital control station, is expected to provide about 500 million kilowatt-hours of electricity to a region that consumes only 800 million kilowatt-hours, meaning it will meet two-thirds of the region's electricity demand.[209]

At Oberlin College (near Cleveland, Ohio), the Adam Joseph Lewis Center for Environmental Studies uses a 60-kilowatt photovoltaic array with both active and

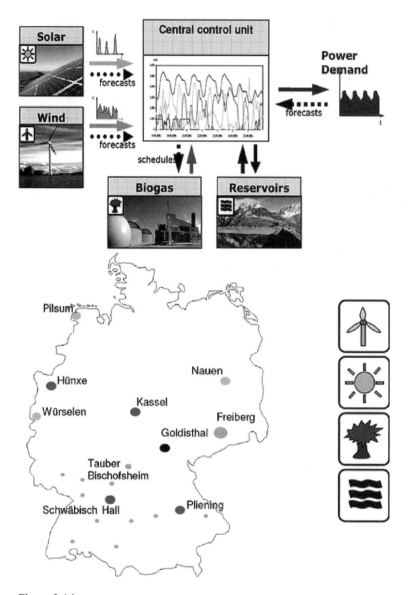

Figure 3.14
Germany's integrated combined-cycle power stations. Source: M. Mendonça et al., *Powering the Green Economy* (Earthscan, 2009).

passive systems to heat, cool, and ventilate. The building uses renewable energy flows and advances in architectural design and energy efficiency to supply local electricity, minimize energy waste, produce food, and restore native vegetation. An ecologically engineered system called "The Living Machine" recycles wastewater directly from the building's toilets and sanitation system, relying on plants, snails, insects, and anaerobic digestion to "treat" waste naturally in a garden-like atmosphere. In days when waste is running low, building operators even pay students to use the toilets through the "Poop Campaign" in order to increase the amount of organic material entering the system.[210]

In Zambia, an interconnected solar-biomass-micro-hydro network will generate base-load electricity for a collection of local villages. The combined system will include one biomass power plant, one micro-hydroelectric station, and a collection of distributed solar panels with a combined output of 2.4 megawatts; and it is expected to begin operation in 2010.[211] In Cuba, a hybrid biomass gasification power plant, four distributed biogas plants, and a wind farm will have a rated capacity of 11 megawatts and will begin generating base-load electricity for the Isla de la Juventud in 2011.[212] In the village of Xcalak, Mexico, 234 solar panels have been integrated with 36 batteries, six wind turbines, a 40-kilowatt inverter to convert DC power to AC, and a sophisticated control system. The system has so far displaced the need to construct a $3.2 million transmission line extension, and in its first year of operation it proved more reliable than the diesel generators that it replaced (although one is still used as a backup).[213]

The development of individual and hybrid renewable electricity technologies has helped increase their potential for commercialization by lowering costs, increasing efficiencies, and improving performance. They suggest that, when integrated in clever ways, the whole can be more than the sum of the parts.

3.6 Synthesis

In view of the magnitude of climate-change and energy-security challenges, it will be necessary to exploit all of the five types of technologies described in this chapter; energy-efficient systems, low-GHG energy production, carbon capture and storage, mitigation of other greenhouse gases, and integrated systems. Numerous "gigaton" solutions are required, each of which will involve transforming and modernizing global energy systems in fundamental ways.[215] For example, one gigaton of carbon emissions reductions could hypothetically be delivered by 1,000 zero-emission 500-megawatt coal-fired power plants with carbon capture and storage. Another gigaton of reductions would result from 50 times the current global capacity of wind or planting energy crops on land areas 15 times the size of Iowa.[216] These solutions in many cases represent more than just the next generation of technology. They will require paradigm shifts in how we generate and use energy and land today as well as acceptance of entirely new, transformational concepts. Optimizing suites of technologies into complex integrated systems is one of these important transformational approaches. As we will see in the forthcoming chapters, however, mitigation of greenhouse gases is not the only option. For example, we can also rely on geo-engineering and adaptation technologies, as discussed in the next chapter.

4

Technologies for Geo-Engineering and Adaptation

Late in the sixteenth century, the Mogul Emperor Akbar the Great decided to build a capital for his empire on the dry sands of what is now Northern India. The Emperor expended vast sums of his resources, hired the finest artisans and crafts persons, and constructed a city of imposing squares and awe-inspiring architecture. In building his capital on such a grand scale, however, Akbar did not anticipate how it would interact with the area's surrounding natural environment. Just 15 years after it was completed, Akbar's capital completely exhausted its water supply and its rulers summarily abandoned it. The residents were forced to disperse into a collection of smaller, more sustainable communities. Practically untouched by the centuries, Fatehpur Sikri still stands today—a testament to the importance of *adapting* to one's environment.

It is common in discussions about energy and climate policy today to commit an error similar to Akbar the Great. It is all too easy to get so lost among the vast array of mitigation options available to policy makers that one forgets to consider the possibility of adapting to climate change. Humans are by nature an adaptable species, yet a number of questions emerge when adaptation is put into the context of climate change. What actions are we likely to take to adapt to climate change? What regulatory and legal challenges will we face as a global society adapting to climate change? How can developed countries help developing countries to tackle their vulnerabilities and increase their adaptive capacity? What actions can be taken to protect other species from the impacts of climate change?

Most of the policy dialog surrounding global climate change has focused on reducing emissions of greenhouse gases as a means of reducing the magnitude of climate change—something often termed "mitigation." However, if the goal is to reduce the harmful consequences of GHG-induced climate change, then there is an array of options extending beyond mitigation, as illustrated in figure 4.1. In addition to mitigation through conservation, energy efficiency, and low-carbon energy, there are two possibilities:

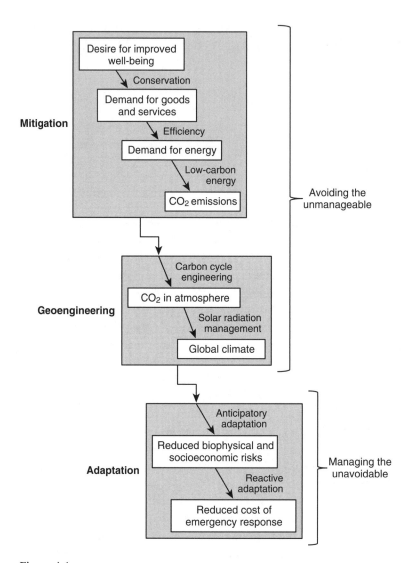

Figure 4.1
Schematic diagram of mitigation, geo-engineering, and adaptation approaches. Adapted from Marilyn A. Brown, "The multiple dimensions of carbon management: Mitigation, adaptation, and geo-engineering," *Carbon Management* 1, 2010, no. 1: 27–33.

• Deliberate actions can be taken to manipulate Earth's environment and thereby offset the negative consequences of climate change. These are called "geo-engineering" approaches, and they include interventions to remove CO_2 from Earth's atmosphere as well as actions to increase Earth's reflection of sunlight. Geo-engineering generally refers to large-scale and rapid manipulation of the Earth's environment intended as a stopgap, beneficial intervention that will allow humans to prevent the most severe impacts of climate change.[1]

• Deliberate actions can also be taken to reduce the vulnerability of humans and ecosystems to the effects of global climate change. These are called "adaptation" approaches, and they can be anticipatory or reactive. Adaptation refers to "changes made to better respond to present or future climatic and other environmental conditions, thereby reducing harm or taking advantage of opportunity."[2] It is sometimes called strengthening "adaptive capacity" or "resilience."

This chapter rounds out our discussion of mitigation in chapter 3 by characterizing the potential roles that geo-engineering and adaptation could play in responding to climate and energy-related challenges. Table 4.1 compares and contrasts the mitigation and adaptation approaches, and also considers how geo-engineering approaches fit into the mix.[3] All three approaches aim to decrease overall biophysical and socioeconomic risks to global climate change. However, mitigation and geo-engineering involve "avoiding the unmanageable" (that is, reducing the pace and magnitude of the changes in global climate being caused by human activities), whereas adaptation involves "managing the unavoidable" (that is, reducing the adverse impacts on human well-being resulting from the changes in climate that do occur).[4]

4.1 Geo-Engineering

Though reducing GHG emissions is the logical first step in the fight against climate change, current mitigation efforts are not sufficient. The effects of climate change are already being felt around the world.[5] Because GHGs last so long in the atmosphere, even immediate reductions in emissions will not mitigate these effects quickly. According to David Keith, one-third to one-half of atmospheric CO_2 will still be present 10,000 years after CO_2 emissions are stopped.[6]

Stabilizing GHGs at levels that will prevent dangerous impacts is possible, but it looks very unlikely before the end of the century. As a result, attention is increasingly focused on what can be done to alter the climate in order to minimize the effects of atmospheric greenhouse gases. The umbrella term for this field of investigation is "geo-engineering."

Table 4.1
A comparative analysis of adaptation, mitigation, and geo-engineering.

	Mitigation	Adaptation	Geo-engineering
Timing	Costs now, benefits delayed	Costs whenever, benefits may be relatively soon after	Costs now, benefits very soon
Temporal incidence	Costs now, benefits to later generations	Benefits mostly to the generation bearing costs	Benefits mostly to the generation bearing costs
Geographic incidence	Local costs, global benefits	Local costs, often relatively local benefits	Local or broader costs, global benefits
Sectoral incidence	Focus on emissions from energy consumption	Very heterogeneous	Only a few options are likely to garner political support
Relation to uncertainty	Must act early despite greater uncertainty	May act later after reducing uncertainty	May act later after reducing uncertainty
Governance issues	Dominated by national goals and international negotiations	Dominated by state and local agencies, but need for coordination is great	International oversight needed because of possible actions of rogue nations and individuals acting on their own

Most scientists concur that geo-engineering should be used only as an emergency response to a climate crisis. However, there is no consensus as to what constitutes such a crisis. Indeed, some have argued that we may already be experiencing dangerous levels of climate change. Others, however, require a more stringent base of evidence before agreeing that geo-engineering should be "triggered." Various high-consequence climate-change impacts might constitute a triggering event, such as the massive melting of summer sea ice in the Arctic Ocean, large-scale droughts and famines, reversal of the thermohaline circulation system impacting deep water and surface ocean currents, or the rapid release of gas hydrates (a particularly gripping scenario because of its positive feedback—more methane in the atmosphere will accelerate global climate change). Disasters like these could constitute the type of situation that would galvanize support for implementing emergency reactions. Because these triggering events tend to be precipitous and irreversible, there is a growing sense that anticipatory rules are needed. Dealing with a climate crisis cannot depend on adaptive evolution. By definition, geo-engineering is one response with the right time constant, but to put it to use quickly requires considerable preparation. Having geo-engineering options available for use is equivalent to having an emergency response plan that could be implemented in the event of a crisis.

Scientists tend to be conservative. Climate scientists deal with 95 percent estimates, and rarely focus on extremes. For example, the Intergovernmental Panel on Climate Change considers the possibility of a 5-centimeter rise in sea level by the end of the century, when a much greater rise is known to be possible with glacial and ice shelf melting.[7] Consistent with this conservatism, the IPCC has not examined the potential role of geo-engineering in its main climate-change documents to date. The economist Marty Weitzman has shown how short-sighted this can be by discussing the "fat tail" of the probability distribution of climate damage.[8] Though turning points and tipping events may be seen as unlikely, they deserve more attention simply because they could be catastrophic. To underscore this point, it is illuminating to explore what a world with 1,000 parts per million of atmospheric CO_2 might look like. According to Steven Schneider (a professor of interdisciplinary environmental studies and biology at Stanford University), such a world would likely face a 10-meter rise in sea levels from the melting of the Greenland and West Antarctic ice sheets, hundreds of millions of refugees from Asian cities at risk to intensifying tropical cyclones, conversion of forests from CO_2 sinks to CO_2 sources as tropical rainforests become vulnerable to wildfire, and the need for global cooperation to sustain geo-engineering for centuries without interruption from wars and economic stresses including unprecedented liability claims.[9] Other studies have noted that a temperature change of greater than 2°C could interfere with the monsoonal rain patterns in India and other parts of Asia, disrupting the farming of rice, or could change evaporation rates in the Amazon, resulting in the collapse of entire ecosystems and releasing large amounts of carbon.[10]

Although there may be a limited number of scenarios in which geo-engineering is seen as acceptable, there is a growing consensus that geo-engineering research should be initiated immediately. There can be great value to knowing earlier, rather than later, what works and what doesn't. In the past 10 years, a few scientists have been patiently developing geo-engineering options and considering the global policy issues surrounding their use. Paul Crutzen ignited a broader discussion of geo-engineering by arguing that the lack of progress on mitigation justifies sustained research on climate engineering.[11] Research and development on geo-engineering options is also advocated by some scientists because of the likelihood that they might be cheaper than mitigation. Granger Morgan estimates the cost of mitigation at a few percent of gross domestic product, whereas geo-engineering approaches are likely to cost only a fraction of a percent of GDP.[12]

There are two types of approaches to geo-engineering. One type—solar radiation management—controls how much solar energy reaches the planet's surface. It

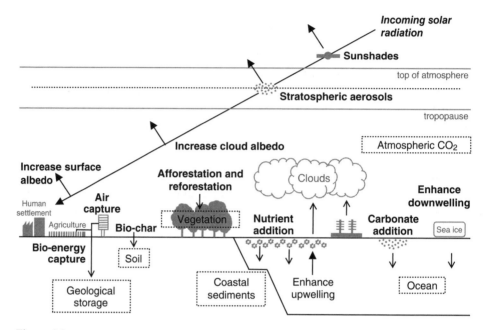

Figure 4.2
An array of geo-engineering options. Heavy arrowheads indicate short-wave radiation; light arrowheads indicate enhancement of natural flows of carbon; gray downward arrows indicate engineered flow of carbon; gray upward arrow indicates engineered flow of water; dotted vertical arrows illustrate sources of cloud condensation nuclei; dashed boxes indicated carbon stores. Source: T. N. Lenton and N. E. Vaughan, "The radiative forcing potential of different climate geoengineering options," *Atmospheric Chemistry and Physics* 9, 2009: 2559–2608 (http://www.atmos-chem-phys-discuss.net/9/2559/2009/acpd-9-2559-2009.pdf).

manipulates the planet's radiation budget to ameliorate the main effects of GHGs (i.e., warming). The other—carbon-cycle management—controls how much heat escapes back into space, which depends on how much CO_2 is in the atmosphere.[13] Put another way, the first approach deals with short-wave solar radiation and the second with long-wave radiation (figure 4.2).

4.1.1   Solar Radiation Management

Controlling the magnitude of solar radiation reaching the Earth can be achieved by many different technology systems; several of these act like sun shades, including stratospheric scatterers such as sulfate particles injected into space, and solar shielding,[14] as well as space-based reflecting mirrors. Solar radiation management can be implemented within the lower atmosphere by increasing the albedo of cirrus clouds. On the ground, there are various ways that the albedo of the Earth's surface can be increased, such as planting more reflective vegetation (for instance, by replacing dark forests with more reflective grasslands) or building more reflective cities (for instance, with whiter roofs and whiter roads). Though it might be possible to increase the albedo of oceans and ice-covered areas, access to these regions are generally more limited because of the harsh conditions and vast distances involved.

By increasing the Earth's albedo, solar radiation management offers a promising method for rapidly cooling the planet. In addition to being relatively fast, it is relatively cheap; however, it is also risky and carries uncertain consequences. Solar radiation management also operates by throwing away the solar energy that could be harvested for renewable power generation. The scattering of light may be good for photosynthesis, but it is not good for solar power production.[15] Also, having less sunlight reach the Earth's surface presents the risk that there will be less evaporation, which may make rain and fresh water scarcer than they are today. Perhaps the most concerning potential unintended impact is the sudden termination of albedo manipulation, which could have worse consequences than the global climate impacts that were avoided because of its abruptness. This approach also does little to manage atmospheric CO_2 and therefore does little to address other climate-change risks, such as increasing ocean acidification.

We do know that oceans are more acidic today than they were in pre-industrial times—perhaps by as much as 30 percent,[16] depending on depth. In addition, oceans are experiencing decreased productivity (e.g., an annual decline of 1 percent of ocean phytoplankton),[17] as well as shifting species distribution and a greater incidence of disease.[18] Coral reefs and shellfish are dying[19]; higher ocean acidity makes it more difficult for crustaceans (including crabs, shrimp and barnacles) and echinoderms (including starfish and sea urchins) to build and maintain calcium carbonate in their shells and skeletons.[20]

Notwithstanding these risks, three geo-engineering options involving solar radiation management have received considerable attention: injecting sulfate particles into the atmosphere, brightening clouds, and building white roads and roofs.

Injecting sulfates into the stratosphere has relatively well-characterized and demonstrated effects and is generally seen as one of the cheapest geo-engineering schemes with a high probability of working as planned. It was proposed as long ago as 1974 by the Russian physicist Mikhail I. Budyko, who envisioned sending airplanes burning high-sulfur fuel into the stratosphere; since then, the theorization of anthropogenic sulfate injection has benefited from the "natural experiments" provided by volcanic activity.[21] The eruption of Mount Tambora in present-day Indonesia in 1816 led to "the year without a summer."[22] More recently, the 1991 eruption of Mount Pinatubo in the Philippines hurled 17 megatons of sulfur dioxide into the atmosphere, forming highly reflective sulfate particles with lifetimes of several years.[23] These sulfur particles produced a measurable reduction in ozone and a global temperature decrease.

Imitating this natural phenomenon in an engineered program is believed to be highly scalable, although the process reaches a point of diminishing returns; its effectiveness tapers off as the atmosphere becomes more saturated with sulfate particles. It has been estimated that a 1 percent reduction in solar irradiance is needed to offset current human forcing, and the sulfate-injection method could provide approximately 0.5 percent, thereby counteracting half of the global warming the world is expected to experience in the next century.[24] Crutzen estimates that delivering the requisite amount of sulfur dioxide to the stratosphere with balloons could cost between 25 billion and 50 billion US dollars (i.e., between 19 billion and 39 billion euros) a year.[25]

Research has also shown that diffuse light effects from the Mount Pinatubo eruption slowed the atmospheric uptake of CO_2 as the result of an increase in photosynthesis (light effect) that was greater than the decrease in respiration (temperature effect). Ozone loss as a result of the additional stratospheric sulfate particles may also be significant, possibly delaying the healing of the ozone hole. In a worst-case scenario, an ozone hole could appear in northern latitudes, which would bathe cities in cancer-causing UV radiation.[26] In addition, rain might also become slightly more acidic and although the acid increase might be small, it could create acid rain in previously pristine areas and further contribute to increased ocean acidification.

Cloud reflectivity enhancement from salt spray, also known as cloud brightening or whitening, is a lower-atmosphere approach to increasing the reflectivity of the sun's radiation. As designed to date, it would involve influencing the Twomey Effect by increasing cloud albedo using a fleet of unmanned sailing vessels that would spray microscopic drops of seawater into the air using pumps and giant "eggbeaters," making clouds whiter and more reflective. It targets low marine stratocumulus clouds that cover much of the ocean. This approach is estimated to provide an increase in albedo similar to sulfate-injection, but its impacts would be geographically more uneven because of the naturally patchy distribution of cloud cover [27] Stephen Salter, Graham Sortino, and John Latham estimate that a fleet of 1,900 ships spraying 8 gallons of seawater a second could increase global albedo by about 1.1 percent, which would alone offset the magnitude of global warming anticipated to occur by the end of the century.[28] The total cost of such an approach, according to Bjørn Lomborg of the Copenhagen Consensus Center, is approximately $9 billion (€7 billion).[29] It is also a relatively quick approach that could be turned on or off in a matter of days, thereby constraining potential risk.

Mechanisms that induce albedo changes are well known, and spraying seawater into the air would appear to be benign, but a better understanding of the impact of reflecting more short-wave radiance into space and reducing skylight at night is needed. And the approach could be disrupted by storms, although distributing seawater spray operations across the world's oceans would strengthen the system's overall resiliency to individual storm events.

White roofs and roads are already being used to increase surface albedo and to cool metropolitan areas. Surfaces made of dark-colored asphalt have lower albedo (e.g., 0.05–0.20 percent) and therefore greater heat absorption, which means that they get much hotter than surfaces made of concrete (with an albedo of 0.10–0.35 percent) or highly reflective roofing materials (with an albedo of 0.6–0.7 percent).[30] Energy stored in roads and rooftops can cause the surface temperature of urban structures to become 10–20°C (18–36°F) higher than ambient air temperatures. This stored heat is then released at night, creating a dome of warmer air over the city that increases energy consumption by encouraging greater use of air conditioning.[31]

Recognizing that white roofs eventually turn gray, becoming unattractive and less reflective, two of the US Department of Energy's national laboratories, the Lawrence Berkeley Laboratory and the Oak Ridge National Laboratory, have been developing "cool colored" roofing materials. Infrared reflective pigments and various coatings and paints are now available for roofs and walls, along with concrete and clay tiles, and painted and stone-coated metals. In cooling dominated climates, such materials could be required in building codes for new construction. Under the leadership of Art Rosenfeld, the California Energy Commission has spent $10 million (€7.7 million) for white "re-roofs" and offers credits for cool roofs in meeting new building standards.[32] To enhance the albedo of land surfaces will require a shift in thinking toward albedo-optimized land-use decision making. This can be done at different scales, but it requires consensus agreement to implement as an approach to urban planning and land-use management.

4.1.2 Carbon-Cycle Engineering

Sometimes referred to as "post-emission carbon management,"[33] carbon-cycle engineering comprises a variety of approaches to removing CO_2 from the atmosphere. These include approaches to capturing CO_2 from air that rely primarily on biological processes and organic compounds such as iron fertilization of the ocean and carbon capture via carbon char burial. Some analysts also include reforestation and afforestation in this category, but they are too slow to qualify as emergency responses. Other approaches to capturing CO_2 from the air can operate more quickly, by relying on chemical processes and organic compounds. These include the industrial capture of CO_2 from ambient air, technologies to accelerate chemical weathering via carbonate or silicate rocks on land or ocean, and spreading crushed silicate rocks over vast areas of land.

As a category, these techniques tend to be slower and more expensive than solar radiation management. However, manipulation of the natural carbon cycle to reduce the buildup of CO_2 also tends to be associated with lower risks, and it can reduce ocean acidification. Four promising geo-engineering options that involve carbon-cycle engineering are described below.

Carbon capture via biochar burial involves creating a charcoal-like substance (called biochar) by pyrolyzing biomass in a closed container with little or no available air, rendering it inert.[34] The biochar is then either put in a landfill or added to topsoils, slightly improving soil fertility and increasing crop productivity, which results in an additional carbon-sequestration benefit.[35] The carbon contained in the soil is therefore unavailable for oxidation to CO_2 and subsequent atmospheric release. As a result, the radiative forcing potential of the avoided CO_2 is removed from the planet's energy balance. The magnitude of carbon sequestration is relatively simple to verify for the purpose of national inventories and to enable farmers to trade their sequestered carbon in carbon markets.

Conversion of biomass carbon to biochar leads to sequestration of about 50 percent of the initial carbon—much higher percentage than is retained after burning (about 3 percent) and several times what occurs with biological decomposition (about 10–20 percent after 5–10 years).[36] Thus, the use of biochar would appear to allow the total organic carbon sequestered in soils to be orders of magnitude larger than is naturally possible.[37] This technique can give $0.52W/m^2$ of globally averaged negative forcing, which is sufficient to reverse the warming effect of about one-third of current levels of anthropogenic CO_2 emissions.

Klaus Lackner, a professor at Columbia University's Earth Institute and the School of Engineering and Applied Sciences, and David Keith, a physicist at the University of Calgary, are both in the process of commercializing systems that capture CO_2 directly from ambient air. At least one of the air extraction devices, in which sorbents capture CO_2 molecules from free-flowing air and release those molecules as a pure stream of CO_2 for sequestration, has met a wide range of performance standards with no physical limit to the capture process. Still, all these types of systems would face similar issues with respect to long-term storage of CO_2 as described in chapter 3—they can be costly and risky, and they may be storage limited.

CO_2 capture from ambient air may, in fact, be more expensive than post-combustion capture from a power plant, because of the low concentration of CO_2 in ambient air. In addition, there are various carbon offset, verification, and pricing issues, as well as questions about the sequestration side of the system. How will the CO_2 be sequestered for posterity? Air capture could be eligible to participate in a carbon credit market because the carbon it sequesters would otherwise be released to the atmosphere. Technical success will require a better understanding of the micro-climate impacts of CO_2-depleted air, along with a research agenda focused on the possible carry-over of small particles of sodium hydroxide (NaOH).

Unlike other techniques, such as carbon capture and storage from power plants, air extraction would allow reductions to take place irrespective of where carbon emissions occur, enabling active management of global atmospheric CO_2 levels. The CO_2-capture device can be located at the point of CO_2 emission, end use, or sequestration, eliminating the current need to match CO_2 sources with sinks. For example, CO_2 emissions equivalent to those from vehicles in Bangkok could be removed from the atmosphere by air-capture equipment located in an oil field in Texas.[38]

Ocean iron fertilization involves adding iron to upper layers of the ocean to accelerate the growth of phytoplankton, which in turn create carbon compounds by removing CO_2 from the atmosphere. When the microorganisms die, a small percentage of their biomass sinks to the ocean's depths, where it is effectively removed from the carbon cycle for decades and perhaps centuries. This approach appears to have vast potential; in a natural experiment involving the iron-rich soil erosion of the Crozet Islands, one study measured a doubling or a tripling of plankton production and CO_2 consumption.[39] In contrast, other studies have estimated that fertilizing the ocean with iron might be able to export only about one gigaton of carbon to the deep ocean, providing perhaps one-sixth the effect of either sulfate injection or promoting the formation of clouds.[40] Since the deep ocean takes in approximately 90 gigatons of carbon a year, this approach may be effective for only a few years, until capacity is reached.[41]

Iron fertilization has some positive potential side effects and some negative ones. The depletion of oxygen from large parts of the ocean as the result of creating giant algal blooms could kill other organisms. Certain phytoplankton produce dimethyl sulfides (DMS), first in seawater and then in the atmosphere. DMS is oxidized to form sulfur dioxide, and this leads to the production of sulfate aerosols, which increase cloud formation and cloud albedo. This complex of feedback effects ultimately leads to a greater reflection of sunlight and a further cooling effect.[42] Iron fertilization creates fundamental changes in the lowest links of the food chains that could provoke huge effects throughout the ecosystem. On the other hand, adding iron filings to dead zones in the ocean could help fisheries recover by starting the ocean food chain in new areas.

Increasing the ocean's alkalinity to accelerate weathering involves removing hydrochloric acid electrochemically from seawater, thereby enhancing the ability of the ocean to absorb CO_2 from the atmosphere.[43] This could be accomplished by building water-treatment plants. The process is similar to the natural weathering reactions that occur among silicate rocks but works at a much faster rate. According to the MIT researcher Kurt Zenz House and his colleagues, "the construction of 100 such water-treatment facilities worldwide could reduce 15 percent of global CO_2 emissions."[44]

Mineral carbonation has been studied for decades. The process involves using calcium containing silicate and magnesium to react directly with CO_2 to produce carbonate minerals, essentially trapping carbon dioxide into a stable, benign geological form. There are places where the process can be examined in the field, because it is a natural system; this could involve simply drilling holes and fracturing rocks. However, there are questions about its scalability. Moving huge quantities of mineral carbonates (i.e., rocks) does not always make sense.

Short-wave reflection and long-wave absorption can have competing effects. Daniel Schrag of Harvard University postulates that compensation of long-wave absorption by short-wave reflection can become less effective as the magnitude of intervention increases. Several different relationships between scale of impact and probability of success are illustrated in figure 4.3, which portrays the seven highlighted geo-engineering approaches: four involving solar radiation management and three using carbon-cycle engineering. The downward-tilting ovals suggest increasing risks and deteriorating effectiveness with magnitude of impact. The upward-tilting ovals reflect economies of scale that can be achieved with large-scale application, increasing the likelihood of success.[45]

4.1.3   Tensions and Complexities

White roofs and roads have only limited potential to combat the effects of global climate change, but they are an attractive element of a geo-engineering plan because

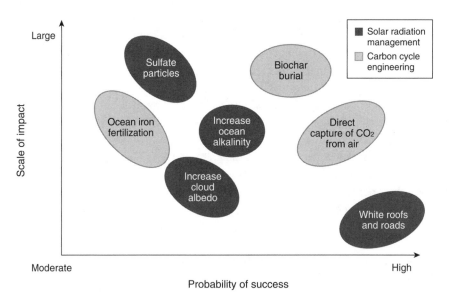

Figure 4.3
Geo-engineering options.

they have a high probability of success and can be implemented quickly. Using sulfate particles and salt spray to enhance the atmosphere's albedo appear to have large-scale potential. They can reverse global warming trends; however, they are fraught with possible unanticipated consequences. The carbon-cycle engineering techniques tend to operate at more modest scales, and even though their effects seem to be more benign, they do have considerable promise. The one exception is ocean iron fertilization, which is a high-risk proposition.

Though the consequences of climate change are worldwide, and geo-engineering aims to provide global solutions, there are still highly variable regional impacts to implementing geo-engineering in practice. As a result, it is important to consider the potential winners and losers. On the one hand, the global effects of climate change can include drought, flooding, extinction of species, displacement of communities, and melting of polar sea ice; on the other hand, it could extend the growing season in northern climates like Siberia and Canada and could also result in opening of the Northwest Passage, providing access to minerals and offering trans-Atlantic transportation benefits in the Arctic.

Unlike global climate-change-mitigation options that require the involvement of a majority of the major emitting countries, geo-engineering can be undertaken unilaterally by single actors: nation-states, corporations, even wealthy individuals. Private companies are already funding experiments on iron fertilization of oceans to prove that it should be worthy of earning carbon offset credits that could be traded in carbon markets (For a discussion of "offsets," see chapter 7.) As David Victor of Stanford Law School states, "it is cheap, easy, and takes only one government with sufficient hubris or desperation to set it [geo-engineering practices] in motion."[46] The United States could choose to save Greenland's ice cap in order to prevent flooding in Florida, or China could choose to preserve its Himalayan glaciers in order to secure supplies of fresh water.

Such unilateral ability to implement geo-engineering systems is disconcerting because of the kinds of problems that could occur if these systems go awry. In *Climate Wars*, Gwynne Dwyer describes such a scenario in which poor countries run an aerosol seeding option and then a war stops the application. The result is cataclysmic. The possible threats include ocean acidification (if solar radiation management is emphasized without compensating strategies to reduce carbonic acid in the ocean), human error (if a system fails, rapid warming and other climatic changes are possible), weaponization (put in the hands of terrorists, geo-engineering systems could be powerful weapons), and reduced solar power (concentrating solar systems may not be able to perform satisfactorily if diffuse light comes to dominate). Such possibilities underscore the need to begin an international dialog on geo-engineering. In addition, research on ethical, social, political, legal, economic, and institutional issues should be initiated quickly, since they are just as likely to limit the use of geo-engineering options as are technological failures.

4.1.4 Geo-Engineering Policy

Geo-engineering creates numerous unprecedented international governance issues. How can we establish a representative global protocol that has the ability to be applied quickly? What burden of proof would be required for global authorization? One of the burdens of proof might have to be that geo-engineering would not become a substitute for mitigation. The moral-hazard problem looms large; to the extent that world leaders deal with geo-engineering, they may be distracted from doing what we know is needed to mitigate carbon emissions.

The UN Security Council has strong authority to set such rules, but it would be a challenge for the Security Council's member countries to come to consensus on the international governance of geo-engineering. Though there is no directly applicable international law, there are, according to University of Maryland Professor John Steinbruner,[47] many suggestive precedents that involved comparatively complex global negotiations. These include the Outer Space Treaty of 1967, the 1974 International Court of Justice Fisheries Case, the Environmental Modification Convention of 1977, and the 1994 Law of the Sea Treaty. Dale Jamieson, Director of Environmental Studies at New York University, recommends the creation of an institutional infrastructure similar to the World Health Organization (which addresses contagious diseases).[48] The shift from intranational to international decision making makes cooperation on geo-engineering cooperation difficult, and the risks and possible repercussions of unilateral action are daunting.

Further, tight administrative control over geo-engineering experimentation and demonstration is needed, because such first steps may otherwise lead inexorably to implementation. The "slippery slope" from the laboratory to large-scale execution concerns many. The slope is slick with enthusiasm for research by the scientific community, donations from beneficiaries to sponsor the R&D, support among political elites, and the potential profitability of vendors. Skewed vulnerabilities complicate each of these factors, as those most vulnerable to climate-change impacts typically have the least resources.

Geo-engineering raises the specter of ecological imperialism: unilateral actions could have worldwide repercussions. Such fears led to a de facto moratorium on ocean fertilization. After the ninth meeting of the UN Convention on Biological Diversity (Bonn, Germany, 2008), the world's governments agreed to such a moratorium (in the Convention on the Prevention of Marine Pollution by Dumping of Wastes and Other Matter 1972, more commonly known as the London Convention), and it was signed into force by 191 countries.

Nevertheless, many in the global climate-change community are becoming proponents of at least initiating a geo-engineering research program. Michael Oppenheimer, Director of the Program in Science, Technology and Environmental Policy at Princeton University, argues that an R&D program is needed for unpredicted emergency climate outcomes, and as a back-stop for political failure. Conditions under which geo-engineering might be justified can be envisioned. Accordingly, some research seems justified.[49]

4.2   Adaptation

Experts generally agree that reducing emissions of greenhouse gases is by far the best way to address climate change. However, the world has not made much progress in reducing such emissions since the signing of the Framework Convention on Climate Change, the international treaty produced at the UN Earth Summit in Rio de Janeiro in 1992. The treaty's aim is to stabilize GHG concentrations in the atmosphere at a level that would prevent dangerous anthropogenic interference with the climate system. Most would agree that this aim has not been achieved, as evidenced by the continuing rise in the atmospheric concentration of GHGs and the mounting scientific evidence of global climate change. Further, there is little reason to think that the world's current emissions trajectory will change dramatically in the near future. Most forecasts suggest that heat-trapping emissions will continue to increase.

No matter how aggressively GHG emissions are reduced, the effects of gases already released will continue to cause some climate change.[50] That is, significant changes in the global climate are, by now, unavoidable. Indeed, there is a meaningful and scientifically plausible chance that the Earth is already committed to catastrophic climate change that even our best efforts at mitigation will not stop. Even with the most progressive mitigation policies, it may not be possible to avoid the impacts of anthropogenic climate change in the decades to come.[51] In that case, adaptation strategies are needed in order to supplement mitigation techniques.

Countries around the world are paying more attention to the need to adapt to changing climates. As Sir Nicholas Stern concluded in the 2006 review of *The Economics of Climate Change,* "adaptation will be crucial in reducing vulnerability to climate change and is the only way to cope with the impacts that are inevitable in the next few decades."[52]

Mitigation, geo-engineering, and adaptation have until recently been treated as distinct responses to climate change. Mitigation has been seen as a priority for developed countries, which have been responsible for most of the GHG emissions to date. In contrast, adaptation is seen as the main concern for developing countries, which are most vulnerable to the effects of climate change.[53] This vulnerability derives from the fact that climate change will affect developing countries disproportionately, and will therefore exacerbate inequities in health status and access to adequate food, clean water, and other resources.[54] The special vulnerability of developing countries is also a function of their resource constraints and their more limited adaptive capacity to reduce the costs of disasters.[55]

Adaptive capacity refers to "the ability of a system to adjust to climate change (including climate variability and extremes), to moderate potential damages, to take advantage of opportunities, or to cope with the consequences."[56] There are individuals and groups within all societies that have insufficient capacity to adapt to climate change. There are also limits and barriers to adaptation, which explains why options to quickly strengthen adaptive capacity and resilience do not always translate into action. As a result, Southern Europe and the United States continue to experience deaths from heat stress because sufficient precautions have not been taken, hurricanes and typhoons cause human loss because construction practices are not robust, and floods continue to take lives because people continue to reside in flood-prone areas.

Mitigation and adaptation are essential components of a comprehensive strategy for responding to climate change.[57] On the one hand, effective adaptation reduces the costs of inadequate mitigation efforts. A resilient, adaptive system is better able to withstand external shocks. As a result, the stabilization level of atmospheric GHG concentrations that avoids "dangerous impacts" is higher in an adaptable world. On the other hand, effective mitigation reduces the need for investments in adaptation, because the magnitude of global climate change is moderated.

Equally important, an integrated approach must include mitigation and adaptation in order to optimize complementarities and avoid conflicts. Mitigation and adaptation strategies can offer synergistic benefits, particularly in the energy sector. For instance, distributed generation (discussed in chapter 3) can deliver electricity and energy services efficiently by allowing waste heat from power generation to be put to productive use in meeting various needs for thermal energy, such as water heating and absorption chilling. Distributed generation also reduces the risk of widespread power loss from severe storms or from peak periods of demand during heat waves. Similarly, advances in the energy efficiency of cooling systems can constrain the growth of GHG emissions and can help to affordably meet the greater need for air conditioning that global warming will bring.

Mitigation and adaptation strategies can also conflict with one another, as was outlined in a recent review of urban planning strategies. In an assessment of approximately 45 adaptation actions spanning biodiversity conservation, sustainable urban form, and alternative energy production, half of the adaptation actions potentially conflicted with mitigation goals. One overarching incompatibility is called the climate-change "density conundrum."[58] On the one hand, the literature on sustainable urban design emphasizes the virtue of high population densities as a means of reducing vehicle-miles traveled and the demand for energy in buildings. On the other hand, adaptation to climate change requires open spaces for storm water management, species migration, urban cooling, urban agriculture, and flood-plain protection.

Finally, an integrated approach could also help bridge the gap between the economic growth and adaptation priorities of developing countries. It could help meet the need to achieve the global engagement required for sufficient emission reductions.[59] In an increasingly interdependent world, the vulnerability of every country is tied to the actions of others. This is true both in terms of climate impacts and security. Conflicts or mass migrations of people resulting from food scarcity, disease, or various ecosystem stresses could threaten the security of countries all over the world.[60]

Societies have a long record of adapting to weather and climate variability through a range of practices. These adaptations have occurred in the past 10,000 years—a period of relative climate stability. They include diversifying crops, installing irrigation systems, and purchasing flood insurance. Planning for adaptation to climate change (rather than variability) poses novel risks with more extreme consequences than have been experienced to date, such as impacts related to drought, heat waves, accelerated glacier retreat, and hurricane intensity.[61] The rapid rate and significant climate change projected for this century will challenge the ability of society to adapt. Failures to adapt have already occurred, as documented in Jared Diamond's book *Collapse,* and more such failures will undoubtedly occur in the future. When changes are rapid, it is particularly difficult and expensive to transform long-lasting infrastructures such as buildings, bridges, roads, ports, and power plants.

4.2.1   Adaptation Practices

Adaptation technologies, systems, and practices are wide-ranging and diverse. Some involve anticipatory actions taken in advance of climate-change events in order to prepare for climate variability or to prevent irreversible damages such as species extinction and coastal land lost to rising seas. In contrast, reactive practices, deployed after climate disasters have stuck, include various types of emergency response, disaster recovery, and migration.

Anticipatory adaptation is an essential part of the optimal response to climate change, as it probably is much less expensive than relying on reactive adaptation only.[62] Though the issue of timing is hotly debated, many argue that the cost of adaptation will be lower if it is planned and implemented in coordination with other public infrastructure investments and as systems evolve naturally.[63] Strategic planning and investment in information infrastructure and in other forms of emergency preparedness can also lower the cost and improve the effectiveness of reactive adaptation. Raising bridges by adding caps to piers (as was done to Alaska's bridge at the Twenty Mile River, for example)[64] reduces the risk of damage from storm surges. Investing in evacuation planning reduces the cost of flooding.

Another way to characterize adaptation practices is according to the climate-related stresses they address. For instance, different actions are needed to adapt to sea-level rise, drought, ocean acidification, permafrost melt, increased intensity of coastal storms, and extreme temperatures.[65] Some of the adaptation initiatives that have been undertaken are listed below by type of stress.[66]

• Sea-level rise—restoring or planting mangroves in coastal embankments to reduce damage from storm surges (Bangladesh),[67] coastal land acquisition programs (United States), building storm surge barriers (Netherlands), building bridges higher to anticipate sea-level rise (Canada), and building seawalls to harden seacoast structures against sea-level rise (Maldives)

• Drought—increased use of traditional rainwater harvesting (Sudan), hydroponic gardening (Philippines), development of drought-resistant strains of wheat (Australia),[68] adjustment of planting dates and crop variety (Mexico and Ethiopia)

• Global warming and extreme temperatures—opening of designated cooling centers (Canada) and heat wave warning systems (United States)[69]

• Permafrost melt—erecting protection dams against avalanches and debris flows stemming from permafrost thawing (Switzerland), railway line construction with insulation and cooling systems to minimize the amount of heat absorbed by the permafrost (Tibet)

In addition to the physical infrastructure oriented adaptation pathways described above, an array of softer adaptation pathways can reduce the cost of global climate change. A taxonomy of three such adaptation pathways is offered here.

• Ecosystem adaptation—Accelerating the adaptation of natural biosystems (e.g., through species transplantation), and establishing and maintaining gene banks to preserve biodiversity

• Strengthening information infrastructures—Improving environmental monitoring, including detection of early warning signals, improving climate prediction and weather forecasting, and improving early warning systems for floods and other natural disasters

• Strengthening institutional infrastructures—Developing agricultural systems and practices better suited to climate variability and change, land-use planning such as limiting development in low-lying coastal, wetland, and flood-plain areas and other actions to limit precarious land uses

As a way of defining and scoping adaptation actions, consider the range of adaptation measures that could be evaluated by government agencies in response to a drought in a traditional agricultural region. Adaptation measures could include the following:

• development of irrigation schemes

• use of organic farming to increase soil density and the ability to maintain moisture through a drought

- change in crops or selective breeding of drought-resilient crops
- a shift to pastoralism and selective migration from the affected region
- enhanced education to retrain for possible alternative occupations
- industrial development to offer residents non-agricultural jobs.

The fact that it is not obvious which of these might be called adaptation underscores the preliminary nature of the discussion surrounding adaptation goals. A narrow "disaster-mitigation" interpretation of adaptation might consider only the first two of these adaptation measures. A more sweeping view of adaptation actions would include the possibility of altering the occupations and education of residents and the possibility of more sustainable industrial development in general.[70]

Adaptation pathways can also be classified by economic sector. For example, Tom Wilbanks and his colleagues at Oak Ridge National Laboratory developed an inventory of 73 pathways that address nine sectors, such as farming, water, health, industry, and energy.[71] These 73 categories blend the infrastructure-oriented approach to climate-proofing with the softer pathways characterized above. Similarly, the Dutch have developed an inventory of 96 adaptation options and have scored them according to effectiveness in avoiding damages, urgency, "no regrets," co-benefits, effects on mitigation, and complexity. After ranking and weighting the resulting scores, three adaptation options were identified as having the highest priority: integrated nature and water management, integrated coastal zone management, and larger catchment areas for water.[72]

New York City has developed a pragmatic mechanism for engaging stakeholders and experts in the evaluation of risks and in prioritizing adaptation measures (figure 4.4).[73] It evaluates the risk of climate change by considering the "likelihood of impact" that a certain climate hazard (e.g., temperature rise or increased storm surge) will result in an infrastructure vulnerability (e.g., buckling of rail lines or flooding of tunnels). In addition, it considers the "magnitude of consequences" (such as capital and operating costs for the infrastructure), the number of people affected, and public-health and worker-safety issues. The resulting 2-by-2 matrix assesses risks as the product of the likelihood of impact and the magnitude of consequences relative to specific types of infrastructure.[74]

Comprehensive assessments of these kinds are being conducted by a number of European countries, by the United States, and by many developing countries in the pursuit of "National Adaptation Plans" to develop strategies for coping with climate change.[75] Some of these plans have been prompted by international climate negotiations, by assessments of the economic costs of inaction, or by the experience of extreme weather events. Implementation of these plans is being hindered by a lack of funding and by a lack of clearly defined responsibilities for different levels of governance.[76]

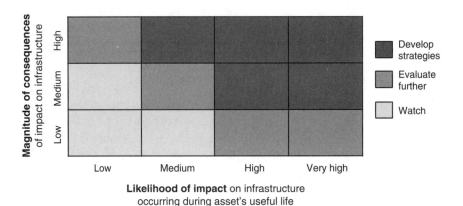

Figure 4.4
The New York City risk matrix. Source: David C. Major and Megan O'Grady, *Adaptation Assessment Guidebook*, 2010 (http://bit.ly/9N78gI), appendix B: 252.

4.2.2 Adaptation Costs

Mitigation policies have a much longer implementation history and government support than adaptation policies, which have only recently gained space in the policy arena. As a result, systematic analysis of adaptation costs and benefits has only just begun. The incipient nature of the knowledge base also contributes to concerns about acting precipitously and locking in "maladaptations" (such as implementing irrigation when the long-term forecast is for more rain, or planting drought-resistant crops, which may be highly vulnerable to disease or pests).

In contrast with many mitigation approaches, few adaptation projects offer opportunities to save resources directly. "Negative-cost" energy-efficiency improvements reduce GHG emissions and save users money, but there are few such examples in the field of adaptation.

Many of the preliminary studies on adaptation continue to focus on the global cost of "climate-proofing" but do not provide estimates of adaptation costs and benefits. At least two studies have estimated that the cost of adaptation is quite large. In 2006, the World Bank estimated that it would be necessary to spend $10 billion–$40 billion a year to "climate-proof" developing countries.[77] In 2007, Oxfam International estimated that the cost was closer to $50 billion per year.[78] In addition, the broader economic impact of adaptation on employment and economic growth is largely unknown.

The level of investment needed to reduce vulnerability to sea-level rise has received considerable attention to date, reflecting the fact that most of the world's population will live in coastal cities by 2030. For instance, one assessment examined the trade-offs between investing in coastal protection and the value of land loss from sea-level rise. The least-cost approach to coastal protection was found to involve variable levels of investment depending on the value of land at risk.[79] Another study found that investments in coastal protection would be quite high for a few Pacific Small Island States such as Micronesia and Palau, but that for the remaining 13 most-affected countries they probably would be just a small percentage of GDP.[80] Put differently, the global protection cost for a one-meter rise in sea level would be approximately $1 trillion (€774 billion).[81]

Evidence suggests that the ability of cities to contend with the impacts of a changing global climate such as sea-level rise is greater if population growth can be limited. This is illustrated in a recent case study of Dhaka, Bangladesh, one of the fastest-growing cities in the world. Because the urban core is almost entirely populated, a low-lying flood plain is being suburbanized. As in most other parts of Bangladesh, flooding is common, and its severity is increasing as sea levels rise, cyclones become more intense, and surrounding forests lose their ability to act as a catchment for water. Dealing with both mitigation and adaptation issues in Dhaka was found to be more successful under scenarios that constrained future growth of the city's population.[82]

In the agricultural sector, there are many low-cost adaptation measures that can enable farmers to better handle droughts, such as changes in planting dates, crop mixes, and the development and adoption of heat-resistant cultivars. In some countries, these measures could be quite cost effective. However, in tropical regions they may not be sufficient to offset the significant damages caused by climate change.[83]

For a 1°C (1.8°F) increase in temperature in 2100, the global benefits from reduced space heating requirements may exceed the cost of increased cooling. For an increase of 2°C (3.6°F) or more, the costs associated with the need for greater space cooling probably will exceed any savings from reduced heating requirements.[84] These adaptation costs include both the capital investment in building equipment to accommodate greater cooling capacity and increased energy expenditures.[85]

The costs of adaptation to permafrost melt are complex. For example, the expense of replacing all ice roads in Canada has been estimated to be as high as $908 million (€703 million), and similar costs might apply to ice roads traversing the Siberian peat bogs. At the same time, the retreat of permafrost would reduce road-building costs.[86] However, permafrost retreat also exacerbates adaptation costs because it releases CO_2 and methane.

An important aspect of adaptive capacity is the ability to innovate. National innovation systems that promote technological advances will have the capacity to deliver new and improved adaptation technologies, such as high-efficiency cooling systems, bio-engineered plants with high drought resistance, and new and improved desalination technologies that could contribute greatly to reducing adaptation costs.

Resources are also critical to success. At the Seventh Conference of the Parties to the United Nations Framework Convention on Climate Change (UNFCCC), held in Marrakech, Morocco, in 2001, several new funds for adaptation were created. These included the Adaptation Fund financed by a 2 percent levy on the Clean Development Mechanism, which may amount to between $100 million and $500 million (€77–€387 million) per year. The Marrakech fund is managed by the Global Environment Facility and may amount to $200 million (€155 million) per year. Relative to the magnitude of the problem and the resources being invested in mitigation, these investments in adaptation are inadequate.[87]

In the meantime, discussions about increasing commitments to adaptation continue at the annual meetings of the Conference of the Parties to the United Nations Framework Convention on Climate Change, led by developing countries. At the eighth meeting, held in New Delhi in 2002, developing countries signed the "Delhi Declaration," which called for urgent action to advance adaptation measures. Since then, the UNFCCC has directed that national communications among signatories should give attention to national adaptation programs.[88] At the 2009 UNFCCC meeting in Copenhagen, US Secretary of State Hillary Clinton promised that the United States would help raise $100 billion (€77.4 billion) per year by 2020 to assist poor countries in coping with climate change as long as China, India, and other fast-growing countries accept binding commitments that are open to international inspection and verification.[89] The next year, at the fifteenth UNFCCC meeting in Cancun, donor nations expanded and reaffirmed their commitments to helping the least developed countries adapt to climate change, bringing the funds to an annual expenditure of $412 million.[90]

4.3   Synthesis

Unlike chapter 3, which focused intensely on mitigation, this chapter has discussed a diverse array of adaptation and geo-engineering options that could enable communities and countries to respond to climate change, and could strengthen energy security. As table 4.1 shows, the three types of action—mitigation, adaptation, and geo-engineering—differ in timing, in effects on future generations, and in geographic and sectoral benefits. Because of their distinct advantages and disadvantages, an optimal, comprehensive strategy probably would involve all three types of action simultaneously. As we shall see in the next chapter, however, all three types of action remain impeded by social, political, economic, and cultural barriers; overcoming of these barriers will require aggressive public policies and government interventions.

5

Barriers to Effective Climate and Energy Policies

In 1992, Al Gore, then vice president of the United States, made a prophetic remark while discussing the causes behind climate change: "We are in an unusual predicament as a global civilization. The maximum that is politically feasible, even the maximum that is politically imaginable right now, still falls short of the minimum that is scientifically and ecologically necessary."[1] Gore feared that the conditions responsible for energy-intensive lifestyles had become so entrenched in American and European society that fighting climate change was politically impossible. The solutions being touted at the time, moreover, touched on the fringes of the issue but did not tackle the main sources of emissions directly. We were, in effect, trying to smother the plumes of a fire without addressing the fire itself.[2]

Almost 20 years later, we have made progress in understanding the forces behind climate change and energy insecurity and assessing available solutions to them, but we still face a mesh of obstacles and impediments to effective policies and actions. The barriers facing climate-friendly technologies are tenacious, interconnected, and deeply embedded in social fabrics, in institutional norms, in regulations and tax codes, and in modes of production.[3] A thorough understanding of these impediments provides a basis for developing effective strategies to shrink the socio-technical gap so that actual social choices draw from the economic-engineering approaches that offer the speed, scale, and scope necessary to tackle climate change and ensure energy security.

For instance, a number of recent US and European studies have documented barriers to innovation for cleaner-energy systems at every stage of the commercialization and deployment process. The Interlaboratory Working Group identified scores of issues relating to misplaced incentives, inconsistent regulations, and information and market failures.[4] Another study has tabled the barriers to renewable energy penetration, highlighting in particular the problem of missing market infrastructure that may increase costs.[5] Yet another assessment has found that subsidies for conventional forms of energy, high initial capital costs, imperfect capital markets, lack of skills or information, poor market acceptance, technology prejudice, financing risks and uncertainties, high transaction costs, and a variety of regulatory and institutional factors impede investments in clean energy.[6] One of us (Sovacool) has interviewed more than 180 experts working for utilities, government agencies, and the national laboratories and identified 38 non-technical barriers to the deployment of distributed generation, renewable energy, and energy-efficiency technologies.[7] Another investigation of the barriers to energy-efficient technologies and energy-efficiency programs has found 80 separate types of barriers.[8]

These studies were recently synthesized and blended with information from expert interviews in an assessment of barriers to the deployment of low-carbon technologies in the United States.[9] This effort resulted in a typology of barriers that were broadly aggregated to include cost effectiveness; fiscal, regulatory, and statutory barriers; intellectual property; and "other" cultural, social, and institutional barriers. After drilling down to greater depths of analysis, these broad categories were subdivided into 20 barrier types, which we present in table 5.1. For example, under the heading of cost effectiveness, the barrier "high costs" is divided into high up-front costs and the high cost of financing. Similarly, "market risks" include the low

Table 5.1
Typology of barriers to GHG-mitigation technologies. Adapted from M. A. Brown et al., *Carbon Lock-In: Barriers to the Deployment of Climate Change Mitigation Technologies*, Oak Ridge National Laboratory Report ORNL/TM-2007/124, 2008.

Cost effectiveness	Fiscal, regulatory, and statutory barriers	Intellectual-property (IP) barriers	Other cultural, social, and institutional barriers
External benefits and costs	Competing fiscal policies	Anti-competitive patent practices	Incomplete and imperfect information
High costs	Fiscal uncertainty	IP transaction costs	Infrastructure limitations
Technical risks	Competing regulatory policies	Weak international patent protection	Industry structure
Market risks	Regulatory uncertainty	University, industry, government perceptions	Policy uncertainty
Lack of specialized knowledge	Competing statutory policies, statutory uncertainty		Misplaced incentives

demand for emerging technologies, uncertain cost of production, the possibility of new competing products, and liability risks. Research undertaken by the Intergovernmental Panel on Climate Change has documented that all the barriers listed in table 5.1 are applicable globally.[10]

Some barriers to deployment affect relatively limited numbers of technologies or limited portions of the economy and the marketplace; others are economy-wide and broad in influence. This range of impacts is illustrated in figure 5.1, which provides a cross-walk between the 20 deployment barriers discussed in this chapter and the 15 technology sectors discussed in chapter 3. A similar assessment of barriers could be completed for the geo-engineering and adaptation options discussed in chapter 4.

Ten barriers were found to have particularly broad impacts, affecting five or more of the 15 GHG-mitigation technology sectors identified by the US Climate Change Technology Program. The most notable among these are the existence of external benefits and costs; the high costs associated with the production, purchase, and use of low-carbon technologies; and technical and market risks. Other widespread barriers include incomplete and imperfect information, lack of specialized knowledge, and infrastructure limitations.

To contextualize these types of barriers, this chapter begins by briefly introducing readers to the concept of market failure, public goods, and policy failure. It then discusses the impediments to effective climate policies and clean-energy technologies according to the typology presented in table 5.1, with a detailed section for each of the four classes of barriers. The chapter ends by discussing the notion of "carbon lock-in" and what the presence of such barriers means for the promotion of climate-friendly technologies and programs.

5.1   Public Goods, Market Failure, and Policy Failure

Let's begin with the concept of a public good. Private goods, such as apples and automobiles, are goods that any individual can own and can exclude others from owning. Public goods, in contrast, are "non-excludable" (that is, once a public good is provided, it is difficult to prevent other people from consuming it). They are also "non-rivalrous" (that is, consumption of public goods by additional parties does not reduce the quantity of the good available to others). The classic example of a public good is national defense: whether you pay taxes or not, you are still "defended," and no matter how secure you are, others can enjoy the same security at no extra cost. Other examples include parks and open space, education, law enforcement, basic research that benefits the human condition, and even Earth's air, water, and land (which can be enjoyed by all citizens of most countries, regardless of an individual's contribution to them).[11]

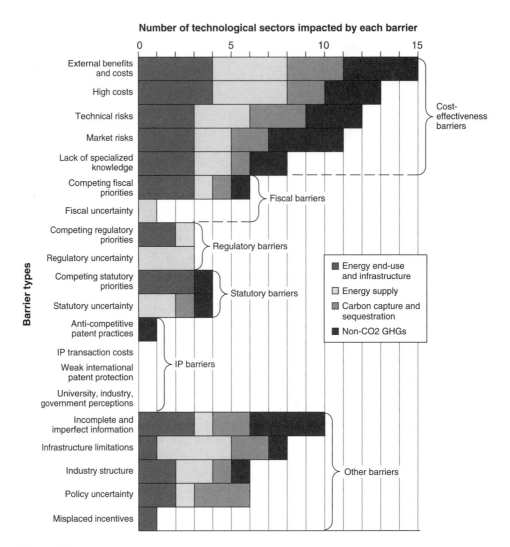

Figure 5.1
Barriers facing various technologies. Source: Committee on Climate Change Science and Technology Integration, *Strategies for the Commercialization and Deployment of Greenhouse Gas Intensity-Reducing Technologies and Practices*, US Department of Energy Report DOE/PI-0007, 2009: 111.

Public goods give rise to the problem of "free riders"—people who benefit from the policies or actions of others without shouldering their fair share of the cost. A man who refuses to be inoculated against smallpox because, if everyone else is inoculated, the risk of smallpox to him is less than the risk of harm from inoculation, is a free rider; so is a woman who refuses to pay for a park even though she uses it, since she believes that others will maintain it. Economic theory posits that rational individuals will optimize their use of a free public good (or commons) until that good is no longer of any value. In other words, people will ride for free unless forced by country or conscience to do otherwise.

Markets, consequently, provide optimal and efficient allocation of private goods, but not of public goods. This fact has serious implications for the energy sector and for climate policy. Companies and industries that import oil from overseas, for example, have little incentive to change the nature of their imports, since the energy-security benefits of doing so—such as less price and supply volatility—are distributed to all companies and importers, including their competitors. The fact that Canada is the biggest exporter of crude oil to the United States, and that at the same time Canada's production of crude oil generates 82 percent more greenhouse-gas emissions than the average barrel of crude oil refined in the United States,[12] highlights another public-good issue: the environmental footprint of imported goods. The issue is particularly acute when applied to pollution that crosses political boundaries. Electric utilities will usually require higher smokestacks on their fossil-fuel power plants as a way to minimize the environmental harm from noxious emissions, shifting the pollution instead to a broader geographic area encompassing other states and countries. In the United States, officials have often located trash incinerators, coal mines, and landfills near state or city borders (when possible), so that some of the harm from leakage and waste are transferred to other localities in a problem known as "state line syndrome."[13] Something similar occurs with climate change. The political scientist Elinor Ostrom put it as follows:

The problem of averting massive climate change—or global public bad—would be a global public good. Millions of actors affect the global atmosphere. All benefit from reduced GHG emissions, but the problem is they benefit whether or not they pay any of the costs. In other words, beneficiaries cannot be excluded from the benefit of cleaner air. Trying to solve the problem of providing a public good is a classic collective action dilemma—and potentially the largest dilemma the world has ever knowingly faced.[14]

Collective-action dilemmas give rise to what economists refer to as "market failure." Table 5.2 shows that perfect markets require at least six interrelated components: information, minimal transaction costs, rational consumers, competition, internalization, and excludability. In order for a market to function perfectly, all six criteria must be satisfied. The violation of just one criterion constitutes a market failure or a

Table 5.2
Six criteria for a functioning market. Source: B. K. Sovacool, *The Dirty Energy Dilemma: What's Blocking Clean Power in the United States* (Praeger, 2008).

	Explanation
Information	All participants in the market must be fully informed as to the quantitative and qualitative characteristics of goods and services (and substitutes to them) and the terms of exchange among them.
Transaction costs	Market exchanges must be instantaneous and costless.
Rational consumers	Consumers must maximize utility and producers must maximize profits; economic actors must be able to collect and process all relevant information, hold rational expectations about prices and products, and make decisions that always promote their self-interest.
Competition	No specific firm or individual can influence any market price by decreasing or increasing supply of goods and services; there must be many buyers and sellers; they must act without collusion; firms cannot use their market power to influence the market themselves; predatory practices by incumbent firms against insurgent firms must be restricted; there must be no barriers to entry and exit.
Internalization	All costs (or positive and negative externalities) associated with exchanges must be borne solely by the participants of the transaction, or internalized in prices so that all assets in the economic system are adequately priced.
Excludability	Those involved in the exchange must be able to exclude others from benefiting.

market imperfection, a circumstance in which the allocation of resources and prices will not yield efficient outcomes without government intervention.

The "market-failure model" that underpins many of today's public-policy debates suggests that markets should be left alone by government unless market failures exist.[15] In competitive and efficient markets, suppliers produce what consumers want and are willing to pay for. As John Donahue of Yale Law School puts it, market failure occurs when "prices lie—that is, when the prices of goods and services give false signals about their real value, confounding the communication between consumers and producers."[16] Market failures are conditions of a market that violate one or more of the conditions that define an ideal fully functioning market, and correcting them often requires some form of external intervention.

The existence of market failures in energy sectors has led to the creation of large-scale government involvement in energy markets.[17] As a result, many deployment policies focus on remedying specific barriers that prevent new technologies and practices from gaining widespread commercial use. In other instances, however, government intervention has resulted in policy failures—policies have become obstacles to the deployment of clean-energy technologies, often as the result of the unintended consequences of well-intended policies.[18] These distortionary policies create confusion in the market for energy technologies. One of us (Brown) identified more than 30 policy failures in the field of clean energy. That study exposed an array of fiscal, regulatory, and statutory policies, and numerous policy inconsistencies across states, over time, between energy resources, and across technologies that are a hindrance to the smooth operation of markets.[19]

A vigorous campaign of policy reform is thus needed to create a consistent, effective, and predictable policy environment in which clear and reinforcing signals encourage the infusion of clean-energy technologies. Perhaps the most obvious justification for such reform is the presence of barriers to "cost effectiveness."

5.2   Barriers to Cost Effectiveness

Commercializing and deploying technologies is largely a private-sector activity to gain market advantage, ultimately leading to increased profits. In the absence of policies or incentives, consumers are not likely to adopt low-carbon technologies and practices that are not cost competitive.

Usually, when a new technology is launched, a niche market, made up of motivated early adopters, begins to use the technology. As awareness of the technology increases and uncertainty decreases, adoption accelerates until it reaches a plateau approaching market saturation; this model is generally referred to as "the S-curve." However, progress along this hypothetical curve is not guaranteed.[20]

Uncertainties associated with the production costs of new products and the possibility that a superior product might emerge are two of the reasons why firms generally focus on their existing competencies and not on alternatives that could make their present products obsolete. Capital investments in firms go preferentially toward perfecting the performance and reducing the production costs of existing products. This "technology lock-in" helps to explain why new enterprises and not incumbent firms are typically the source of radical innovations that displace existing dominant designs. Lock-in is also reinforced by financial institutions, which prefer to make loans to companies with collateral and the ability to repay debts—characteristics of successful firms within the existing network.[21]

The cost-effectiveness barriers impeding diffusion and market penetration include unpriced externalities, high costs, and technical and market risks. Externalities can make it difficult for clean-energy technologies to compete in today's market, in which GHG emission reductions have only limited market value. Clean-energy technologies also often have inherently higher up-front costs, because additional features and subsystems are needed to achieve GHG reductions. Such embellishments can increase the ratio of capital to operating expense. For example, SF_6 is a high-GWP gas used in the magnesium industry as a "cover gas." SO_2 is being considered as an alternative, but it is more toxic and therefore requires additional monitoring (and cost) to deal with the health and safety issues. There are no simple drop-in substitutes. Similarly, discounted cash-flow analysis makes capital-intensive solar-electric projects with high up-front costs appear less attractive than conventional power generators that are expense intensive and have high fuel costs.[22]

Costs can also be artificially inflated by import tariffs, a barrier to free trade that is imposed by countries to protect their domestic industries from international competitors. Countries all over the world impose tariffs on importing such clean-energy products as wind turbines, solar panels, and gas turbines. (See table 5.3.) Russia has particularly high import tariffs for solar panels (20 percent), gas turbines (10 percent), and wind turbines (5 percent); India is not far behind, with tariffs of 15 percent for solar panels and 7.5 percent for wind turbines and gas turbines. The import tariffs of the European Union and the United States are much lower: 2.7–4.1 percent for the EU, 1.3 percent for the US.[23] The World Trade Organization espouses opposition to all tariffs on the presumption that market efficiency is greatest in the absence of interventions to a freely operating market.

The efficient operation of energy markets is also compromised by the existence of *external benefits and costs.* Externalities occur when important societal benefits and costs are "external" to, or not priced in, the market. "Think of externalities as a second price tag on every product we consume," notes the University of Richmond law professor Noah Sachs, "representing the real costs of disposing the product and the environmental impacts directly flowing from the existence of that product. The

Table 5.3
Import tariffs on clean-energy goods. Source: Peter Evans, "Clean energy technology innovation and deployment," presented to Council of Foreign Relations, Washington, 2010.

Wind turbine tariffs		Solar panel tariffs		Gas turbine tariffs	
Mexico	10.0%	Russia	20.0%	Nigeria	10.0%
China	8.0%	India	15.0%	Russia	10.0%
S. Korea	8.0%	UAE	5.0%	India	7.5%
India	7.5%	Colombia	5.0%	S. Korea	6.3%
Russia	5.0%	Brazil	3.8%	Canada	4.8%
UAE	5.0%			EU	4.1%
EU	2.7%			China	3.0%
US	1.3%			US	1.3%

price tag may be less than a cent for some products and several dollars for others, but because this price is never actually 'paid' by consumers or producers, the price becomes externalized as a social cost."[24] A pack of cigarettes would cost $3.43 (€2.65) more if it included the cost of lost wages, house fires, and greater health-care expenditures associated with smoking,[25] and a hamburger at McDonald's would cost $2–$3 (€1.55–€2.32) more if its price included the risks associated with high cholesterol and heart disease.[26] Energy technologies are no different: most power plants and cars are like big cigarettes that spew toxic elements into our air, land, and water.

Indeed, clean-energy technologies may be difficult to deploy (without public intervention) if their principal benefits are entirely societal and external to the market.[27] Across most of the United States, and in many areas around the world, low-carbon technologies for GHG mitigation are not currently governed by explicit regulatory legislation, and are not rewarded by the market for their public goods.[28] When the developer of a low-carbon technology cannot capture all the benefits that might accrue to society, the result is under-investment in its development and a sub-optimal supply of the technology. Because polluters do not pay for their societal damages, the "free rider" makes it difficult for the higher-priced clean-energy technologies to compete. In general, goods generating positive externalities are under-produced and goods generating negative externalities are over-produced.[29]

Specific examples of externalities in the energy and climate sectors are striking. In the electricity industry, the generation costs from a coal power plant may appear low, but they do not include the costs of coal mine dust that kills thousands of workers each year; black lung disease that has imposed at least $35 billion (€27.1 billion) in health-care costs; and coal emissions that cause acid deposition, smog, and global warming and also contribute to asthma, respiratory and cardiovascular disease, and premature mortality. These external costs would easily double the price of coal if they were incorporated into its price.[30] To recap the numbers presented in chapter 2, the negative externalities associated with electricity generation overall amount to about $13.46 (€10.42) per kilowatt-hour or more than $2 trillion (€1.55 trillion) in global damages each year. The global chemical industry would have to spend an amount equal to 8 times its annual profits (more than $20 billion, €15.5 billion, in the 1990s) to pay to incinerate the waste from its top 50 products.[31] In the transport sector, vehicle crashes are the leading cause of injury-related deaths in the United States for people between the ages of one and 65, causing 40,000 deaths, 2 million injuries, and $150 billion (€116.1 billion) in economic losses each year that are not reflected in the price of a new vehicle. Gasoline is cheap because its price does not incorporate the cost of smog, acid rain, and their affects on health and the environment. Charles Wheelan, author of *Naked Economics*, put it this way:

The problem is not that we like cars; the problem is that we do not have to pay the full cost of driving them. Yes, we buy the car and then pay for maintenance, insurance, and gasoline. But we don't have to pay for some of the other significant costs of driving: the emissions we leave behind, the congestion we cause, the wear and tear on public roads, the danger we pose to drivers in smaller cars. The effect is a bit like a night on the town with Dad's credit card: We do a lot of things we wouldn't do if we had to pay the whole bill. We drive huge cars, we avoid public transportation, we move to far-flung suburbs and then commute long distances. Individuals don't get the bill for this behavior, but society does—in the form of air pollution, global warming, and urban sprawl.[32]

These types of negative externalities, along with countless others that have not been discussed, have five main consequences apart from their social and environmental damage:

• They do not align price signals properly. The practice of ignoring price and masking the true cost of goods and services is the stuff of which historians write epitaphs for entire civilizations, and it deludes us into thinking that we are much richer than we are.[33]

• They incentivize energy companies and firms to emit GHGs and other pollutants. The more a company can externalize its cost of doing business, the greater return on capital it will receive in the short term.

• The more firms externalize costs, the more difficult it is to attribute blame. The more factories that pollute the air and water, the more liability is negated. The more liberally acceptable levels of pollution are set and the greater number of smokestacks and discharge points through which pollution is emitted, the lower the residual probability that a culprit can be made responsible, resulting in what the sociology professor Ulrich Beck has called the "organized irresponsibility" of the modern economy.[34]

• Externalities are wholly undemocratic and exclusionary. The most affected parties—often the poor or disenfranchised—are under-represented in the market, and have external costs imposed upon them,[35] particularly air pollution and toxics.[36] Yet on a per capita basis, lower income households consume less energy and hence contribute less to GHG emissions and other pollutants. The difference between price and cost is a matter of fair accounting and a reminder that, sooner or later, someone will pay (although this "someone" is often in the future and not the person imposing the cost).

• Negative externalities become normalized and accepted so that consumers learn to tolerate them. Surveys of social norms and rules in the United States have found an odd discrepancy between willingness to pay to prevent pollution and a willingness to accept it. Consumers asked about the damages from climate change, for example, have stated they would ask an electric utility be charged as much as $750 (€580) per kilowatt if it wished to emit carbon dioxide into the air but would only pay $10

(€7.70) to prevent that kilowatt from being emitted themselves—implying that they believed externalities should be priced *unless* they were the ones doing the paying.[37]

High costs mean that some combination of the capital cost of the technology, its cost of operations, or other aspects of a project that employs the technology yields a product that costs too much relative to other options that perform essentially the same function. High costs of a technology deter investments and thus make it difficult to justify providing capital to the high-cost technology or financing the use of its outputs in the absence of deployment assistance.

High costs are impacted by market and technical risks associated with commercialization or commercial deployment of a technology. The high cost barrier is a function of endogenous costs (e.g., the nature of the fabrication process and its materials requirements), but it also reflects fiscal and regulatory uncertainties—the intermittent nature of tax credits and the lack of approved permitting procedures for off-shore wind development in the United States being two examples. Infrastructure limitations can also contribute to high costs, as with wind generation in the upper Midwest, which requires investment in transmission lines to reach major metropolitan markets.

Although high costs in and of themselves are not a barrier that requires government intervention, they have recently been exacerbated by the global financial crisis. The United Nations Environment Program collected data on investment trends for renewable energy from 26,000 organizations covering 15,000 projects in more than 50 countries in 2008.[38] The study found that investment in the second half of 2008 was down 17 percent, whereas it had been growing by 200–400 percent from 2005 to 2007. On a regional basis, investment in Europe rose only 2 percent, and it fell 8 percent in North America. The study warned that many banks had shortened periods of loan repayment, in some cases up to 5 years or less, placing projects at risk of refinancing. It noted that small-scale project developers and independent producers were most at risk and were having difficulty raising capital, with a majority of the funding being lent to corporate conglomerates with strong balance sheets and with whom banks already had close relationships.

Technical risks are associated with unproven technology; they are particularly troublesome during the early stages of innovation diffusion when there is limited validation of technology performance. Insufficient operational experience can create an environment of uncertainty that the innovation will be able to perform to specifications. Fear of excessive production or operational downtime, lack of standardization, and inadequate engineering leads to a reluctance to invest.

Once the R&D process resolves these issues, the technology often transitions into problems of scalability as it moves into mass production. Solar photovoltaics clearly demonstrate this issue as laboratory efficiencies of 19 percent for crystal silicon and 10–17 percent for thin films were not duplicated in modules (11–14 percent and 4–8 percent) or in the field (9–13 percent and 3–7 percent).[39] Years of advances in processes have led to increases in efficiencies across the board, but the gap between performances in the laboratory versus the field remains today.

The term *market risks* refers to uncertainties associated with the cost of a new product relative to its competitors, and with the likelihood of the new product's acceptance in the market. It includes the risk of long-term demand that could fall short of expectations, lower-than-anticipated prices for competing products, the possibility that a superior product could emerge, rising prices for inputs including energy feedstocks, lack of long-term purchase agreements for outputs, and transportation constraints for inputs and outputs. Market risks may be particularly high under certain industry structures: fragmented industries are generally slow to adopt innovation (in part because smaller firms have less access to capital), and industries characterized by monopolies are often sluggish, preferring to aggressively defend the incumbent technologies that have gained them so much profit and market control.

The history of the electric car in the United States in the 1990s illustrates the monopoly effect, which is also chronicled in the 2006 documentary film *Who Killed the Electric Car?*[40] In 1990, the California Air Resources Board passed a zero-emissions-vehicle (ZEV) mandate, and General Motors (GM) responded by introducing a car called the EV1. After lawsuits from auto manufacturers, the oil industry, and the George H. W. Bush administration, the mandate was ended, and production of the EV1 ceased in 1999. The demise of the EV1 was the result of oil companies who feared the loss of large-scale profits from their transportation fuel monopoly, and GM, which was afraid of the losses associated with launching a new vehicle. (Interestingly, about 20 years after first introducing the EV1, GM is mow marketing the plug-in electric Chevy Volt, perhaps recognizing consumers' increasing demand for alternative-fuel vehicles, stoked partly by the British Petroleum *Deepwater Horizon* oil spill in the Gulf of Mexico in 2010.)

In general, the cost of technology declines as risks are addressed.[41] Actions such as operating experience gained by first adopters, increased scale of production by manufacturers, greater confidence in technology by regulators and insurers, and experience gained by project engineers and construction companies can all lower cost. Technological improvements drive costs down as well, and early commercial deployments can stimulate these improvements.

Skilled workers' *lack of specialized knowledge* about how to install, operate, maintain, and evaluate technology is generally considered to be a product of inadequate or unavailable training programs. In other cases, organizations may simply lack the capacity to implement novel energy or climate-protection technology.

In the building industry, few small enterprises have access to sufficient training in new technologies, new standards, new regulations, and best practices. Local government authorities tend to face this difficulty too: many building officials lack the skills necessary for maintenance and installation of technologies that increase efficiency. And those who repair and service automobiles and trucks have limited knowledge of advanced power-train designs, alternative-fuels such as biodiesel and ethanol, and electric vehicles.

In the electricity sector, many smaller utilities and rural electric cooperatives have difficulty promoting demand-side management and installing and operating renewable energy systems. Only about 10 percent of municipal utilities and less than half of cooperatives in the United States promote energy efficiency, versus 80 percent of investor-owned utilities.[42] The explanation appears to be that rural communities tend to be poorer than urban ones, lack access to capital, and have little experience managing marketing or rebate programs. Rural areas tend to lack retail business centers that can deliver a full array of renewable energy or energy-efficiency products. Small towns usually have no single large industrial customer, fragmenting the electricity market among residential and commercial users.

The solar photovoltaics industry lacks not only trained workers but also adequate purchasing channels—consumers cannot find complete systems or get them installed or maintained. Photovoltaic technology would benefit from (and is well suited to) being purchased, installed, and serviced by nationwide retailers. The industry should move toward simpler and more reliable systems that are easy to maintain, perhaps even systems that mimic other industries' "plug-and-play" designs.[43] In the United States, the Interstate Renewable Energy Council found that only eight states had PV system providers in 2006. Similarly, there are no clear one-stop vendors for integrated gasification and combined cycle or supercritical pulverized coal technology as exists for conventional pulverized coal.[44]

Proponents of nuclear power are concerned about the availability of qualified construction and fabrication talent. Many of these craftsmen (welders, boiler-makers, heavy equipment operators, and others) go through multi-year apprenticeships, and there are doubts that the current supply of craftsmen and engineers trained in nuclear fields would be sufficient to meet the expected demand for fission plants. The nuclear industry is also anxious that the limited availability of materials and heavy machining capacity may stall a nuclear renaissance. Some parts, such as reactor vessel heads, can be fabricated by only one company in the world.[45]

Finally, the knowledge barriers that business managers face are exacerbated by the absence of motivation to obtain credible information and by the absence of trusted sources of knowledge.[46] In addition, industrial managers have very specific distillation, annealing, forging, and other process technologies for which no off-the-shelf improvements are available. To complicate matters, facility managers often distrust hired experts (such as energy services companies specializing in energy efficiency), because these companies do not always have industry-specific knowledge.[47]

5.3   Fiscal, Regulatory, and Statutory Barriers

Unfortunately, just as markets can fail, so can public policies designed to correct market failures. Government action has its own set of problems, too. Public policies can provide broad societal benefits that increase overall economic welfare, for example, but can also inadvertently disfavor certain segments of the economy, including, in some cases, inhibiting the commercialization and deployment of clean-energy technologies. Many competing priorities result from policies established years ago for a public purpose that could be better addressed in other ways today.

Competing priorities also arise as a result of legal inertia. For example, regulations take a long time to adopt and modify; as a result, they can be slow to adapt to technological advances, and therefore they can inhibit innovation. Similarly, environmental standards that propelled the large-scale reduction of acid rain in the 1980s now enable the continued operation of some of the most polluting power generators in the United States far beyond their normal life spans and disincentivize investing in plant upgrades.[48] Competing policies attributable to outdated fiscal rules include the tax depreciation schedules that the Internal Revenue Service put into place more than 20 years ago as part of the IRS Tax Reform Act of 1986. These rules have not kept up with technology breakthroughs, and they inhibit the advancement of some modern low-carbon technologies. For example, back-up generators (which provide reliability at the expense of energy efficiency and clean air) are depreciated over 3 years, whereas a new combined heat and power system (providing both reliability and energy efficiency) is depreciated over 20 years.[49]

Some *tax policies* unequally impact markets in which a technology is expected to compete. Examples include distortionary taxes that favor conventional energy sources and high levels of energy consumption; tax treatment favoring operating versus capital expenses that slow the pace of capital stock turnover; outdated tax depreciation schedules that reward older energy facilities; standby charges, buyback rates, and uplift fees that make distributed generation more expensive; and lack of marginal cost pricing and time-of-use rates.

The use of traditional rules of thumb for allocating tax dollars and regulated revenues can also create conflicting priorities that impede low-carbon technologies. Allocating tax revenues on the basis of level of activity is a traditional principle of public finance, but it can promote inefficient use of resources. For example, apportionment of resources from the Federal Highway Trust Fund on the basis of vehicle-miles rewards increased use of energy for transportation. Similarly, utility companies' profits in traditionally regulated electricity markets are a function of how much electricity they sell to their customers. Under traditional rate-of-return regulation, utility companies' profits are based on the total amount of capital invested in selected asset categories (such as transmission lines and power plants) and the amount of electricity and natural gas sold. A utility company's rates are set on the basis of an estimation of costs of providing service over some period of time (including an allowed rate of return) divided by assumed sales of electricity and/or natural gas over that period. If actual sales are less than projected sales, the utility will earn a smaller return on investment and in fact could fail to recover all of its fixed costs. Thus, financial incentives favor increasing energy sales and traditional utility-scale supply-side infrastructure and discourage this important set of stakeholders from promoting energy efficiency and distributed generation—including rooftop solar photovoltaics and cogeneration at commercial and industrial sites.[50] As of late 2010, only nine US states had implemented policies to "decouple" electricity profits from sales (figure 5.2).

Similarly, it is common for state constitutions to limit the obligation of public revenues to the current fiscal year and prohibit multi-year contracts that would obligate funds in advance of annual appropriation cycles. In many states, these

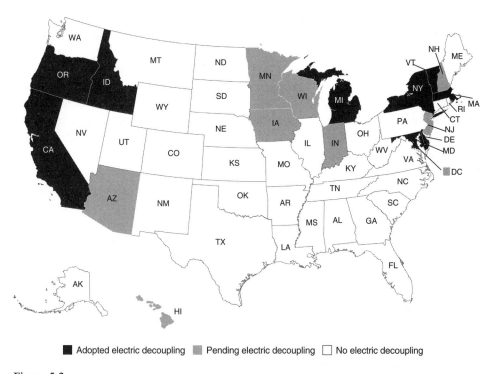

Figure 5.2
Decoupling of electricity sales from profits in the United States. Drawn from data published in Database of State Incentives for Renewable Energy (www.dsireusa.org).

administrative rules effectively prohibit financing by energy services companies to upgrade the energy efficiency of government-owned buildings.[51]

In developing countries, in Eastern Europe, and in the former Soviet republics, energy subsidies are strong deterrents to clean-energy technologies. Energy subsidies in the 20 largest non-OECD countries hit $310 billion (€240 billion) in 2007, exacerbating environmental effects. In most non-OECD countries, at least one fuel or form of energy continues to be subsidized, most often through price controls that hold the retail or wholesale price below the true market level. Widespread fuel poverty has been the driver of energy cost subsidies in Russia.[52] Other forms of direct financial intervention by government, such as grants, tax rebates or deductions, and soft loans, are commonplace. Indirect interventions also occur, such as the free provision of energy infrastructure and services. Most of these countries have announced their intention to phase out consumption subsidies eventually, though at varying rates.[53]

Fluctuating and sporadic fiscal incentives lead to uncertainty and to abandonment of initiatives before their potential can be realized. This is particularly the case for capital-intensive improvements and technologies that require a large investment for an uncertain return. One example of this is the US renewable production tax credit (PTC), which provides a tax credit for each kilowatt-hour of electricity generated by qualified technologies.[54] These tax credits were initially made available for the first 10 years of operation for all qualifying plants that entered service from 1992 through mid 1999. The subsidy was later extended to 2001, then to 2003, then, with the Energy Policy Act of 2005, to the end of 2007. In 2006, the provisions were extended for an additional 2 years. In 2009, the American Recovery and Investment Act extended the production tax credit through 2012, ending many years of on-again, off-again subsidies. Because planning and permitting for new wind turbines takes about 2 years, expirations of the PTC contribute to investment downturns even if it is reauthorized shortly afterward. PTC reauthorization stimulates market activity, and PTC expiration is promptly followed by declines in capacity additions.[55] The tax credit has created a seller's market resulting in increased competition for electricity from wind energy, with many projects delayed by at least a year, causing price inflation of as much as 50 percent in some areas (and causing the national average for wind projects to climb to $1,600 per kilowatt). Further, with the sporadic nature of the tax credits, developers have an incentive to maximize electricity production in the short term only, instead of optimizing production over the long run by investing in longer-term facility needs, systems, and personnel training.

Local variations in tax credits, deductions, and subsidies also add another layer of geographic diversity and sometimes inconsistency. The US Department of Energy counted nearly 500 different types of policies and programs attempting to reduce GHG emissions in the United States in 2009 (figure 5.3). As of early 2009, policy

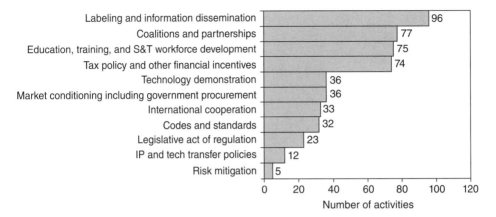

Figure 5.3
National policies and activities to reduce greenhouse gases in the United States, by type of policy and measure. Source: Committee on Climate Change Science and Technology Integration, *Strategies for the Commercialization and Deployment of Greenhouse Gas Intensity-Reducing Technologies and Practices*, US Department of Energy Report DOE/PI-0007, 2009.

targets for renewable energy existed in 73 countries, including all 27 members of the European Union, 33 of the 50 US states, and 9 of the 10 Canadian provinces. At least 64 countries had energy policies mandating use of renewable energy.[56] At least 37 countries had some form of price-based support mechanism, 49 states and countries had portfolio standards, and more than 30 countries had tax credits, tax exemptions, and/or public financing.[57] One recent study noted that 55 different types of policy mechanisms were currently in use for supporting renewable energy resources around the world.[58] Another study, after surveying hundreds of experts in Asia, Europe, and North America, identified 30 favored mechanisms.[59] Yet another investigation of subsidies for renewable energy found that policies can take the form of direct financial transfer (grants), preferential tax treatment (tax credits, exemptions, accelerated depreciation, and rebates), trade restrictions (quotas), financing (low-interest loans), and direct investment in energy infrastructure, research, and development.[60] The sheer magnitude of the types of energy and climate policies makes the international market for renewable energy somewhat chaotic and very complex, and it can contribute to uncertainty and risk when policies change.

Competing regulations can also slow the adoption of new technologies. A regulation is a legal restriction promulgated by government administrative agencies through rulemaking supported by a threat of sanction or a fine. Regulations are imposed in pursuit of the public good to produce outcomes that might not otherwise occur, but they can become impediments to innovation and competition. Common examples of regulation include attempts to control market entry, prices, wages, pollution, and standards of production and performance. Regulatory barriers that disadvantage clean-energy technologies and impede efficient market functioning include a range of environmental performance standards, power plant regulations, rules pertaining to the use of combined heat and power, parts of the federal fuel economy standards for cars and trucks, and certain codes and standards regulating the building industry. Burdensome and underdeveloped regulations and permitting processes can also inhibit low-carbon technologies.[61]

Consider the New Source Review (NSR) program, which was established as part of the 1977 US Clean Air Act and its amendments.[62] To implement this program, the US Environmental Protection Agency issued New Source Performance Standards to dictate how much pollution a new stationary source may produce.[63] These standards are intended to promote use of the best air-pollution-control technologies, taking into account the cost of such technology and any other non-air quality, health, and environmental impact and energy requirements. However, these standards apply only to electricity-generating units that have been constructed or modified since the proposal of the standard. This "grandfathering" has enabled the continued operation of some of the most polluting and highest-CO_2-emitting electricity generators in the country far beyond their normal life spans,[64] and some contend that it has resulted in the underutilization of newer power plants.[65] As one legal scholar put it, the New Source Review program "imposes pollution controls where they are least needed and artificially inflates the value of the dirtiest plants."[66]

Consider also the International Energy Conservation Code (IECC). As of August 2010, only 32 of the 50 US states had residential codes that met or exceeded the 2006 IECC, and 10 states either had no statewide energy codes for residential construction, or they pre-dated 1998. Evidence regarding building code compliance also suggests a need for verification by a third party. In a survey of code officials, one in ten reported that inspections did not occur in their jurisdictions.[67] Other studies have confirmed a considerable lack of building code enforcement and compliance.[68]

Fluctuating, variable, and unpredictable regulations can undermine the market's efficiency by introducing policy uncertainty. These barriers include uncertainty about future regulations of GHGs, uncertainty about the disposal of spent nuclear fuels, uncertain siting regulations for off-shore wind, lack of codes and standards for hydrogen use, and interconnection standards in the electric industry.

For example, the ability to legally connect equipment for distributed generation to the grid depends on national, state/provincial, and local rules and regulations. In the United States, the legal right to connect to the grid is provided in federal laws such as the Public Utilities Regulatory Policies Act of 1978 and by states' net metering statutes.[69] Connection standards are designed to prevent unnecessary fluctuations in the electric system caused by improperly functioning or out-of-phase electric generators. These standards keep the electric system safe from fires, surges, brownouts, and blackouts; however, in some cases, their application can be seen as onerous, rather than as due diligence. State-to-state variations in net metering policies cause confusion in the market and raise the cost of completing distributed generation projects.

More than 40 of the US states have net metering laws, which allow a two-way flow of electricity between the distribution grid and customers with their own generation. But state-to-state variations in regulations impose significant burdens on project developers. One study examined the adoption rates of net metered distributed generation systems at hospitals, schools, and other commercial facilities in Illinois. It noted that organizations considered regulatory complexity as an obstacle when deciding whether to embrace distributed generation technologies.[70]

Competing statutes include environmental permitting and building codes, variance in rate-based recovery mechanisms for energy-efficiency investments, bans on private wires crossing public streets, and state laws that prevent energy-saving performance contracting. The universal ban on private electric wires crossing public streets is a good example of competing statutes. This ban was established to maintain safety on roadways by preventing the introduction of low-hanging wires. By forcing would-be power entrepreneurs to use their competitors' wires, often at a high cost, this ban penalizes local generation, which offers the potential for high-efficiency power delivery.[71] In today's mature market for electricity, specifications could be designed to permit private wires.

The expectation of a stream of immediate and future benefits drives most investment and consumption decisions. *Uncertainty* is a deterrent to investment. It is particularly problematic when new clean-energy technologies are being launched into a market in which codes and standards have not been developed, policies are expected, statutes fluctuate over time, and "the rules" vary from place to place. For example, there are pressing property-rights issues surrounding the legal right to harvest wind and the ownership of coalbed methane.[72]

For carbon capture and sequestration (CCS), two critical areas of uncertain property rights pertain to the surface injection of the CO_2 and the sub-surface reservoir of CO_2.[73] When CO_2 is injected underground, it is not clear who is to be paid and who has the right of refusal. If the injected fluid goes beyond the surface boundaries, the floor space a mile deep in adjacent lots may or may not be available to that well, for example. The ownership of CO_2 storage rights (reservoir pore space) will have to be resolved before CCS approaches can proliferate.[74]

5.4   Intellectual-Property Barriers

Generally, intellectual-property law is intended to stimulate innovation, entrepreneurship, and technology commercialization. However, its application can also impede the innovation process. For example, patent filing and other transaction costs associated with strong patent enforcement and protection, as well as the anti-trust challenges related to technological collaboration and patent manipulation, can be serious barriers to technology diffusion. Anti-competitive business practices can also play a part in impeding cleaner energy systems.

Patent warehousing, suppression, and blocking are *anti-competitive practices* undertaken by incumbent firms that impose barriers to technological change.

Patent warehousing is a form of patent manipulation that involves owning the patent to a novel technology but never intending to develop the technology.[75]

Patent suppression involves refusing to file for a patent so that a novel process or product never reaches the market. For example, in 1977, Tom Ogle developed, for the Ford Motor Company, a system that used a series of hoses to feed a mixture of gas vapors and air directly into the engine. Ford built a small number of prototypes that averaged more than 100 miles per gallon at 55 miles per hour (2.35 liters per 100 kilometers), but the technology was ultimately suppressed.[76]

Patent blocking occurs when firms use patents to prevent another firm from innovating. For example, though Ford has used Toyota technology (in the Ford Escape), Ford has resisted purchasing Toyota technology for hybrid vehicles because of hefty licensing fees. Likewise, Honda has not been able to successfully negotiate a license to use nickel metal hydride batteries in its hybrid vehicles. General Electric has used its patent on variable-speed wind turbines to prevent Mitsubishi (a Japanese firm) and Enercon (a German firm) from entering the US market.[77]

The high cost of filing a patent application can serve as a financial impediment for inventors and firms with scarce capital including many small businesses. Though the *transaction costs* associated with patent filing will vary depending on the type of technology and breadth of the patent, it has been estimated that typical costs range from $10,000 (€7,740) to hundreds of thousands of dollars per patent.[78] These expenses do *not* include the costs associated with patent continuation, maintenance, and enforcement against infringement. It has been calculated, on the basis of the uncertainties of the patent searching process and the number of amendments and drawings that may be required, that inventors and firms can expect to expend an additional $20,000 (€15,478) for each foreign country in which patent protection is sought. Furthermore, the patent filing process typically takes between 24 and 36 months.[79]

Inconsistent or nonexistent patent protection in developing countries and emerging markets can deter innovation, as firms believe they would be at a competitive disadvantage to distribute their technology. Also, many companies do not want to collaborate with overseas partners because participation may attract those that have the most to gain and the least to contribute, risking an asymmetrical relationship in which sharing is uneven between firms. Moreover, host companies in developing countries may be reluctant to purchase or acquire technology that they believe competitors could freely copy in their own markets.

Weak international patent protection affects both the supply component and the demand component of technological diffusion, especially in the cases of efficient industrial boilers, fluidized bed combustion, coal gasification, and various end-of-pipe pollution-abatement technologies (such as carbon capture and storage). For example, weak protection of intellectual property has discouraged American and European companies from developing more advanced clean coal technologies (such as more efficient coal washing processes, advanced combustion turbines, and carbon capture and storage systems). Intellectual-property concerns connected with clean coal systems are cited as one of the most significant impediments toward diffusing such technologies to China, Indonesia, and other developing countries—especially in cases in which new technologies could be reverse engineered or copied.[80]

Changing relationships among universities, national laboratories, and industry leaders may have reduced the motivation for collaboration. A shift in universities' focus to include more patenting and enforcement of intellectual property in collaborations with businesses and national laboratories has created a discussion among scientists and academics about the role of the university.[81]

Disagreements between faculty members and university administrators seeking to commercialize academic research can become more than just theoretical debates. Disputes between faculty members who prefer an open or accessible license for discovery (which would maximize the breadth of knowledge dissemination in their mind) and entrepreneurial universities (which want a more lucrative, more exclusive license) can turn into protracted legal battles.[82] A 1997 case in California illustrates the financial danger of such conflicts. Two professors won multi-million-dollar settlement in a lawsuit against the University of California at San Francisco after claiming that the university had defrauded them by licensing their patents to other companies at a discount in exchange for sponsorship of research by those companies.

5.5   Other Cultural, Social, and Institutional Barriers

Many additional barriers inhibit the deployment of GHG-mitigation technologies in ways that are not captured by the categories discussed thus far. These barriers stem in part from the cultural traits that affect individuals' behavior and their choices. The influence of lifestyle and tradition on energy use is most easily seen by means of cross-country comparisons. For example, in China clothes are traditionally washed in cold water, whereas in the United States and in Europe hot water is commonly used.[83] Similarly, there are international differences in how lighting is used at night; in the preferred temperatures of food, drink, and homes; and in the operating hours of commercial buildings.[84] Cultural differences also appear to help explain regional variation in the market penetration of ENERGY STAR appliances and ENERGY STAR homes in the United States (figure 5.4).

The *provision of information* is subject to a classic public-goods problem having to do with completeness and quality. If one person generates useful information, it creates a positive externality because it provides knowledge to others. Those who have information may have strategic reasons to manipulate its value, self-interested sellers have incentives to provide misinformation about their products, and often well-distributed misinformation can overpower the distribution of unbiased and more accurate information.

The information problem relates not only to situations in which people are forced to base their decisions on incomplete information, but also to situations in which one party to a transaction knows more than another party. The economist George Akerlof demonstrated that the existence of "lemons" in the automobile market revealed the issue of asymmetric information; bad cars sell at the same price as good cars, since it is difficult for a buyer to determine which cars are good and which are

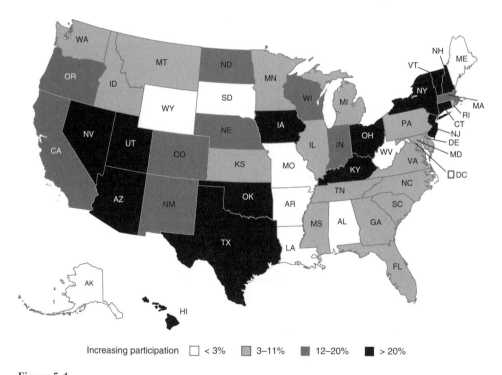

Figure 5.4
Market penetration of ENERGY STAR homes. Source of data: Database of State Incentives for Renewable Energy (www.dsireusa.org).

"lemons."[85] A supplier of air conditioners will have better information than the buyer and thus can deceive customers, which makes consumers reluctant to trust even an honest seller's claims about a certain air conditioner's efficiency. Climate-change negotiations are also vulnerable to asymmetric information. One study compared the wide-ranging climate-change impacts forecast for Norway, Russia, and the Middle East and concluded that because these differences were not broadly understood, bargaining undermined international negotiations.[86]

The economist George Stigler argued that prices change with varying frequency in all markets, and that unless a market is completely centralized no one will know all the prices that various sellers quote at any time. Price dispersion is indeed a manifestation and a measure of ignorance in the market.[87]

Individuals and firms are also limited in their ability to use, store, retrieve, and analyze information. The social scientist Herbert Simon hypothesized that human decision makers are only as rational as their limited computational capabilities and incomplete information permit them to be. He characterized this condition as "bounded rationality."[88] Simon argued that, in searching for decision alternatives, people look for satisfactory choices instead of optimal ones. "Bounded rationality" is an understandable reaction to a complex and uncertain world in which humans must calculate consequences, resolve uncertainties, and pursue courses of action that are sufficient, rather than truly self-maximizing.

It is costly to evaluate all the possible alternatives and futures for any particular decision. Relying on heuristics, or a simplified set of assumptions, allows people to make quick judgments based on just a few aspects, such as color, brand, or availability. With durable goods purchasing experiments, a majority of consumers (58 percent) were found to rely on screening—getting rid of what they like least; far fewer relied on choice (12 percent)—picking what they like most, and the remainder used a combination of screening and then choosing.[89] Copying others is another type of heuristic; it reduces risks and uncertainty by making things more familiar, and it is also approval seeking. Copying is such a strong motivator of behavior that it is principle number one of behavioral economics.[90]

In view of the immaturity of many advanced energy technologies, acquiring or proving information about their performance can be a major obstacle. Diffusion of solar photovoltaics, cogeneration (combined heat and power) systems, resource efficiency, substitution of materials, recycling, changes in manufacture and design, and fuel switching remains impeded by high transaction costs for obtaining reliable information.[91] Information collection consumes time and resources, especially for small firms, and many industries prefer to expend their human and financial capital on other investment priorities. Neal Elliott, Industrial Program Director of the American Council for an Energy Efficient Economy, attributes the insufficient investment in industrial efficiency to a lack of trusted information.[92] Since many industrial operations don't have in-house engineering resources, they have difficulty sorting through all the advice they are given.

It is often hard to determine the performance and costs of energy- and carbon-efficient technologies because the benefits are bundled and difficult to disaggregate. For example, residential consumers get a monthly electricity bill that provides no breakdown of individual end uses, which makes it difficult to assess the benefits of efficient appliances, televisions, and other products. This situation contributes to making energy savings "invisible," and makes energy-use patterns and load profiles hard to understand and difficult to link to energy-bill savings. Similarly, the price paid for different levels of vehicle fuel economy is buried in base prices or in the price of complete subsystems such as engines. Further, efficiency differences between vehicles are coupled with substantive differences in other attributes that are important to consumers, such as acceleration, level of luxury, and handling. Fuel-economy ratings (such as those available at www.fueleconomy.gov), appliance energy labels, and ENERGY STAR branding all help, but studies have shown that many consumers do not understand them.[93]

The accumulation of these cognitive limitations and information asymmetries perpetuates misunderstandings about energy and climate issues. A survey undertaken by Southern California Edison in the 1970s asked thousands of customers where electricity came from, and most people replied "out of the plug in the wall"; others even said "lightning" and "static electricity."[94] In another poll, this one taken in the 1990s, only 39 percent of those interviewed had ever heard of GHG emissions, and more than half of those who had did not know where they came from.[95] In 2004, 41 percent of respondents in a Kentucky survey identified coal and oil as "renewable resources."[96] A national survey in 2006 found that 80 percent of electricity customers were unable to name a single source of renewable energy.[97] A separate study found that nearly 70 percent of owners of flexible-fuel vehicles (i.e., automobiles capable of running on gasoline and/or bio-based ethanol) were unaware that they owned such a vehicle.[98] A survey of 1,300 residents of the US found that 80 percent thought that idling a vehicle's engine for more than 30 seconds was better than restarting the engine, when in reality idling wastes fuel and emits more carbon dioxide per year than the entire American iron and steel manufacturing sectors.[99] Another study of drivers found that no respondents analyzed vehicle fuel costs in a systematic way, almost none tracked gasoline costs over time, and few considered transportation fuel costs in their household budgets; the study concluded that drivers "lack the basic building blocks of knowledge" necessary to make intelligent decisions about driving.[100] People in the United States are so uninformed about energy that a mail-order scam consisting of offering a "solar clothes dryer" for $30 and then mailing the buyer a piece of rope and some clothespins has been profitable for decades.[101]

In Europe, somewhat novel efforts to make energy visible—such as a "Power Aware Cord" that glowed with greater intensity the more energy that was flowing through it—resulted in increased energy consumption because people found the colors aesthetically pleasing.[102] Almost two-thirds of people in Belgium do not know how much they pay for heating, and half do not know how much they pay for electricity.[103] In the Chinese province of Liaoning, an assessment of electricity use in homes found that only 2 percent of respondents reported understanding ways to save energy and 40 percent indicated they did not remember their electricity bill.[104] A global assessment of patterns of energy use and transportation choices among 17,000 consumers in a total of 17 countries confirmed that overall "literacy" in regard to energy issues is low.[105]

Technologies that are otherwise expected to be successful may still have difficulty penetrating the market because of *infrastructure limitations*. These limitations include a wide range of supply-chain shortfalls, ranging from inadequate physical systems and facilities and shortages of supporting technologies, to insufficient supply and distribution channels and inadequate operation and maintenance (O&M) support. These infrastructure limitations hinder the adoption of many mitigation technologies in developing countries. For example, poor power quality can interfere with the operation of sensors, controls, and other electronic devices needed to operate energy-efficient devices in developing countries.[106] To facilitate investment in infrastructure, developing countries usually require assistance with capacity building. Such assistance may include enhancing knowledge and management skills, creating appropriate institutions and networks, and acquiring and adapting hardware and infrastructure.[107] Access to information is particularly problematic. The creation of energy service companies and traditions of energy performance contracting could help to manage the perception of market and technical risk by lenders; however, an informed workforce and user population are still critical to success.[108]

Potential types of US energy infrastructure needed to support a transition to clean energy are illustrated in figure 5.5. One of the most critical needs is to strengthen the US transmission grid. Additional capacity is needed to reach the country's large but often remote wind resources. At the same time that demand for electric power is increasing rapidly in the US, the high-tech economy is placing greater requirements on increasing levels of power quality and power conditioning. Despite these trends, investment in the transmission and distribution infrastructure has declined with the restructuring of the electric industry, and evidence of grid congestion is increasing. The National Electric Transmission Congestion Study identified two critical congestion areas already existing in 2006 and additional areas in which future congestion is likely to occur.[109]

Biofuels also face geographical incongruence with the current system. Today ethanol is most economically produced within about 100 miles (161 kilometers) of

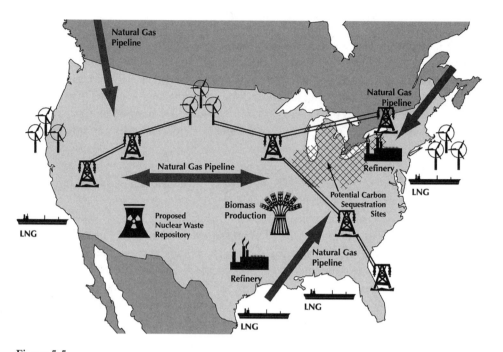

Figure 5.5
Major elements of the national energy infrastructure of the United States. Source: National Commission on Energy Policy, *Ending the Energy Stalemate*, 2004.

where the corn feedstock is grown, because corn's biomass density is so low. Historically, biomass-to-ethanol plants have been built in the rural Midwest, which is not coincident with large concentrations of fuel consumption. To realize a significant displacement of petroleum, the produced bioethanol must be shipped long distances to consumption centers. The current pipeline infrastructure is unidirectional from the coasts to demand centers, so this economic transport solution cannot be utilized; in addition, ethanol is not compatible with existing petroleum pipelines. Waterways would be desirable for transport, but their orientation is inconsistent with the necessary routes from producers to existing blenders or refiners. Currently, long-distance shipments of bioethanol go via rail or roadways; hauling fuel on the roads is costly, and rails have limited expansion capacity.[110] For bioethanol, the distribution infrastructure is clearly insufficient; a wholly new infrastructure will have to be built, or perhaps the US will have to consider alternatives to transporting bioethanol (i.e., replacing petroleum with bioethanol only near production centers or more concentrated blends with petroleum).

Two other GHG-reducing technology areas have large infrastructure issues: capture and storage of hydrogen and of carbon. Hydrogen presents a unique challenge for transmission, storage, and distribution because of its chemical nature; these infrastructure design issues are still undergoing research. Carbon capture and geologic storage, working together, will require significant expansion of CO_2 transmission from the points of emission to underground storage sites; it is estimated that 11 trillion tons of CO_2 storage capacity is available within geologic formations in the United States, but detailed characterization of this potential infrastructure has only begun.[111]

There are two very different barriers to innovation related to *industry structure*: the natural monopoly of some energy industries, which hinders competition, and the fragmented nature of other industries, which slows technological change, obstructs coordination, and limits the availability of investment capital.

Natural monopolies occur when average cost declines over the relevant range of demand. With declining average costs, a single firm can produce the necessary output at lower cost than any other market arrangement, including competition. The resulting market power and economic inefficiencies associated with natural monopolies are limited when close substitutes for the product are available. In the case of electricity generation, in which natural monopolies have been dominant, close substitutes do not exist. As a result, small-scale competition is difficult in the electric power sector.[112] The existing utility monopoly attempts to maintain control over the production and distribution of power—examples of utilities and municipalities opposing small-scale renewable electricity projects have occurred across the US.[113] This opposition often is exerted through utility pricing and tariff policies such as standby charges, uplift charges, and interconnection requirements. Tariff barriers to distributed generation include excessive standby charges and utility buyback rates that do not provide credit for on-peak electricity production.[114] Standby charges are fees levied by a utility service company for the potential use of the company's electricity in the event that a combined heat and power project goes down. These fees are often based on worst-case scenarios assuming the need to maintain backup capacity to support distributed generation units during times of peak demand.

Economic sectors that are diverse and fragmented, such as agriculture and forestry, impact technologies related to terrestrial sequestration, methane recovery from livestock and poultry operations, and nitrous oxide emissions. Industry fragmentation slows information dissemination and technological change, complicates coordination efforts, and limits investment capital. The agricultural industry is significant for terrestrial sequestration efforts as well as for reducing emissions of other gases, including methane and nitrous oxide. Many thousands of actors operate in this industry, largely autonomously. In addition to making widespread change in the farming and forestry industries more complicated, the fragmented nature of these industries also limits access to capital.

Construction is another fragmented industry that illustrates these barriers to rapid technological diffusion. The building industry has a record of being slow to adopt innovations. For example, 1.06 million new homes were purchased in 2006. The largest homebuilder (D. R. Horton) was responsible for only 53,410 (5 percent) of these, and the five largest homebuilders for only 20.2 percent.[115] This fragmentation means that a large number of firms and individuals must act in concert to produce a significant collective response, because those engaged in building design and construction generally have little interaction with each other.[116]

"Policy uncertainty" refers to the unknown future legal status of greenhouse gases. In the United States it is not yet clear if GHGs will be taxed, capped, or regulated under the Clean Air Act; if credit will be given for past actions; and which emissions might be treated as "offsets" for CO_2 emissions. Investors, electric utilities, and other stakeholders who deal with fuel futures must decide what to build as a next generation of power plants and transportation fuels without knowing if CO_2 and other GHGs will remain uncontrolled by policies, or if regulations are imminent. When the basis for estimating long-term operating costs and competitive advantage is so uncertain, how are consumers to make "rational" choices about the purchase of new energy-using systems? How are producers to decide whether or not to invest in alternative energy technologies? All the uncertainties associated with future and current GHG treatment are impediments to positive action.[117]

More and more US companies have been participating in voluntary GHG emission reduction and registries to prepare for eventual federal regulations.[118] But whether or not these early actions will receive credit in any future GHG cap-and-trade program depends on future congressional legislation. To add further complexity to this already uncertain situation, the existing GHG emissions reduction registries in the United States differ in ways that could affect the provision of credit under future federal legislation.[119] These uncertainties contribute to a "wait-and-see" attitude among many GHG emitters.

Misplaced incentives occur when the buyer or owner of a technology is not its consumer or its user—a phenomenon that is referred to in the economics literature as the principal/agent problem. In general, this problem occurs when one party (the agent) makes decisions in a particular market and a different party (the principal) bears the consequences of those decisions. An assessment done for the International Energy Agency found such market failures to be significant and widespread in many energy end-use markets in the United States and in other International Energy Agency countries.[120] In many market situations, buyers purchase equipment on behalf of consumers without taking their best interests into account. The following are a few examples of misplaced incentives inhibiting energy-efficient investments in low-carbon technologies:

• In many countries, the energy bills of hospitals are paid from central or federal funds but investment expenditures must come from either the hospital itself or from a local government agency.[121]

• Architects, engineers, and builders select equipment, duct systems, windows, and lighting for future building occupants who will be responsible for paying the energy bills.

• Landlords purchase the air conditioning and major appliances and decide on major renovations, whereas tenants often pay the energy bills.

• Industrial buyers choose technologies that manufacturers use in their factories.

• Specialists write product specifications for military purchases.

• Vehicle fleet managers select the cars and trucks to be used by drivers.

• New car buyers determine the pool of vehicles available to buyers of used cars.[122]

The involvement of intermediaries in the purchase of energy technologies limits the ultimate consumer's role in decision making and leads to an under-emphasis on life-cycle costs. Nearly one-third (32 percent) of US households rent their homes,[123] and 51 percent of privately owned commercial buildings are rented or leased.[124] In these segments of the market, landlords have a powerful influence over the energy efficiency of buildings and their equipment.[125] The landlord-tenant relationship is a classic example of misplaced incentives. If a landlord owns the energy-using equipment but the tenants pay the energy bills, the landlord is not incentivized to invest in efficient equipment unless the tenants express their self-interest. Thus, the circumstance that favors the efficient use of equipment (when the tenants pay the utility bills) leads to a disincentive for the purchase of energy-efficient equipment. The case that favors the purchase of efficient equipment (when the landlord pays the utility bills) leads to a disincentive for the tenants to use energy wisely.[126] About 85 percent of all households in multifamily buildings are renters, which makes misplaced incentives a major obstacle to energy efficiency in urban housing markets.[127]

5.6   Toward Carbon Lock-In

As the sections above imply, the unique barriers faced by different types of technologies highlight the fact that specific deployment policies and programs may be required. At the same time, economy-wide actions may be more efficient in addressing common barriers in a broad, systematic fashion in ways that could significantly accelerate the uptake of clean-energy technologies. This tension between highly specific versus general policy interventions requires careful consideration.

Many GHG-reducing technologies and practices involve novel and sometimes radical departures from earlier practices. As such, they must overcome a wide range of technical and market risks to gain widespread commercial use. Risks must be minimized because success requires displacing the market shares of established and mature incumbent technologies with demonstrated performance records.

A host of technologies relating to energy end-use infrastructure, energy supply, carbon capture and storage, and high-GWP gases are currently available and technically feasible. However, various barriers impede progress across the complete spectrum of clean-energy technologies and at every stage of commercialization and deployment.

Some obstacles are broad in scope; others are narrower. Some appear to be amenable to policy solutions; others may not be. The numerous barriers described in this chapter have led analysts to suggest that consumer and household actions should be a focal point of public policy, since behavioral modifications can reduce energy consumption with available technology, at a low cost, without appreciable lifestyle changes, in a cumulative manner. Indeed, some of our colleagues have argued that the technology gigaton solutions to climate change discussed at the end of chapter 3 should be expanded to include a behavior gigaton, because the impacts of lifestyles are that large.[128] Information programs would appear to hold great promise for delivering such a gigaton solution, but could necessitate significant economic and political costs.

Other barriers are attributable to existing regulations, statutes, and fiscal policies that treat technologies for mitigating climate change unfavorably.[129] In addition to reforming these existing policies, policy makers should consider the traditional policy instruments as well as the novel climate policies being launched in smaller test beds all over the world. Designing policies to address the numerous and specific deployment barriers impeding low-carbon technologies will lead to fuller realization of the great economic and technical potential of climate-change-mitigation solutions.

Broadly speaking, barriers hinder low-carbon technology commercialization and deployment in different ways: by locking in incumbent technologies, by escalating the business risks of innovation, or by increasing transaction costs associated with change. These powerful and restraining influences reinforce one another. Systems of positive feedback between government, financial institutions, suppliers, and existing infrastructure support and sustain status-quo technologies even in the face of superior substitutes. Inventions and innovations face an array of obstacles in the market, and since many clean-energy technologies are relatively new, these obstacles can strongly impact them. Costs associated with gathering and processing information, developing patent portfolios, obtaining permits, and designing and enforcing contracts can all be prohibitive during the early stages of a technology's deployment.

Unfavorable policy environment

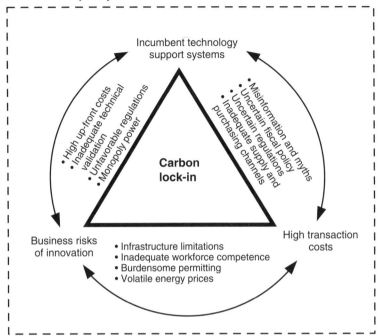

Figure 5.6
Causes of "carbon lock-in." Source: M. A. Brown et al., *Carbon Lock-In: Barriers to the Deployment of Climate Change Mitigation Technologies* (Oak Ridge National Laboratory report ORNL/TM-2007/124, 2008).

Further reinforcement of incumbent technologies is provided by the policy environment that tends to support the status quo. Clean-energy technologies are often subjected to unfavorable treatment by fiscal, regulatory, and statutory policies, and they are impacted by policy uncertainty that causes market inefficiencies and a reluctance to innovate. Taken together, these barriers create "carbon lock-in"; that is, they "lock" societies into carbon-intensive modes of energy production and use (figure 5.6).

Tackling these systematic forces requires comprehensive forms of intervention. For example, overcoming lock-in of incumbent technologies suggests that government organizations should be decoupled from the systems that support mainstream technologies, whereas overcoming business risks of innovation requires reduction of costs and of financing hurdles. Many options available to address the numerous barriers that impede the progress of clean-energy technologies are described in the following chapters.

6

Overcoming Barriers to Effective Climate and Energy Policies

Denny Ellerman, a leading authority on emissions trading and energy economics at the Massachusetts Institute of Technology, compares today's environmental problems to an earlier problem dealing with a common pool resource—land. Much like air and water today, land was once free for the taking, but population growth made it scarce and property rights emerged to allocate the scarcity. Today, practically no one questions that allocation of property rights, even though the initial allocation may have been "coercive and unfair." Ellerman notes that the "ancient act" of switching to property rights is "lost in the mists of history and no one really cares now, even though a significant portion of everyone's lifetime income is devoted to acquiring the right to call a small piece of the earth home."[1] Today, the question is whether a similar private property-rights system can be devised to manage current common pool energy and climate problems. Just as people thousands of years ago questioned the need to purchase land, we now find it hard to understand the need to purchase the right to pollute. It is particularly challenging to explain the need to purchase the right to emit invisible greenhouse gases.

One overarching lesson for those wanting to promote cleaner and more secure forms of energy infrastructure, energy supply, and energy use is that relying on the market alone is insufficient. As chapter 5 documented, robust government policy is needed to reduce market risks, provide information, and accelerate technological breakthroughs. Because of the numerous interconnected and interlocking barriers to a low-carbon society, intervention is necessary if the future is to differ from the past. As a review of the political landscape shows, government intervention can take a variety of forms—from putting a price on carbon emissions and creating performance standards to financing and procuring cleaner technologies or incentivizing more efficient agricultural, waste, and water practices. As the need for technological change accelerates, how should local, national, and global bodies choose among the available policies and mechanisms?

This chapter argues that there is an urgent need for policies that combine speed (i.e., actions sooner rather than later), scope (i.e., broad participation across

economic sectors and GHGs), and scale (to achieve the depth of impact and economies of mass production) to meet energy and climate targets. That petroleum remains
unrivaled as the world's dominant transportation fuel nearly 40 years after the
1973–74 Arab oil embargo shows clearly that the speed of existing responses is inadequate. When 71 percent of the world's carbon emissions come from countries
without CO_2 reduction obligations (as figure 6.1 shows), it is difficult to achieve
scale economies, prevent spillover effects, and minimize free riding. Similarly, when
unregulated sectors of the economy account for a large proportion of the world's
CO_2 emissions, planet-destabilizing changes in climate become almost impossible to
avert. Efforts to address these deficiencies in speed, scale, and scope are mounting.
Whether those efforts are sufficient remains to be seen.

To better understand the bases of different types of public policies for energy
and climate, this chapter begins by contrasting two underlying paradigms behind
policy formation: the precautionary principle and the more common "risk paradigm." The chapter then offers a typology of public-policy mechanisms and
summarizes different methods of selection, including cost-benefit analysis, cost-
effectiveness analysis, and various hybrids. The chapter finishes by exploring the
potential and pitfalls of putting a price on carbon, the dynamics of carbon cap-and-

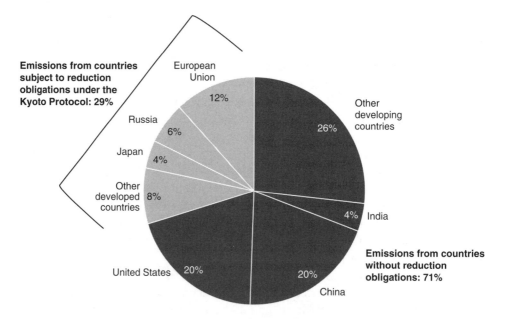

Figure 6.1
Global CO_2 emissions from fuel combustion in 2006. Redrawn from data published in table 1 of CO_2
Emissions from Fuel Combustion, 2008 Edition (International Energy Agency).

trade schemes, and how pricing carbon can be complemented with other policies in the electricity supply, transport, agriculture and forestry, and waste and water management sectors.

6.1   The Risk Paradigm versus the Precautionary Principle

Let us begin by talking about two paradigms—underlying thought structures—behind forming climate and energy policies. The precautionary principle is an elegantly simple practice of informed prudence. Most definitions of the principle have two main elements:

• A need to anticipate harm before it occurs. Within this element lies an implicit reversal of the onus of proof: under the precautionary principle it is the responsibility of possible polluters to establish that their proposed project is not likely to result in significant harm.

• An obligation to prevent or minimize harm when a risk of significant harm exists, even when the absence of scientific certainty makes it difficult to predict the likelihood and magnitude of the harm.

The precautionary principle emerged as an explicit basis of policy formulation during the 1970s in West Germany as the *Vorsorgeprinzip* (foresight principle) of German water-protection law. At the center of this conception was the belief that society should seek to avoid environmental damage through careful planning. The principle was introduced internationally in 1984 at the First International Convention on Protection of the North Sea.[2] It has been slowly assimilating into various facets of policy in the United States, in both criminal law and in medical practice.[3] In US criminal law, for instance, a person is considered innocent until proven guilty beyond a reasonable doubt. Society has decided to place an extremely high value on precaution against preventing the execution of innocent people by the state, even if it risks letting some guilty individuals go free. Similarly, in medical practice if a baby is born to a mother with a mild fever but exhibits no symptoms of infection themselves, doctors will still require that the baby undergo intravenous antibiotic treatment as a precaution. The course of action is taken not because doctors know the child has an infection, but because an infection during the first days after birth can be fatal.[4]

Perhaps the best-known definition of the precautionary principle is contained in Principle 15 of the UN Framework Conference on Environment and Development (1992)[5]:

Where there are threats of serious or irreversible damage, lack of full scientific certainty shall not be used as a reason for postponing cost-effective measures to prevent environmental degradation.

Though this may not be obvious, the precautionary principle is a radical departure from the "ordinary" way of calculating risks and identifying problems. Risk is generally defined as the impact of some adverse event multiplied by the probability of its occurrence. Thus, similar risks can result from a modest impact with a virtual certainty of occurrence and a catastrophic event with a low probability of occurrence. Yet the magnitude of the harm (e.g., that caused by an oil embargo or by a rise in sea level) and the likelihood that it will occur may both be largely unknown, so there is "deep uncertainty." Nevertheless, policy analysis utilizes mechanistic decision tools based on the following:

• placing the burden on the public to prove that actions will be harmful, and explaining potential harm in terms of causality and probability

• separating science from social issues, with an emphasis placed on closure and scientific consensus

• measuring direct harm by a few core variables at the human or the molecular scale, often in the laboratory

• treating uncertainty as resulting from a lack of scientific data.[6]

Conventional policy making is still trapped in this "risk paradigm," which is also sometimes called the "acceptable discharge," "pollution control," or "technocratic" approach to risk calculation.[7] Risks are, by definition, quantifiable probabilities of things that either do or do not happen; the word thus captures the fact that risks refer to only those types of damage that can be expressed in a narrow and numerical way, such as cancer rates or the probability of birth defects.

Take the example of laws that manage toxic pollution in many countries by setting limits on "acceptable" discharges and then ensuring that releases do not exceed "minimum" standards of acceptable contamination. The paradigm assumes that ecosystems have an assimilative capacity to absorb and degrade pollutants. It also assumes that organisms can accommodate some degree of exposure with no or negligible adverse effects so long as the exposure is below the "threshold" at which toxic effects become significant. Regulations determine acceptable exposures by working backward from acceptable exposure level to calculate the maximum release rate that will ensure that this level is not exceeded; industries comply with these limits by installing pollution-control devices such as filters, scrubbers, and evaporators that capture pollutants at the end of the smokestack and then move them to a different place. Excluded from the paradigm are impacts that are difficult to quantify (such as immunosuppression, altered behavior, and reduced fertility) and risks that scientists do not yet fully understand. In the risk paradigm, lack of data is misconstrued as evidence of safety.

The risk paradigm also says something about the politics of the entire system. In common parlance, a risk is endured voluntarily in expectation of some benefit, as when one bets on the stock market, boards an airplane, or eats fatty foods. People can reduce the risks they take, but they can never eliminate them entirely. The paradigm implies that the only way to live risk-free is to do nothing at all.[8] The risk paradigm also tends to set the goal of pollution control as reducing the quantity of emissions per unit of product. Yet production is usually increasing, so pollution increases too.

The risk paradigm has come under fierce attack for being "utterly ill suited" to address long-term global threats from things such as pollution and climate change. The inadequacy of the paradigm begins with the very concept of "acceptable" discharges or emissions. Some chemicals and pollutants (such as mercury or benzene) persist in the bodies of living organisms and build to higher and higher concentrations, so that acceptable levels may ultimately reach unacceptable limits.[9] Others, such as carbon dioxide, accumulate in the atmosphere over very long periods. Recent research in toxicology and biology has shown that for many effects, including cancer, developmental impairment, and neurotoxicity, there is no clear threshold of harm. Similar studies in climatology have warned about dangerous and unknown "tipping points" in the atmosphere that, once crossed, cannot be recovered from, such as a permanent absence of ice during summers in the Arctic Ocean. This irreversible change is expected because of the positive feedback between warming and sea-ice melting, which reduces Earth's reflectivity.[10] Abrupt and threatening changes in human and ecological system—especially systems experiencing other forms of stress, such as over-population, an economic recession, or the introduction of invasive species—may also be triggered by climate change. Most people are used to thinking of climate change as like the turning of a dial, but a better metaphor is the flicking of a switch. Very real laws bind the natural world, and these limits are full of surprises, irreversible changes, and non-linear interactions. Though we may be uncertain about what these limits are, nature is an accountant that makes no errors. Any exposure, no matter how small, can contribute to the incidence of severe disease, to functional impairment of the global climate.

Pollution-control devices and carbon offsets may appear to reduce local pollution or emissions when in fact they merely shift its location (and do not reduce overall rates of contamination). Furthermore, such technologies and practices seldom perform as well as they are supposed to and are always susceptible to human error, poorly maintained equipment, and fluctuating environmental conditions. Landfill liners decay and leak, incinerators undergo explosions, trees planted to sequester carbon burn down, and chemical spills release contaminants into waterways and farms.[11]

The risk paradigm thus locks us into a situation like that of the Red Queen in *Alice's Adventures in Wonderland*, who had to run faster and faster just to stay in the same place. Companies must spend more and more on pollution-control devices or on technologies for capturing and sequestering carbon just to maintain a constant rate of production. The paradigm offers no way to address the burden of total pollution, and it is insufficient as a single decision-making paradigm for policies about the Earth's climate, which is a highly unpredictable, densely interconnected, hierarchical, and incompletely characterized complex system.[12]

The precautionary principle, by contrast, calls for the following:

• placing the burden on the proponents or producers to prove their actions will be safe, with explanations given in terms of patterns and associations
• multidisciplinary and interdisciplinary approaches that promote open-ended dialog instead of closure and consensus
• measuring direct harm by the disruption of biological, ecological, or social systems rather than laboratory experiments
• treatment of evidence and data as experimental, highly case-specific, and inclusive of both quantitative and qualitative differences
• accepting that uncertainty will always exist.

The precautionary principle implies that ecosystems and organisms should be viewed as complex and dynamic systems in which innumerable parts are connected in webs of interdependency with multiple causal and feedback loops. It suggests that humans can never truly predict or diagnose the impact of pollutants or emissions on a particular natural system, and that, faced with so many unknowns, we cannot afford to take risky bets on predictions or wait to know for certain if a particular activity is harmful. We should instead avoid practices that have the potential to cause damage in the absence of scientific proof that they do not. The precautionary principle, simply put, calls for us to err on the side of caution when the potential impacts of mistakes are serious, widespread, irreversible, and/or incompletely understood.[13] The precautionary principle provides strong additional justification for government actions to reduce the impacts of climate change.

Unfortunately, from a practical perspective, precaution does not provide the basis for balancing among competing goals when resources are limited or when an action supports one goal but undermines another.[14] For example, using the precautionary principle, oil companies might argue for reducing the risks of petroleum dependence by developing tar sands, oil shale, and liquid fuels from coal, while environmentalists might argue for addressing the threat of climate change by reducing the consumption of fossil fuels. In turn, fiscal conservatives might argue against any limits on carbon emissions in order to maintain economic growth. Like risk analysis, the precautionary principle is a tool to better understand vulnerabilities and consequences; therefore, like other tools, it can be used either properly or improperly.

With an appreciation for risk analysis and precaution in mind, the next section explores which types of policies seem best attuned to lowering GHG emissions and improving energy security, and how to select among them. If we recognize the value of precaution and the need to consider risks, an over-arching approach of "iterative" or "adaptive" governance appears judicious, admitting that no single set of judgments at one point in time is sufficient, but rather ongoing assessment, action, reassessment, and response is needed. Iterative governance emphasizes that many energy and climate-related decisions are fraught with deep uncertainties; as a result, the policy environment must be continuously monitored. That way, new knowledge about global climate change and energy security can be assimilated into decision making, along with information about advances in technology solutions and understanding of decision making and social systems. This approach builds on the work of Bryan Norton, a professor of public policy at the Georgia Institute of Technology who has written about the virtues of adaptive management of environmental resources. His approach does not evaluate static policy interventions, but rather iteratively evaluates "development paths"—paths unfolding into the future based on choices made in the present.[15]

Without reassessment and updating of policies, outmoded regulations can persist. For example, the US Clean Air Act exempts old and inefficient power plants, while inferior technical solutions (e.g., corn-based ethanol in the United States, coal-fired electricity in Indonesia and China) continue to be supported. At the same time, effective policies must be sufficiently steadfast to endure constant attacks from interest groups with stakes in their demise.[16]

6.2   Types of Public Policies and Ways to Evaluate Them

How should governments intervene? No single typology of public policies has gained dominance in the literature.[17] Table 6.1 (which is based on our research) offers an inventory of policy mechanisms organized by the principal objectives supported by each policy. The mechanisms listed are applicable to both industrialized and emerging economies. This typology does not prescribe, but rather illustrates the mosaic of diverse policies that currently guide nation-states. Further complicating matters, different methodologies exist for evaluating policy mechanisms—even after a particular goal has been declared.

6.2.1   Cost-Benefit Analysis

Cost-benefit analysis compares policies by quantifying their total monetary costs and total monetary benefits, then selecting the policies and projects that have the largest net benefits after discounting for time and risk. It can be used to recommend one policy over another for future implementation or to evaluate policies that have

Table 6.1
Energy and climate policy objectives and illustrative mechanisms.

Policy objectives	Illustrative policy mechanisms
Security of energy supply and diversity of energy resources	Portfolio standards, targets and quota obligations; tax incentives; regulated vs. open markets; negotiated agreements between producers and government; voluntary agreements; liability protection; administrative reforms to cut permitting time and costs
Economic growth, job growth, innovation, expansion of markets	R&D cost sharing and subsidies; grants for demonstrations and pilot programs; loan guarantees and zero-interest loans
Environmental protection, reduced GHG emissions	Renewable energy funds; green and white certificates; command-and-control regulations; air quality standards; investment tax credits; production tax credits; feed-in tariffs; eco-taxes; voluntary green electricity schemes; equipment and appliance standards; performance labels
Energy affordability and availability	Subsidies and tax incentives; power pricing policies; privatization and restructuring
Political posturing	Presidential directives and other statements by political leaders

been implemented.[18] For example, a coastal community considering an investment in a sea wall will assess the cost of building and maintaining the wall and will also assess any adverse impacts the wall might cause. As benefits, the community might consider the avoidance of storm damages. If benefits exceed costs, cost-benefit analysis recommends taking the action. Challenges to cost-benefit analysis include the need to monetize all costs and benefits and the need to estimate future costs and benefits accurately.

6.2.2   Cost-Effectiveness Analysis

Cost-effectiveness analysis is used in policy debates in which it is difficult to place a value on the benefits being generated, such as programs that promote public transit, public education, and drug prevention. It measures costs in monetary units, and effectiveness in units of goods, services, or some other valued outcome. In the context of energy and climate, cost-benefit analysis would require weighing the benefits of more secure energy and slowing climate change, "which introduces highly contentious issues in damage valuation, dealing with extreme climate risks, and intergenerational discounting."[19] Cost-effectiveness analysis avoids the problem of measuring benefits in purely monetary terms. As a result, it is more easily applied than cost-benefit analysis and is commonly applied to discussions of energy and climate policy.

Cost-effectiveness analysis is most feasible when policy makers have decided upon specific objectives or targets. Setting targets is generally viewed as a sign of good governance because it can help communicate policy goals, motivate appropriate action, and enable retrospective evaluation. For example:

• During his 2006 State of the Union Address, President George W. Bush declared that "America is addicted to oil" and set a goal of replacing more than 75 percent of US oil imports from the Middle East by 2025.

• On September 27, 2006, Arnold Schwarzenegger, governor of California, signed a bill (Assembly Bill 32) pledging to cap its emissions of greenhouse gases at 80 percent below 1990 levels by 2050.[20]

• In 2007, Shinz Abe, prime minister of Japan, announced the "Cool Earth 50" plan, which called for Japan and other industrialized countries to commit to reducing greenhouse gas emissions 50 percent by 2050.

• In March 2009, the Group of Eight countries endorsed a target of limiting the rise of global mean surface temperatures to 2°C (3.6°F) above pre-industrial levels.[21]

• In December 2009, China pledged to reduce its carbon dioxide emissions per unit of gross domestic product by 40–45 percent from 2005 levels by 2020.[22]

Most of these proposals are aspirational and not binding.[23] In general, agencies have not been assigned the responsibility of achieving them, and rewards and penalties have not been set.

In addition to considering the cost effectiveness of alternative policy mechanisms, analysts typically consider other criteria, including these:

• Effectiveness, meaning whether an alternative mechanism results in achievement of the policy objective. If the policy mechanism offers only a moderate level of assurance that a goal can be achieved, an alternative approach may be advisable. (For instance, it may be necessary to activate geo-engineering or adaptation if mitigation goals are not met.)

• Efficiency, meaning the amount of effort required to produce a certain level of effectiveness. This criterion is often calculated in costs per unit or volume of a product or a service (for example, dollars per gallon of alternative fuel, metric tons of carbon dioxide emissions avoided, or 100 gallons (379 liters) of municipal water per dollar expenditure).

Policies can also be evaluated by hybrid measures such as the following:

• Equal-cost analysis—that is, maximizing effectiveness within a set level of available resources. (For instance, a municipality may have only a set amount for promoting mass transit, so city planners must maximize ridership levels achievable by that investment.)

• Equal-effectiveness analysis—that is, minimizing costs while achieving a fixed level of effectiveness. (If public transit must serve at least 100,000 people per year, the problem is to identify those alternatives that will achieve this fixed level at least cost.)

• Variable-cost/variable-effectiveness analysis—that is, maximizing the ratio of effectiveness to costs (for example, choice of an optimal budget to maximize the attainment of agency objectives).

• Equal-cost/equal-effectiveness analysis, in which analysts are limited by the requirement that costs not exceed a certain level and also by the constraint that alternatives satisfy a predetermined level of effectiveness—for instance, public transit must serve a minimum of 100,000 people per year, but costs cannot exceed $100 (€77) per person.

With two goals that must be met, solutions using this last approach may be unachievable. While announced targets and objectives for energy security and climate change focus on achieving a single goal, the limited resources available to address them may move them into this double-constrained category, making them unachievable.

6.2.3   Equity and Distributive Justice

Last, but not least, issues of equity must be considered. An equitable policy is one in which effects (e.g., units of service or benefits) or effort (e.g., costs) are fairly or justly distributed. An inequitable or unjust policy might result when those most in need do not receive services in proportion to their numbers, when those who are least able to pay bear a disproportionate share of costs, or when those who receive most of the benefits do not pay any costs. Equity is related to the competing concepts of distributive justice (or fairness) and maximization of social welfare. Climate change is laden with issues of equity. Data on per capita CO_2 emission show a close correlation with affluence, yet, as was previously mentioned, the costs associated with climate change are likely to be much greater in developing countries.

Indeed, there are different ways to maximize individual welfare and social welfare. One way to include distributional values in a policy analysis is to consider the costs and benefits for relevant groups, such as income classes, geographic regions, and racial or ethnic groups.

• Maximizing individual welfare simultaneously is often sought, but it is difficult to achieve because there are usually trade-offs between winners and losers of a public policy.

• Protecting minimum welfare is an alternative that involves increasing the welfare of some persons while still protecting the positions of the worst off. The goal of this approach is to achieve a Pareto optimum, meaning that it is not possible to make one person better off without also making another person worse off. The goal of a Pareto optimum is seldom applicable, since most policies involve tradeoffs.

• Maximizing net welfare involves increasing total benefits less total costs, assuming that the resulting gains could be used to compensate losers. According to the Kaldor-Hicks criterion: one social state is better than another if there is a net gain in efficiency and if those who gain can compensate losers and still be better off.[24]

• Maximizing redistributive welfare focuses on selected groups in society—for example, the racially oppressed, the poor, or the sick. It reflects a principle that the American philosopher John Rawls has outlined: one social state is better than another if it results in a gain in welfare for members of society who are worst off.[25]

When applied to energy and climate, concepts of equity reveal that policies that raise the price of energy to consumers can have important distributional consequences and numerous tradeoffs. Energy-intensive industries and operators of coal-fired power plants will see their costs rise and may have difficulty competing in international markets. Since energy expenditures represent a much higher fraction of a low-income household's income (12 percent vs. 4 percent for the lowest versus highest-income quintiles), prices that raise energy costs will be highly regressive.[26] In the United States, the carbon emissions of a high-income household are 4 times those of a low-income household; at the same time, any harm that high-income households might experience from climate change is likely to be moderated by their greater adaptive capacity. The equity issues related to climate change between developed economies (which are responsible for a majority of the world's emissions) and developing economies (which are responsible for a small amount of emissions, but in which a majority of climate-related impacts will occur) were discussed in chapter 2. In evaluating a US policy that involves a cap-and-trade system aimed at reducing CO_2 emissions by 15 percent and allocating carbon allowances, Terry Dinan from the Congressional Budget Office and Diane Rogers from the Joint Economic Committee estimate that the households in the lowest-income quintile would be worse off by approximately $500 (€387) per year, whereas households in the upper-income quintiles would realize a net gain of approximately $1,000 (€770) per year. The gains experienced by members of the highest-income group are attributable to increased stockholder wealth that more than compensates members of that group for higher energy costs.[27] The distributional impacts of emissions-control policies can be quite variable across income, ethnic, and racial groups, across regions of a country, and across countries, and can penalize or subsidize particular economic sectors.

6.2.4 Political Feasibility and Critical Stakeholder Analysis

Closely linked to issues of cost effectiveness and equity is the issue of political feasibility—that is, whether or not a recommended policy could be implemented. In democratic societies, political feasibility can be assessed by conducting a "critical stakeholder analysis," which provides a framework for understanding allies and competitors whose involvement can affect the success or failure of a project or a policy. The word "stakeholder," first used in 1708 to refer to a person who held a stake or stakes in a bet or a game of chance, now refers to a group or an individual that can affect, or is affected by, an organization or an event. Critical stakeholder analysis identifies relevant stakeholders for a specified project or policy, maps out their relative power, influence, and interests, and assesses the broader context in which they interact.[28,29]

Critical stakeholder analysis brings three important benefits. First, it can facilitate dialogue and discussion among previously disconnected actors, which makes it an important component of democratic decision making; and it also can reveal power asymmetries between stakeholders.[30] Second, the process of identifying stakeholder interests can promote common understanding of agendas and help incentivize collaboration; conversely, it can identify zero-sum tradeoffs and incommensurable or irresolvable views among stakeholders that must be resolved for consensus to occur.[31] Third, by making some stakeholders and their power relations more visible, critical stakeholder analysis can improve social responsibility and force desirable change. Numerous examples abound of organizations and stakeholders being forced to alter their practices—especially those related to the destruction of the environment and to human-rights abuses—in response to broader stakeholder participation.[32]

By collecting and analyzing data on stakeholders, one can develop an understanding of—and possibly identify opportunities that influence—how decisions are made in a particular context. According to William Dunn, author of the well-known book *Public Policy Analysis*, stakeholder analysis involves four steps:

- identify relevant stakeholders to include for analysis
- assess each stakeholder's objectives, interests and beliefs
- characterize stakeholder resources
- elucidate the strategies and venues stakeholders use to achieve their objectives.[33]

In the remainder of this subsection, to give readers a sense of how to perform their own stakeholder analysis and of what the application of stakeholder analysis can reveal, we have depicted the stakeholders associated with three energy projects: renewable electricity in the state of Georgia, a natural-gas pipeline network in Southeast Asia, and an oil pipeline in the Caspian Sea and Eastern Europe.

For Georgia, we consider the institutions and organizations that have stakes in a statewide renewable electricity standard (a law that requires utilities to provide a minimum level of renewable electricity). The result of this analysis, presented in table 6.2, suggests that passage of a law setting a standard for renewable electricity in Georgia would require a considerable infusion of resources to realign stakeholders, since the most heavily resourced organizations would strongly oppose the policy.

Table 6.3 sets forth a slightly different approach, illustrating the underlying interests behind a sample of four stakeholders involved in two multi-billion-dollar pipeline projects in Asia, one dealing with natural gas (the Trans-ASEAN Gas Pipeline—TAGP) and one with oil (the Baku-Tbilisi-Ceyhan, or BTC, Pipeline). The exercise reveals how stakeholders can view pipelines as mechanisms of resource extraction that transfer wealth from developing countries to developed ones, or as systems of segregation that separate negative externalities from energy production from the positive attributes of energy consumption, or as symbols of national pride and modernity, or as components of large plans to promote economic stability and guaranteed returns on investment in the wake of economic and political crises. The Association of Southeast Asian Nations (ASEAN) has envisioned the TAGP as a mechanism to distribute energy, raise standards of living, and fight poverty, whereas the government of Thailand envisions it as a way to bolster its competitive advantage by becoming an "energy hub" where the government can take in cheap natural gas and transform it into expensive commodities and exports (even if it worsens poverty elsewhere). The government of Myanmar views the TAGP as a way to solidify its political control and build military strength, at times to be used against the country's own population, whereas Petronas (an oil and gas company owned by the government of Malaysia) views it as a way to cultivate the political and economic legitimacy of Muslims and Malaysians. The World Bank conceives of the BTC pipeline as a tool for legitimizing its approach to economic development, whereas British Petroleum and its consortium of companies see it as a way to convert the stranded oil reserves of the Caspian Sea into revenues and profits. Environmentalists see the BTC pipeline as a system of environmental segregation that enables the European Union and other importers to receive oil "free" of its negative upstream environmental impacts, whereas leaders in Azerbaijan interpret it as means of cultivating a strong, economically robust, independent state. Thus, various actors and stakeholders have differing reasons for becoming invested in such pipelines, and these reasons can interact to create social and political environments that largely dictate the pace of a project's development.

6.3 Putting a Price on Carbon

Just as important as evaluating public policies is deciding which ones ought to be specifically implemented to improve energy security and minimize climate change. One of the simplest actions countries and international institutions such as the United Nations could take to provide an equity-increasing and welfare-maximizing

Table 6.2
Stakeholder analysis of a Georgia renewable electricity standard.

Stakeholders	Interests and objectives	Expected position	Access to resources	Strategies and venues of influence
Georgia Power Company	Maintaining a reliable electricity system and generating profits	Oppose	Significant financial resources; lobbyists	Lobby for maximum time to meet requirements; lobby for all possible grants and funding to allay initial high costs
Renewable Energy Equipment Suppliers	Maximizing profits and growing businesses	Support	Limited financial resources; lobbyists	Lobby for funding to increase capacity; lobby to incentivize RE expansion
Fossil fuel Suppliers	Maximizing profits	Oppose	Significant financial resources; lobbyists	Lobby for legislation to minimize negative impact on profits
Ratepayers	Lowering electricity rates and maintaining reliability	Mixed	Influence Georgia Public Service Comm.	Elect candidates who minimize retail costs
Shareholders	Rising Georgia Power/Southern Company stock values; corporate responsibility	Mixed	Shareholder votes; influence corporate boards of directors	Influence Georgia Power's corporate leadership
Georgia General Assembly	Represent the interests of constituents	Mixed	Voting privileges; staff resources	Craft, introduce, and vote on legislation
Georgia Public Service Commission	Getting reelected; ensuring fair and sensible policies	Oppose	Regulatory authority over ratemaking	Craft and approve rules and regulations believed to be sensible
Federal Energy Regulatory Commission	Ensuring fair and legal interstate energy trading	Support	Regulatory authorities; staff and R&D resources	Craft and approve laws to maintain fair trade in developing market

Table 8.3
Stakeholder analysis of the Trans-ASEAN Gas Pipeline Network (TAGP) and the Baku-Tbilisi-Ceyhan (BTC) oil pipeline. Sources: B. K. Sovacool, "A critical stakeholder analysis of the Trans-ASEAN Gas Pipeline (TAGP) network," *Land Use Policy* 27, 2010, no. 3: 788–797; Sovacool, "Exploring the conditions for cooperative energy governance," *Asian Studies Review* 34, 2010, no. 4: 489–511; Sovacook, "The interpretive flexibility of oil and gas pipelines," *Technological Forecasting & Social Change* 78, 2011, no. 4: 610–620.

Actor	Type	Pipeline	View of project	Underlying interest
Association of Southeast Asian Nations	International Governmental Organization	TAGP	Delivery system for natural gas	To use the natural gas flowing from the TAGP to equalize economic disparities between ASEAN members and other developed countries as well as fight poverty and grow economies within Southeast Asia
Myanmar	Government	TAGP	Tool of political control	To use natural gas revenues facilitated by the TAGP to purchase military weapons and maintain political authority
Thailand	Government	TAGP	Mechanism for competitive advantage	To transform Thailand into an "energy hub" and also accrue revenue for state-owned enterprises
Petronas	State-owned Energy Company	TAGP	Component of national identity	To promote a unique brand of Malaysian economic legitimacy
World Bank Group	Multilateral Development Bank	BTC	Validation of organizational strategy	To affirm the organization's style of development assistance and prove that extractive industrial projects can benefit communities and prevent environmental degradation
British Petroleum/BTC Company	Multinational energy corporation	BTC	System of wealth transfer	To transform the "stranded" oil reserves of the Caspian Sea into billions of dollars of revenue and profits
European Union	Intergovernmental organization	BTC	Facilitator of environmental segregation	To "segregate" the environmental impacts associated with oil exploration, production, and distribution with use so that the European Union (and other importers) receive "clean" high quality crude oil while the environmental costs are distributed to Azerbaijan, Georgia, and Turkey
Azerbaijan	Government	BTC	Harbinger of national revitalization	To assert Azerbaijani sovereignty and independence from Russia and demonstrate the country's technical, economic, and political sophistication

response to climate change is to provide a market price for GHG emissions and to charge emitters for the cost of climate-change-mitigation technologies. Currently, GHGs can be emitted into the atmosphere for free in many countries, but the impacts of these emissions impose real costs on other societies. In contrast with the many barriers that are specific to individual technologies or sectors, this single obstacle is economy-wide. A carbon cap-and-trade system, carbon tax, or other policy mechanism for internalizing externalities in energy prices could help address cost-effectiveness barriers connected to unpriced costs and benefits related to carbon emissions. Such an approach would increase the competitiveness of energy-efficiency technologies and low-carbon fuels and power, and it would place greater value on carbon capture and sequestration projects and technologies for reducing non-CO_2 GHGs. Implementation of such mechanisms would also help to address the policy uncertainty that has become an important barrier to the domestic deployment of low-carbon technologies. Energy markets face numerous uncertainties even when operating within a stable policy framework.

Putting a price on GHG emissions can be accomplished with various policies, including energy and carbon taxes and cap-and-trade systems. An extensive academic literature suggests that macroeconomic efficiency favors a carbon tax with socially productive revenue recycling over other forms of regulation.[34] Energy-related taxes and standards have been high in many European countries and in other members of the Organization of Economic Cooperation and Development, including Japan (where energy audits for industrial facilities are mandatory) and Singapore (where vehicle use and ownership is tightly controlled), and the result has been energy-efficient practices in industry, buildings, transportation, and in the energy-supply sector. Raising energy taxes in the United States to comparable levels, however, has proved politically infeasible. However, cap-and-trade programs focusing on carbon and other GHGs have taken hold in several regional programs and have been incorporated into draft federal legislation, including the American Clean Energy and Security Act of 2009 (the "ACES Act").[35]

It has been argued that the choice of a policy instrument is less important than having an effectively designed instrument.[36] In a cap-and-trade program, sources of GHG emissions covered under the program receive sanctioning the amount of emissions they can produce. On the basis of that amount, sources of emissions can design their own control strategy, using any of several reduction options. Among the options are adopting new technology, purchasing offsets, and trading in the emissions market. This flexibility provides numerous advantages. Because emissions trading uses markets to determine how to deal with the problem of pollution, cap-and-trade is often touted as an example of effective free-market environmentalism.[37] Markets encourage low-cost solutions rather than mandating specific technologies. Though the cap is usually set by a political process, individual companies are free to choose whether and how they will reduce their emissions. In theory, firms will choose the least costly way to comply with the pollution regulation.[38]

In the framework provided by the National Commission on Energy Policy (NCEP),[39] design features of a cap-and-trade program pertain to the following:

- emission targets (What are the program's emission-reduction targets?)
- point of regulation (Where in the chain of energy production and consumption will the program be implemented?)
- price ceiling and floor (Will there be mechanisms to provide cost certainty?)
- offsets (How will credit be given for sequestration, agriculture and forestry projects, projects in other countries, and so on?)
- banking and borrowing of allowance credits (How are emissions reductions in the past or future treated?)
- allocation of allowances (Will allowances be auctioned, or given away? Who will receive them?)

6.3.1 Emissions Targets

In the Kyoto Protocol, emission-reduction goals were set individually by different countries. Australia, for instance, set a CO_2 emissions target of 8 percent above 1990 levels, Russia's target is to return to 1990 levels by 2012, whereas Japan's target is 6 percent below 1990 levels.[40] The US goal, which was debated but not ratified, would have capped US emissions at 7 percent below 1990 levels by 2012. US bills debated recently have called for reducing economy-wide global warming pollution by 80 percent of 2005 levels by 2050.

6.3.2 Point and Scope of Regulation

The greater the number of sources and gases covered, the greater the number of opportunities there are for low-cost reductions. The Kyoto Protocol covers a "basket" of six GHGs that represent the vast majority of global-warming gases: carbon dioxide (CO_2), methane (CH_4), nitrous oxide (N_2O), and industrial gases such as hydrofluorocarbons (HFCs), perfluorocarbons (PFCs), and sulfur hexafluoride (SF_6). The ten-state Regional Greenhouse Gas Initiative in the United States restricts its regulation to just one gas (carbon dioxide) and one sector (power plants). Though broader coverage of sectors and gases increases the options for low-cost reductions and increases the environmental benefits of the program (including many small sources and minor gases), it increases administrative complexity and transaction costs.

Cap-and-trade systems can regulate emissions at upstream facilities where fuels enter the economy, at a midstream point, or at downstream locations where emissions occur. The European Union's trading program exemplifies the midstream approach to regulating four broad sectors: iron and steel, minerals, energy, and pulp and paper. This approach covers only 46 percent of the EU's CO_2 emissions. If emissions were regulated further upstream, some of this "leakage" could be prevented.[41]

Downstream programs are expected to be more costly because they require the regulation of more numerous sources of emissions.[42] At the extreme, it would simply be infeasible to monitor emissions from vehicles, home heating fuels, and small-scale industrial boilers in a downstream system.[43] It has been estimated that regulating carbon dioxide at upstream facilities would involve less than 3,000 entities in the United States.[44] For example, there would be about 200 petroleum refineries and refined product importers, 500 natural-gas processing plants and importers of liquid natural gas, and 800–900 coal mines.[45] As a result of these high transaction costs, no major cap-and-trade program has taken a downstream approach to date.

6.3.3   Price Ceiling and Floor

Safety mechanisms involve setting a predetermined price ceiling at which the government will sell additional CO_2 allowances, thereby effectively capping the price of emissions allowances and constraining compliance costs. This can be achieved by having government make available an unlimited quantity of allowances at a predetermined price. If allowance prices rise above the prescribed limit, government allowances could then be purchased by emitters, providing "price certainty." This "safety valve" gives companies the regulatory certainty needed to optimize long-term investment decisions. However, it compromises "environmental certainty": to the extent that a price cap is activated, the cap on greenhouse gases will be exceeded. One way to reduce environmental uncertainty is to have the initial price rise from time to time. For example, increasing the price cap 5 percent per year would gradually strengthen the trading program's market signal and encourage emissions reductions.

When regulators establish a floor to allowance prices, investors in mitigation technologies can be assured of some level of consistency in financial incentives. Boom-bust cycles have become detrimental to the development of alternative transportation fuels (such as bioethanol and hydrogen for use in fuel cells) in the United States, to the development of solar photovoltaic panels in Germany and Japan, and to the development of wind turbines in Denmark and China. By providing greater assurances of market stability, allowance price floors would lead to a more orderly development of low-carbon technologies. Together, a price ceiling and a price floor create a "price collar" that has the potential of reducing long-term abatement costs.[46]

6.3.4   Offsets

Offsets are credits for emission reductions from sources outside of a regulated area. They allow entities outside the capped countries, sectors, and gases to participate and reduce emissions. To be eligible, offset reductions should be real, additional (that is, deliver emissions reductions beyond business-as-usual projections), verifiable, and permanent. As a result of these stringent requirements, verifying offsets can be difficult and costly. Reforestation appears to be one of the largest opportunities for offsets worldwide; the UN's Collaborative Programme on Reducing Emissions from Deforestation and Forest Degradation in Developing Countries is helping to organize these offsets. That program will help to ensure that offsets can contribute to worldwide cap-and-trade programs while helping to attract revenues to developing countries that will need financial and other assistance with programs to adapt to threats posed by global climate change.

Most cap-and-trade proposals limit the permissible volume of allowable offsets (e.g., 10–15 percent of allowances). Such quantitative limits on the use of international offsets help ensure that domestic emission reductions do not fall too short of the goals. To reach the 1 billion tons of offsets necessary for a climate bill in the United States by 2015, it would be necessary to review and approve thousands of projects in the first 3 years of the cap-and-trade program.[47]

6.3.5   Banking and Borrowing

Under cap-and-trade systems, the volatility of emissions prices can also be moderated by permitting the banking and borrowing of allowances. Banking mechanisms allow sources to carry forward surplus allowances into future years, thereby creating a cushion that can hedge price spikes in the future. Most existing cap-and-trade programs, including the federal SO_2 cap-and-trade system and the regional CO_2 programs in the United States and the European Union's CO_2 program, contain banking provisions.[48]

Borrowing mechanisms allow sources to use allowance allocations from future years to meet current obligations to reduce emissions, thereby mitigating price spikes. Some proposals with this mechanism have included interest rates on borrowed allowances. A variation on borrowing is the use of a multi-year compliance period, which allows a source to use future reductions in emissions to qualify for the current year without penalty. A three-year compliance period is used in the Regional Greenhouse Gas Initiative, and a two-year compliance period is proposed in the American Clean Energy and Security Act.[49]

6.3.6   Allocating Emission Allowances

In many trading programs (for example, the EU Emission Trading Scheme and the US cap-and-trade programs for SO_2 and NO_x), most allowances have been given away to regulated entities. In an upstream GHG trading program, however, regulated entities (fuel producers and processors) will pass most of the allowance "costs" downstream to fuel consumers. Giving away allowances to these regulated entities thus creates a "windfall profit" for them, since they receive higher prices for the fuels they sell, but they do not have to use these revenues to purchase allowances. This is happening in the European power sector as a result of free allocation in the EU Emission Trading Scheme. It has been estimated that the UK power sector received a windfall of about $1 billion (€774 million) in 2005 as a result of the allowance allocation.

Allowances should be allocated in a manner that recognizes and roughly addresses the disparate costs imposed by the program. In addition to regulated entities, other groups that could receive allowance allocations include electricity suppliers, energy-intensive industrial sectors (steel, chemicals, etc.), and low-income households. Some portion of the allowances could also be auctioned to raise revenues for technology research and deployment programs.[50]

Some schemes extend beyond the electricity and energy-supply sectors to cover transport and aviation. In 2008, for example, the European Union agreed to expand its Emission Trading Scheme to include aviation, an industry responsible for about 3.5 percent of global climate change when atmospheric chemical reactions are taken into consideration. The EU system caps carbon dioxide equivalent emissions from both European and foreign airlines, and permits them to buy and sell pollution credits on the European market. The scheme requires airlines traveling to any of the current 27 member countries to buy 10 percent of the requisite permits under an auction system, and to distribute the rest for free.[51] Though the scheme will not be fully implemented until 2012, the International Civil Aviation Organization has already begun developing tools to measure carbon emissions from aircraft operators, and in 2009 the Arab Air Carriers Organization (consisting of eleven national airlines in the Middle East) were already tracking their emissions and formulating reduction plans.[52]

Though cap-and-trade approaches are most common among existing carbon pricing systems (e.g., the Kyoto Protocol and the EU Emission Trading Scheme), carbon taxes are still actively being debated in some countries. Tax systems can be similarly integrated with cap-and-trade schemes operating elsewhere in the world by authorizing Certified Emission Reductions from the Clean Development Mechanism to serve as tax offsets; firms could also be allowed to sell offsets to international buyers. Allocating and trading allowances domestically, however, would not be part of a tax system.[53]

6.3.7   Economic Impacts and the Regulatory Alternative

The economic impacts of cap-and-trade programs depend in part on how allowances are allocated; they also are a function of the emission-reduction targets and numerous other variables. Several different types of models are available for evaluating the economic cost of cap-and-trade systems. At one extreme, "top-down" computable general equilibrium models focus on capital dynamics, demand responses, and factor substitution, but tend to have limited technology characterization. At the other extreme, "bottom-up" engineering-economic models tend to have detailed representation of technologies and can characterize technological innovation but are more limited in modeling macroeconomic effects. Between these extremes are several hybrid models that have been developed to evaluate energy and climate policies.[54] Partly as a result of these modeling variations, analysts have produced divergent conclusions about the cost of implementing cap-and-trade systems. When credible studies reach different findings, many policy makers find it difficult to endorse change at all and thus maintain the status quo.

A version of the status quo in the United States is to regulate greenhouse gas emissions under the US Clean Air Act. In 2007, the Supreme Court ruled (in *Massachusetts v. EPA*) that the EPA has the authority to regulate heat-trapping gases. Indeed, the Supreme Court stated that the EPA cannot sidestep its authority to regulate GHGs unless it can provide a scientific basis for its refusal. With this backing from the Supreme Court, EPA has deemed greenhouse emissions threats to public health under the Clean Air Act. In December 2010, the EPA announced a schedule for setting GHG standards for power plants and oil refineries over the next two years. That means the agency can require emitters to reduce their GHG emissions. Regulating GHGs under the Clean Air Act has a number of relative weaknesses. Because such a rule raises no revenue, the government cannot easily compensate consumers for the disparate costs imposed by the regulations, including energy-intensive industries and low-income households. Because it depends on the policy preferences of the president, new administrations might move to weaken the policies, leading to the type of regulatory uncertainty that frustrates business today. Finally, and perhaps most important, because it relies on federal mandate, it may not put the US on the cheapest path to sustainable energy production.

6.4 Complementary Policies

Putting a price on carbon is a critical "core" policy because it addresses the principal market failure that has prevented individuals and firms from responding effectively to the damages precipitated by GHG emissions. Some have argued that putting a price on carbon is all that is needed. Evidence is mounting, however, that complementary policies are needed to ensure the speed, scope, and scale of response required to "avoid dangerous anthropogenic interference in the Earth's climate" as called for in the UN Framework Convention on Climate Change.

The first argument for complementary policies is based on the insufficiency of current cap-and-trade programs and proposals, with their reliance on offsets, price ceilings, "gifted" allowance allocations, and protracted phase-in periods. Policy makers do not appear to be willing to act quickly to cut emissions deeply. There are many factors in this reluctance, including skepticism about the reality of the climate threat, anticipated costs to emitters, possible impacts on competitiveness, challenges to enforcers including measurement difficulties, and concerns about risks of gaming and cheating. Emissions-pricing policies can also have deleterious distributional effects that complementary policies can ameliorate.

Another persuasive line of argument suggests that the tendency for the obstacles facing GHG-reducing technologies to be technical as well as political and economic means that policy instruments should be similarly multidimensional. Despite billions of dollars spent on R&D, government procurement, tax incentives, tax credits, subsidies, standards, and financial assistance, the impediments to more sustainable forms of energy supply and use are at least partly social and cultural. Until these remaining cultural barriers are targeted in the same way that engineers and scientists tackle technical impediments, the promise of new climate-proof systems will remain unfulfilled. Consumer attitudes, values, beliefs, and expectations are just as important as improved tires, better fuel economy, longer-lasting batteries, and tougher and lighter wind turbines in explaining why people embrace some forms of technology but not others.

For example, if consumers were rational, the adoption of energy-efficient technologies and practices would be commonplace, because energy costs money and thus prevents the consumption of other goods. (If consumers properly understood money and risk, Las Vegas would still be only a desert). In a perfect world, consumers would always choose the more efficient product, all other characteristics being equal. Responding to consumer choices, suppliers would improve the efficiency of their products, with demand pulling supply.

In reality, psychological elements can challenge the rational actor:

• Deliberation costs. Consumers reduce the cost of evaluating alternatives by relying on heuristics, copying others, and defaulting to the status quo. Brand loyalty, stock availability, the persuasiveness of sales people, and other screening devices are used to simplify the selection process. The ultimate tendency to default to the status quo means that many energy decisions are actually *non*-decisions, simply the result of ingrained or repetitive behavior.[55]

• Aversion to losses. People strongly prefer avoiding losses to acquiring gains—perhaps by as much as a 2:1 ratio. The "sunk cost" fallacy of protecting investments already results in high discount rates and inertia, contributing to the "lock-in" of old energy-intensive and carbon-intensive choices. Daniel Kahneman received a Nobel Prize in Economics for work in this area (done with Amos Tversky).[56]

• Experience. People put greater weight on options grounded in recent experience and personalized information. Providing feedback can improve decision making by connecting decisions with outcomes. This is particularly apparent in energy and water consumption when we consider how quickly people forget the details of their monthly bills. Smart meters that provide real-time information to customers and display panels that communicate fuel consumption to drivers show great promise to improve the efficiency of energy use.[57]

How government policies can best address these cultural and behavioral impediments is not obvious and is the subject of a sizeable body of research by social scientists. Evidence to date clearly indicates that the strongest influences on behavior are often contextual (e.g., lack of local availability of a new technology, or inability to retrofit it to a factory, a home, or an office building because of characteristics of the current structure). The weaker such contextual constraints, the stronger the influence of personal factors [58]

After a great deal of analysis, dialog, and debate, the Intergovernmental Panel on Climate Change concluded that a few complementary policy measures had proved successful in multiple countries.[59] We have adapted the policies proposed by the IPCC into our table 6.4. It is unlikely that this list of options is the exact policy remedy for any single country, in view of the divergence of contexts; however, they represent an excellent list for consideration. Organized by sector, they largely mirror the five challenges discussed in chapter 2. They are divided into two types. The mandatory regulatory standards ("sticks") deliver fairly predictable outcomes (if enforced) but may suffer from issues of equity and efficiency; the voluntary incentives ("carrots") subsidize and motivate choices that contribute to energy security and to climate-change mitigation, and include support for R&D, but their effects are less predictable.

6.4.1   Energy-Supply Options

Policy options for supplying energy include "renewable energy obligations," such as renewable portfolio standards and real-time pricing for electricity, as well as reducing fossil-fuel subsidies and passing feed-in tariffs. Some of these instruments have become quite popular, with 85 countries having some type of policy target for renewable energy in 2009, a jump from only 45 in 2005.[60] Europe's target of 20 percent of final energy by 2020 is predominant among countries belonging to the OECD, Brazil is targeting 75 percent renewable electricity by 2030, China 15 percent final energy by 2020, India 20,000 megawatts of solar by 2022, Kenya 4,000 megawatts of geothermal by 2030.[61] In early 2010, no less than 50 countries and 25 states and provinces had some type of feed-in tariff, and 46 countries were home to renewable portfolio standards for electricity. A number of towns and municipalities around the world—including Güssing (Hungary), Dardesheim (Germany), Moura (Portugal),

Table 6.4
Selected sectoral policies, measures, and instruments.

Sector	Policies, measures, and instruments shown to be environmentally effective
Energy supply	**Mandatory Regulatory Standards** Renewable energy obligations* Net metering and improved electricity pricing systems **Voluntary Policies/Incentives** Producer subsidies* Reduction of fossil-fuel subsidies* Feed-in tariffs for renewable energy technologies*
Transport	**Mandatory Regulatory Standards** Mandatory fuel economy and biofuel blending* Limiting the number of driving days—CO_2 standards for road and aviation transport* **Voluntary Policies/Incentives** Influence mobility needs through land use regulations, and infrastructure planning* Investment in attractive public transport facilities and non-motorized forms of transport* Incentives for energy services companies*
Buildings	**Mandatory Regulatory Standards** Appliance standards and labeling* Building codes and certification* **Voluntary Policies/Incentives** Demand-side management programs* Public sector leadership, including procurement* Incentives for energy services companies*
Industry	**Mandatory Regulatory Standards** Performance standards* **Voluntary Policies/Incentives** Subsidies, tax credits* Voluntary agreements—Provision of benchmark information*
Agriculture and forestry/forests	**Mandatory Regulatory Standards** Land use regulation and enforcement* Harvest quotas and limits on timber extraction—Renewable energy obligations or incentives for anaerobic digestion or landfill gas capture* **Voluntary Policies/Incentives** Financial incentives and regulations for improved land management, maintaining soil carbon content, efficient use of fertilizers and irrigation*

Table 6.4
(continued)

Sector	Policies, measures, and instruments shown to be environmentally effective
	Financial incentives (national and international) to increase forest area, to reduce deforestation, and to maintain and manage forest*
	Payments for ecosystem services
Waste management and water	**Mandatory Regulatory Standards**
	Waste management regulations*
	Volumetric water pricing—Renewable energy incentives or obligations for anaerobic digestion or landfill gas capture*
	Voluntary Policies/Incentives
	Financial incentives for wastewater treatment and recycling
	"Cradle to cradle" design—Extended producer responsibility

*According to the Intergovernmental Panel on Climate Change (*Climate Change 2007: Mitigation of Climate Change*, Cambridge University Press, table SPM.7), these policies have been shown to be environmentally effective in the respective sector in at least a number of countries. R&D investment in low emissions technologies have proved to be effective in all sectors.

Varese Ligure (Italy), Samsø (Denmark), Thisted (Denmark), Frederikshavn (Denmark), Bourk (United Kingdom) and Rock Port (Missouri) have already achieved 100 percent renewable energy sectors or will achieve them by 2015.[62] Table 6.5 provides an overview of these policies at the national level around the world.

As table 6.5 demonstrates, policy makers have adopted numerous other types of policies to promote renewable energy, many in combination. Direct capital investment subsidies, grants, and rebates are offered in 45 countries; tax credits, import duty reductions, and other tax incentives are offered in more than 30 countries; net metering laws now exist in 10 countries and in 43 US states.

In one innovative program, Ellensburg, a city in the state of Washington, began promoting virtual net metering to incentivize residents to invest in a municipal-scale community solar photovoltaic system. The city built a 36-kilowatt solar array in 2006 and asked interested residents to contribute to its capital cost; in return, participants receive a credit on their electricity bill apportioned to their level of investment. The municipal utility purchases electricity from the solar PV system (currently at the wholesale rate of 3.5 cents (2.7 euro cents) per kilowatt-hour but soon to be 15 cents (12 euro cents) per kilowatt-hour once qualified under the state's production tax credit), and then divides the total amount among the system's contributors. The use of virtual net metering is intended to overcome the high capital cost of

Table 6.5
Enacted policies for promoting renewable energy as of 2010.

	Feed-in tariff	Renewable portfolio standard/quota	Capital subsidies, grants, rebates	Investment or other tax credits	Sales tax, energy tax, excise tax, or vat reduction	Tradable re certificates	Energy production payments or tax credits	Net metering	Public investing, loans, or financing	Public competitive bidding
European Union										
Austria	×		×	×		×			×	
Belgium		*	×	×	×	×		×		
Bulgaria	×		×						×	
Cyprus	×		×							
Czech Republic	×		×	×	×	×		×		
Denmark	×		×	×	×	×		×	×	×
Estonia	×		×		×		×			
Finland	×		×		×	×	×			
France	×		×	×	×	×			×	×
Germany	×		×	×	×			×	×	
Greece	×		×	×				×	×	
Hungary	×		×	×	×				×	×
Ireland	×		×	×			×			×
Italy	×	×	×	×	×	×		×	×	
Latvia	×				×				×	×
Lithuania	×		×	×	×				×	
Luxembourg	×		×	×	×					
Malta			×		×			×		
Netherlands			×	×	×		×	×		
Poland		×	×		×	×			×	×
Portugal	×		×	×	×				×	×
Romania		×			×	×			×	
Slovakia	×			×	×				×	
Slovenia	×		×	×	×	×			×	×
Spain	×		×	×	×	×			×	
Sweden		×	×	×	×	×	×		×	
United Kingdom	×	×	×		×	×			×	

Table 6.5
(continued)

	Feed-in tariff	Renewable portfolio standard/quota	Capital subsidies, grants, rebates	Investment or other tax credits	Sales tax, energy tax, excise tax, or vat reduction	Tradable re certificates	Energy production payments or tax credits	Net metering	Public investing, loans, or financing	Public competitive bidding
Other developed/transition countries										
Australia	*	×	×		×	×			×	
Belarus									×	
Canada	*	*	×	×	×			×	×	×
Israel	×				×					×
Japan	×	×	×	×		×		×	×	
Macedonia	×									
New Zealand			×			×			×	
Norway			×		×	×			×	
Russia			×							
Serbia	×									
South Korea	×		×	×	×				×	
Switzerland	×		×		×					
Ukraine	×									
United States	*	*	×	×	*	*	×	*	*	*
Developing countries										
Algeria	×			×	×					
Argentina	×		×	*	×		×		×	×
Bolivia					×					
Brazil				×					×	×
Chile		×	×	×	×				×	×
China	×	×	×	×	×		×		×	×
Costa Rica							×			
Dominican Republic	×		×	×	×					
Ecuador	×			×						
Egypt					×					×
El Salvador				×	×				×	
Ethiopia					×					
Ghana			×		×				×	

Table 6.5
(continued)

	Feed-in tariff	Renewable portfolio standard/quota	Capital subsidies, grants, rebates	Investment or other tax credits	Sales tax, energy tax, excise tax, or vat reduction	Tradable re certificates	Energy production payments or tax credits	Net metering	Public investing, loans, or financing	Public competitive bidding
Guatemala				×	×					
India	*	*	×	×	×	×	×		×	
Indonesia	×			×	×					
Iran				×			×			
Jordan					×				×	×
Kenya	×			×						
Malaysia									×	
Mauritius			×							
Mexico				×				×	×	×
Mongolia	×									×
Morocco				×	×				×	
Nicaragua	×			×	×					
Pakistan	×							×		
Palestinian Territories					×					
Panama							×			
Peru				×	×		×			×
Philippines	×	×	×	×	×		×	×	×	×
Rwanda									×	
South Africa	×		×		×				×	×
Sri Lanka	×									
Tanzania	×		×		×					
Thailand	×				×				×	
Tunisia			×		×				×	
Turkey	×		×							
Uganda	×		×		×				×	
Uruguay		×								×
Zambia					×					

*In these countries, some states or provinces have policies but there is no national-level policy.

purchasing a solar array, and was designed to offer a middle ground between voluntary green power programs (in which electricity customers can voluntarily contribute $5 (€3.87) to $10 (€7.74) per month to promote renewable energy) and the massive $20,000 (€15,478) to $30,000 (€23,217) investment required to place a solar PV system on one's property. More than 70 utility customers (including a number of departments at Central Washington University) initially contributed about $120,000 (€92,868) to the community solar PV system, the Bonneville Environmental Foundation (a nonprofit organization) contributed 40 solar modules worth $45,000 (€37,147), and the city pledged to cover the expense of operations and maintenance. The program was so successful that the city is collecting additional contributions and expects to expand the PV system.[63]

6.4.2   Transport Options

Transport policies include mandates for biofuel blending along with investments in alternative transport and carbon dioxide standards for cars or airplanes. Biofuel blending mandates exist in 41 states and provinces and 25 countries as of 2010, with most requiring a blending of 10–15 percent ethanol with gasoline or 2–5 percent biodiesel with diesel fuel.[64] Biofuels targets exist in more than 10 countries plus the European Union, and exemptions for fuel taxes and production subsidies are also common.

The city of Vauban, Germany, creatively limits the use of automobiles through coordinated restrictions on parking, building standards, road construction, and driving. A former German army base with only 5,500 residents, Vauban was never meant to accommodate private cars and has an abundance of narrow roads. It has forbidden street parking and the parking of automobiles in driveways and home garages in most areas. Likewise, it has mandated that the streets be completely free of cars, excepting one main thoroughfare and a few streets on the edge of the district. Though automobile ownership is still permitted, drivers can park only in a large garage at one end of the city (for $40,000 a space) or at their home. As a result, 70 percent of families residing in Vauban do not own a car, and 50 percent sold their car to move there.[65] Residents of Vauban boast that life there has a slower pace. People get around largely by walking and bicycling. When people want to take a trip outside of the city, they mostly rely on public transport (such as the tram; in rare cases, they pool funds to purchase cars, which they share, or they rent from a car-sharing club).[66]

Mexico City has taken a different approach to improve the flow of traffic and reduce pollution coming from automobiles. The city has strictly limited the days on which particular automobiles may operate in the city through a government program known as Hoy No Circula (meaning Day Without a Car). Started in 1989, the program bans drivers from using a vehicle with a certain number plate one weekday per week. (Taxis are excluded.) The restrictions apply from 5 a.m. to 10 p.m. During its first year, the program kept 460,000 vehicles off the streets per day. It now keeps about 1.6 million cars off the streets per week.[67]

In Israel, the government has started an ambitious program to promote plug-in hybrid electric vehicles (PHEVs). The government has teamed up with automobile and battery manufacturers to distribute PHEVs, construct recharging facilities, and create service stations that can quickly replace depleted batteries. Renault and Nissan provide the cars (at prices lower than those of gasoline vehicles thanks to an Israeli subsidy), and Project Better Place provides lithium-ion batteries that are capable of traveling 124 miles (200 kilometers) per charge. The government provides the infrastructure needed to keep the cars going, such as small "plugging stations" on city streets and at service stations and highways. When the batteries in one's car no longer perform well, one can visit a car-wash-like station and have them replaced in a few minutes. To get drivers interested, the government offered generous tax incentives. It has also invested $200 million in public funds in electric vehicle

Figure 6.2
An electric car being recharged in Israel. Source: Marilyn A. Brown.

infrastructure. Drivers get an electric vehicle at a greatly reduced price and then pay a fixed monthly fee for mileage for the electricity they use.[68] Since in Israel gasoline prices exceed 6 US dollars per gallon, and since 90 percent of car owners drive fewer than 45 miles per day, PHEVs are gaining market share. In one survey, 57 percent of Israeli drivers reported being interested in a PHEV as their next car, and 28 percent indicated they would *only* consider purchasing a PHEV. A few hundred PHEVs had already hit the road as of early 2009, and by 2011 the government expects to have half a million PHEVs distributed throughout the country.[69]

6.4.3   Options for Buildings and Homes

Policy options for buildings and homes are numerous, including regulatory approaches (such as appliance standards and labeling and building codes) and voluntary policies (such as demand-side management programs operated by electric and gas utilities and incentives for energy services companies). Cities and local governments around the world are especially becoming involved in setting building standards that require the installation of renewable energy. For example, in 2008, Spain became the first county to mandate solar water heating. And in Jiangsu, one of the most populous provinces in China, all new residential buildings with no more than twelve stories must use solar water heating.[70]

In the early 1990s, the city of Minneapolis and CenterPoint Energy initiated an innovative neighborhood-based energy program. Employees for the city trained volunteers to serve as block captains and to invite their neighbors to "energy workshops." The workshops focused on providing information about energy-use habits (e.g., how daily routines such as washing clothes and showering affect energy bills), on the energy efficiency of electric appliances and furnaces (e.g., by providing a list of how much electricity or natural gas these devices used), on weatherization techniques that could be implemented quickly and at low cost (e.g., caulking or adding insulation), and on large-scale retrofits and upgrades that could drastically reduce energy consumption (e.g., replacing windows or installing geothermal heat pumps). Participants were given $50 (€39) worth of materials and then trained to take low-cost actions independently. Those who then desired to pursue major, more capital-intensive projects were eligible for a free energy audit and were offered low-interest financing from the city. Participants electing to choose this latter option were also given a list of city-certified contractors and free infrared thermographs to help them spot areas where insulation could be improved and heat losses minimized. The entire program cost about $80 (€62) per household, and one assessment of energy savings found a reduction in total energy use of 4.3 percent, an average payback time of 1.9 years, and block participation averaging 35–40 percent.[71]

Japan has been especially successful at promoting appliance standards. Minimum energy performance standards, which went into effect for refrigerators and air conditioners in 1983, were later expanded to virtually all appliances, including electric toilet seat warmers. The appliance standards effectively reduced electricity consumption within a short period of time. Average electricity use of refrigerators, for example, declined by 15 percent from 1979 to 1997, even though the size of the average refrigerator increased by 90 percent. Japanese regulators also applied their performance standards to imported automobiles, televisions, air conditioners, and computers, demanding that new imported products be as efficient as the most efficient Japanese counterparts and in some cases requiring energy-efficiency improvements of more than 50 percent.[72]

Electric and natural-gas utilities all over the world have implemented demand-side management programs to encourage their customers to use energy more efficiently. Recognizing that cost barriers hinder consumer investments, these programs often provide rebates, low-interest loans, or other kinds of financial assistance. Most recently, on-bill utility financing has gained popularity, because it enables energy-efficiency improvements without up-front capital costs to the building owner and it addresses two important behavioral barriers: deliberation costs and loss aversion.[73] In these programs, a utility's billing system is used to collect a charge that is attached to the electric or natural-gas meter as a special tariff to repay the cost of energy improvements. Typically, the monthly charge is less than the expected savings from the efficiency improvements, so bills do not rise, thus minimizing risk to the consumer. Because the payment is tied to the meter and not the homeowner, if the current occupant moves, the next occupant must accept responsibility for repayment; this minimizes the risk to the utility. Such a program has been successfully operated by Manitoba Hydro since 2001. That Canadian utility has signed more than 41,000 loans, a particularly high level of participation considering that the utility has some of the lowest residential electric rates in the world. The most popular retrofits have been installation of energy-efficient windows, upgrades to doors, and improvements in heating systems. The Manitoba Hydro program reviews applications within a day. Its loan approval rate is high (94 percent of total applications), and its default rate is impressively low (0.2 percent).[74]

6.4.4   Industry Options

Policy options for industry include mandatory performance standards or audits for manufacturers along with voluntary agreements and the provision of benchmarking information. For example, the Netherlands has taken a proactive stance on industrial energy efficiency, beginning in 1992 with the Long Term Agreements on Energy Efficiency with industry. These agreements were established through an understanding by the industry that the government was observing energy consumption closely and would not initiate strong regulations so long as industry met the agreed targets. A second phase of this program, launched in 2000, entails "benchmarking" the most energy-intensive industries to comparable industries worldwide. The affected industries must be the best in their class in energy efficiency; in return, the government promises not to implement additional stringent climate-change policies.[75]

As is true of most developing countries, India's industrial makeup is dominated by small and medium-size companies. To achieve the ambitious goal of reducing their energy intensity by 5 percent each year, the government of India has introduced an energy efficiency trading program. It is expected that this market will be worth $15 billion (€11.6 billion) and will cover nine sectors of the economy by 2015. This approach is very similar to other markets for efficiency credits, but India mandates reductions similar to those in a cap-and-trade program. This is a unique approach for a developing country, with the expected outcome of more rapid deployment of efficient technologies throughout the Indian economy.[76]

From 1980 through 2000, China reduced its national energy intensity by 65 percent as the result of process and technological changes and structural shifts throughout Chinese industry. Rapidly developing countries typically see an increase in energy intensity; China was able to reverse this trend through a series of policy reforms allocating capital toward energy efficiency and developing energy service conservation and energy management centers, which act similarly to energy services companies. However, China has recently seen an increase in energy intensity for the first time in decades. Late in 2005, noting worsening energy efficiency conditions, the Chinese government announced a mandatory reduction in energy intensity of 20 percent by 2010. Initial responses, however, were not sufficient to reverse the trend. In an attempt to meet the mandate, the Energy Conservation Law was revised, tax policy was modified for export products, tax credits for efficiency investments were granted, and a "Top-1000 Energy Consuming Enterprises" program was initiated to promote energy efficiency among large industrial energy users. These policies have placed China on a path toward reaching its mandates and again reducing its energy intensity.[77]

6.4.5   Agriculture and Forestry Options

Agricultural and forestry options include regulation of land use, harvest quotas for timber, financial incentives for improved land management and to increase forest area, and payments for ecosystem services.

One policy used effectively in the state of Tamil Nadu, India, provides financial incentives for organic fertilizer. Owners of tea plantations have grown and distributed bio-organic fertilization to replenish degraded land, restore soil fertility, and improve productivity. After decades of excessive application of chemical fertilizers and pesticides had depleted the soil's fertility and crippled the productivity of tea plantations (in some cases resulting in crop losses as high as 70 percent of ordinary yield), plantation managers collaborated with university researchers and a fertilizer company to use natural methods to restore the land. Researchers placed vermicultured earthworms in trenches between rows of tea plants. They used tea prunings and high-quality organic and horticultural waste from nearby farms to create organic fertilizer, which they then distributed to six large tea estates. The combination of earthworm trenching and organic fertilization increased tea yields by 76–239 percent, and profits rose significantly.[78]

In Costa Rica, payments for ecosystem services are distributed to the owners of forests and forest plantations in exchange for their preservation and management of the land. The Costa Rican program, passed as part of that country's 1996 Forest Law and termed the Private Forest Project, recognizes four services provided by forests—protection of biodiversity, sequestration and fixation of carbon, erosion prevention and water purification, and scenic beauty—and uses revenue from activities that threaten those services to pay land owners for their services. A 5 percent tax on gasoline collects about $16 million per year to enhance biodiversity protection, the sale of carbon credits (called Certifiable Tradable Offsets) helps to pay for carbon sequestration, and donations from private hydropower companies sponsor hydrologic services. During the first 2 years of the program, more than 1,000 land owners signed contracts to receive payments averaging $120 per hectare per year for plantations, $60 for forests, and $45 for forest management and reforestation. The combined taxes and donations now produce about $16 million–$20 million a year and generate 4 million tons of carbon credits to be brokered on the international market.[79]

In Malaysia (which has less than 0.25 percent of the world's forests but 10 percent of the world's plants and 7 percent of its biodiversity), regulators passed a National Forest Act in the 1980s to classify forests and set limits on harvesting and deforestation. The rules mandated that only trees of a certain length and age could be felled (protecting both young and old trees), prohibited harvesting of timber and wood within an extensive network of reserves, set strict quotas, and relied on surveillance (now done by satellites) to track compliance. In 2007, the maximum harvest quota was 50,000 cubic meters, and newer standards require that those forests that have been harvested undergo regeneration and restoration efforts. Under such policies, forest area increased from 58.7 percent of land area in 2000 to 63.6 percent of land area in 2005.[80]

6.4.6   Waste and Water Options

Policy options for waste and water include waste-management regulations, volumetric water pricing, incentives for waste incineration, anaerobic digestion, cleaner production processes, and establishment of the principle of extended producer responsibility.

Policies that promote "design for recycling" are becoming increasingly popular. They minimize waste streams by producing products that can be reused and recycled back into the manufacturing process at the end of their useful life.[81] Designing products for "closed-loop supply systems" is an extension of this concept and is a relatively new field. It draws upon decades of related experience from design for serviceability and maintainability, and incorporates the newer concepts of remanufacture and recycling. Policies to promote the design for closed-loop supply chains are complicated by the fact that post-consumer returns can occur years after the product was designed in a world in which technology and business conditions have changed.[82] A related promising approach is the use of "cradle-to-cradle" designs that require the recycling of products and materials back into the manufacturing process at the end of their useful life.

The European Union has begun to address the pollution coming from discarded products through the principle known as extended producer responsibility (EPR).[83] First legislated in Germany and then expanded into an EU directive in 2001, EPR assigns long-term responsibility for the environmental impacts of many products (including lawnmowers, household paints, computers, batteries, and cellular telephones) to their manufacturers. It requires that manufacturers take back their products or charge consumers a small fee to pay for collection and recycling.[84] EU member countries have implemented the EPR directive differently, the four most prevalent types of implementation being economic implementation (which requires manufacturers to pay all or a portion of end management and disposal or recycling costs), physical implementation (which requires manufacturers to take possession of discarded goods to ensure that materials and components are recycled), informative implementation (which requires manufacturers to publish information about where consumers can recycle their product), and legal implementation (which makes manufacturers liable for the environmental damage resulting from their products, including costs for remediation, cleanup, and disposal).

The central premise of extended producer responsibility is that a manufacturer should be made responsible for its goods. As a result of EPR legislation in Europe, manufacturers have designed products to be more recyclable and/or decreased the use of environmentally damaging raw materials, have improved their efforts to collect discarded goods, have incorporated recycled components and materials back into their production processes, have adopted modular designs that are easier to disassemble, and have unified and harmonized standards for various types and grades of materials and plastics.[85]

6.5   The Need for Synergy

Unfortunately, complementary policies such as undertaking research and development, adjusting subsidies, internalizing externalities, promulgating standards, and improving information will not work in isolation. Changing R&D practices without removing subsidies for carbon-intensive technologies, for instance, would have to swim against the strong current created by existing incentives and momentum. Removing subsidies without promoting public information and education will ensure that consumers remain uninformed about other options and the inefficiency of their current practices. Some energy services fulfill social functions independent of cost, so people will ignore price increases until a price becomes prohibitive. Consumers want to preserve their lifestyles and often do so until costs become significant, and manufacturers will protect their current practices against any changes that might threaten to disrupt productivity. Policy makers and regulators must design policy mechanisms that match the technical-economic-political-socio-cultural nature of the barriers facing cleaner systems and technologies. They must consistently pursue a variety of policy mechanisms that simultaneously alter R&D practices, fine tune subsidies, price externalities, and better inform the public if they are to affect consumer demand and promote sustainable energy practices at the speed, scope, and scale required.

The next two chapters show precisely how such comprehensiveness can be obtained. Chapter 7 discusses the importance of polycentric scales of action, and chapter 8 presents eight case studies of effective polycentric solutions that have been implemented in a variety of countries around the world.

7

The Case for Polycentric Implementation

At this point in the book, readers could rightly ask why it is that, despite all the policy and technology options available to cities, states, countries, and global leaders, greenhouse gases continue to be emitted and many components of energy security continue to deteriorate. Although this book does not provide an ultimate single answer, recent advances in psychology and academic professionalism may provide a partial explanation.

Some research has shown that human beings develop psychological protective mechanisms to shelter them from the full force of many of their daily actions. Psychologists have tended to describe this function as "diffusion of responsibility" or the "passive bystander effect." The classic example is the 1964 murder of Kitty Genovese, who was attacked for almost an hour in a densely populated New York neighborhood where 38 people heard her screams but never called the police or came to her assistance.[1] The lesson appears to be that when individuals know that others have the potential to address a common problem, they often make the assumption that someone else will intervene and do nothing themselves. Such inaction is enhanced when individuals are anonymous members of a large group, or when situations seem so complicated that individuals worry that they may be misinterpreting them and don't want to suffer embarrassment from acting.[2] The longer no one acts, the more the desire to do nothing is reinforced. When put into the context of pressing global problems such as climate change and energy security, "diffusion of responsibility" suggests that—counter to intuition—the most pressing conflicts will be those in which people and institutions have the greatest incentive *not* to act.

Something similar occurs within academic and scientific disciplines. Alongside the deterioration of natural resources and the Earth's climate, some theorists have posited that an "intellectual commons" is developing. The major problems of modern society are perpetuated—allowed to intensify and develop—through the insulated and highly specialized nature of academic disciplines. The natural sciences are able to recognize that some problems are not technically solvable and relegate them to the "nether-worlds" of politics, whereas the social sciences recognize that some problems have no current political solution and postpone a search for solutions while they wait for new technologies. As one influential study published in *Science* concluded,

[Academics] can thus avoid responsibility and protect their respective myths of competence and relevance, while they avoid having to face the awesome and awful possibility that each has independently isolated the same subset of problems and given them different names. Thus, both [the natural sciences and the social sciences] never have to face the consequence of their respective findings.[3]

The ultimate result of such specialization and insularity, the study noted, is that many of society's most critical problems "lie in limbo" while the specialists involved go on to less critical problems for which they can utilize readily available technical and political solutions.

One of the thorny or wicked complications about effective climate and energy policy is that it must fight both of these tendencies in order to succeed. It must prevent "diffusion of responsibility" by getting all major stakeholders in a certain region to agree to reduce their GHG emissions, and it must encourage approaches that avoid the trap of falling into specialized and often parochial disciplines. The costs and benefits of improving energy security and fighting climate change are distributed across geographic and temporal scales, from those as small as a household to as large as the entire planet, and from the momentarily fleeting to the geologic time frame.[4]

Because of this weird spatial and temporal multi-dimensionality connected to energy and climate problems, this chapter argues that similarly multi-dimensional scales must be utilized to implement policies that respond to them.[5] Table 7.1 offers 20 novel and successful examples of different actors (some of them highlighted in earlier chapters) attempting to address different energy-related challenges at different scales. Although these examples could be replicated at different levels (by firms, trade organizations, non-governmental organizations, communities, states, regions, countries, and global alliances), relying on a single scale in isolation—whether it is local and from the bottom up or global and from the top down—usually brings with it unique costs and benefits. Polycentric approaches, those that blend scales and engage multiple stakeholder groups, can capture the benefits of operating at each scale, while minimizing costs.

This chapter describes the substance and the advantages of polycentric approaches. It begins by discussing five benefits of global, centralized, homogenizing action: consistency, economies of scale, equity, mitigation of spillovers, and minimization of transaction costs. It then discusses five benefits of local, decentralized, heterogeneous action: diversity, flexibility, accountability, simplicity, and positive contagion (table 7.2). Next, the chapter explains the benefits behind polycentrism, or how properly designed policies can capture most of the benefits of both global and local action

Table 7.1
Successful "traditional" policies and the barriers they address.

Sector and type of program	Scale	Policy description	Barrier addressed	Result
Transport				
Parking fees and road restrictions	City	The German city of Vauban charges $40,000 per permanent parking space and prohibits motorized transport on most roads	External benefits and costs	Nearly three-fourths of residents no longer own a car and rely on walking, cycling, mass transit, and car sharing or car rental to satisfy their transport needs
Limited driving days	City	Mexico City bans drivers from using their automobiles one weekday per week	External benefits and costs	About 1.6 million vehicles stay off the city roads per week
Plug-In Hybrid Electric Vehicle Tax Breaks and Charging Stations	Country	The government of Israel subsidizes the cost of purchasing a PHEV to make it equivalent to conventional vehicles and has committed $200 million to build recharging and maintenance stations throughout Israel	High costs, technical risks, market risks, and infrastructure limitations	A few hundred PHEVs were in operation as of early 2009 with the number of charging stations expected to jump to 500,000 by 2011 and one-third of drivers claiming their next automobile purchase will be a PHEV
European Union—Emissions Trading Scheme (EU-ETS) for Aviation	Multi-country (regional)	Brussels imposes a binding cap and a system of trading on carbon dioxide equivalent emissions associated with air travel to any of the 27 member states of the European Union	External benefits and costs and industry structure	Full implementation will occur by 2012 and the International Civil Aviation Organization and Arab Air Carriers Organization have already begun programs to measure and reduce emissions

Table 7.1
(continued)

Sector and type of program	Scale	Policy description	Barrier addressed	Result
Electricity				
Virtual Net Metering	City	Ellensburg, Washington accepted contributions from the community to fund an array of solar panels, sell the solar electricity to the municipal utility, and then apply a credit to the contributors' electricity bills	High costs	73 customers donated $150,000 to fund the system and the City is currently accepting contributions to expand the program by a factor of 10
Financing Initiative for Renewable and Solar Technology	City	Berkeley, California used low-interest loans repaid through property taxes and secured by a lien on property to encourage owners to invest in solar energy or energy-efficiency improvements	High costs, lack of specialized knowledge	The pilot program funded $1.5 million worth of investments spread across 40 properties and generated more than 1,300 inquiries from other cities about how to copy the initiative
Neighborhood Energy Workshops	Multi-city	In Minneapolis and Minnegasco, Minnesota, volunteers held free workshops to educate and inform homeowners and the city provided a small amount of free materials, information, low-cost financing, free energy audits and thermographs	Incomplete and imperfect information and lack of specialized knowledge	Reduced household energy consumption by 4.3% and enrolled more than one-third of city residents

Renewable Energy and Energy Efficiency Partnership (REEEP)	Global	A central secretariat in Vienna, along with 17 regional secretariats, collects contributions from governments, banks, businesses, and civil society groups and then funds small-scale energy efficiency and renewable energy projects in the developing countries	High costs, lack of specialized knowledge, regulatory uncertainty, and infrastructure Limitations	In 2008 REEP managed more than 145 projects worth a total cumulative investment of €65 million and was expanding to include 37 new projects
Agriculture and forestry				
bio-organic fertilization	State (local)	In Tamil Nadu, India, managers of tea plantations place earthworms in trenches and use organic fertilizer to restore degraded farmland and improve soil fertility	Technical risk	Tea yields increased 76%–239% and profits per hectare increased by more than $5,500
Payment for Ecosystem Services	Country	In Costa Rica, the Private Forest Project collects a 5% tax on gasoline, revenues from the sale of carbon credits, and donations from hydropower companies and then distributes funds to encourage plantation owners and forest managers to preserve or afforest their land	High costs and external benefits and costs	More than 1,000 property owners signed contracts within the first two years of the program and $16 million–$20 million are currently dispersed per year to protect forests

Table 7.1
(continued)

Sector and type of program	Scale	Policy description	Barrier addressed	Result
Harvest quotas and forest reserves	Country	In Malaysia, legislation sets limits on the types of trees that can be felled, prohibits harvesting in reserves, sets quotas on overall harvest rates, and requires degraded forests to be restored	External benefits and costs	Forests have grown from occupying 58.7% of Malaysia's land area in 2000 to 63.6% in 2005
Reducing emissions from deforestation and land degradation (REDD)	Global	The United Nations Development Programme verifies carbon emissions reductions associated with improved forest management and provides compensation to build capacity to preserve tropical forests	External benefits and costs and lack of specialized knowledge	Once fully implemented REDD would reduce carbon dioxide emissions at a price of $2.31 per ton and the program could reduce GHG emissions associated with deforestation 66% by 2020
Water and waste				
"Cradle to cradle" design process	Factory	Wädenswil, Switzerland, relies on a simplified production process utilizing only 16 non-toxic dyes and natural textiles and threads to produce biodegradable fabric and upholstery that is "safe enough to eat"	External benefits and costs	The plant's production process cleans the water it uses rather than polluting it, produces compost for local strawberry farms, and has reduced overhead costs by 20%

Anaerobic digestion	Factory	Jurong, Singapore, utilizes anaerobic digestion to transform organic food waste into electricity and compost	High costs, technical risk	Produces revenue from electricity and compost sales, carbon credits, and tipping fees and has extended the lifespan of local landfills by 20 years
Landfill gas capture	Factory	In São Paulo, Brazil, the Bandeirantes Landfill Gas-to-Energy Project captures methane that would otherwise escape from a large landfill and converts it into 20 MW of electricity	High costs	Supplies electricity to more than 400,000 homes and has prevented the release of about 800,000 tons of carbon dioxide equivalent per year
Extended producer responsibility	Multi-country (regional)	EU standards require manufacturers to pay for disposal or recycling costs, take possession of discarded goods, provide information about recycling, and fund environmental clean up efforts for their products	External benefits and costs and industry structure	Manufacturers have designed products to be more recyclable, less environmentally damaging, and more easily collected
Climate change Various measures	City	In Woking, UK, city planners have reduced energy consumption and GHG emissions through coordinated policies regulating electricity use, heat, water, waste disposal, and transport	External benefits and costs, infrastructure limitations, misplaced incentives, competing statutory priorities	The city has reduced energy consumption by 44% and carbon dioxide emissions by 72% between 1990 and 2005

Table 7.1
(continued)

Sector and type of program	Scale	Policy description	Barrier addressed	Result
Various measures	City	In Beijing, city planners have relied on combined heat and power, energy efficiency, fuel substitution programs, electric bicycles, and ring roads to reduce emissions	External benefits and costs	80% of all traffic is now served by ring roads and emissions have grown at a rate of only 3.9% while the economy has grown at 15%
Zero-net-emissions goal	Local electric utility	The municipal electric utility Seattle City Light pledged in 2001 to be GHG emissions neutral by 2005 and achieved their goal by using renewable energy, promoting energy efficiency, and purchasing carbon credits	External benefits and costs and misplaced incentives	The electric utility releases no net GHG emissions at a cost of only $2 per customer per year
Kyoto Protocol	Global	Requires all industrialized (Annex 1) countries to reduce their GHG emissions 5.2% below 1990 levels by 2012 and permitted global emissions trading, Joint Implementation, and Clean Development Mechanism projects	External benefits and costs	Ratified by 183 countries and helped create a €116 billion market for global carbon credits

Table 7.2
Costs and benefits from local and global scales of action.

Criteria	Local	Global
Favors global policy		
Consistency	Building national markets for technology solutions is difficult when policies vary; local efforts can sometimes be "captured" by major emitters or incumbent energy firms	Standardization minimizes transaction costs and policy uncertainties
Economies of scale	Inefficient because of redundancies of R&D efforts and data collection systems	Better matched to promote economies of scale and avoid redundancies
Equity, race to the bottom	Scope and effectiveness of climate and energy policy differ between jurisdictions, some of which may decide to do little or nothing	Ensures that at least a minimum level of climate and energy security is reached for all communities
Spillovers, dilemmas of collective action	Vulnerable to free ridership and emissions leakage	Minimizes free ridership and emissions leakage
Transaction costs and the regulatory commons	Coordination and negotiation is time consuming, uncertain, and costly	Harmonization involves fewer actors and lower transaction costs
Favors local policy		
Experimentation and innovation	Encourages innovation and diversity in designing policy	Stifles innovation and experimentation
Flexibility	More flexible and able to adapt to local conditions; promotes administrative efficiency	More uniform and rigid; tends to fail to account for local conditions
Accountability	Allows for closer fit between policies and preferences and affords option to sort between jurisdictions	Depresses self-determination and prevents people from expressing policy preferences
Simplicity	Simpler and enables speedier responses	More complex and often slower to change
Positive contagion	Can sometimes compete in a race to the top to craft better policies	Can preempt healthy competition by setting uniform standards

while avoiding their disadvantages. The chapter concludes by focusing on for the requisite conditions for successful polycentric approaches.

7.1   The Benefits of Global Action

Many strong arguments exist in favor of promoting energy and climate policies at the global or national scale. Action at this level often facilitates uniformity and consistency, creates economies of scale, ensures equity, prevents spillovers and a "race to the bottom," and minimizes transaction costs associated with coordination and negotiation.

7.1.1   Consistency

Simply put, global or uniform standards and actions offer consistency. Having one top-down climate or energy policy can engender a more efficient regulatory regime than a multiplicity of state and local standards, which tend to heighten barriers between individual states and lead to inefficiencies.[6] Global policy creates consistent and predictable statutes that manufacturers and industries can anticipate and deal with. A single national currency, for example, is cheaper to use than scores of separate local currencies, and a uniform gauge for railway tracks is undeniably sensible. "Sometimes," the law professor Steven Calabresi noted, "variety is not the spice of life; as to some items it may be a downright nuisance and an expensive one at that."[7] In terms of energy and climate policy making, it may make sense to have single, harmonized standards instead of multiple, confusing rules and regulations.

To consider what happens when there is too much variation, take the example of climate policy. Policy variations and fragmentations exist across countries, regions, states, and localities. International corporations today are operating in a patchwork of markets, some with strong carbon constraints and others without any carbon regulations. Within the United States, numerous fractionated policies exist. Two of these (residential building codes and net metering rules) were discussed in chapter 5 and are illustrated in figures 7.1 and 7.2.

Figures 7.3 and 7.4 provide two additional examples of geographic disparities. The first pertains to regional carbon cap-and-trade initiatives. Ten states in the northeastern US (Connecticut, Delaware, Maine, Maryland, Massachusetts, New Hampshire, New Jersey, New York, Rhode Island, and Vermont) are currently participating in the Regional Greenhouse Gas Initiative (RGGI), which will reduce emissions of carbon dioxide from power plants by 10 percent in 2019, but more than half of the US states do not even have GHG-reduction goals. This mosaic of divergent policies is particularly challenging to entrepreneurs who are striving to develop national markets. The second example illustrated in figure 7.4 pertains to

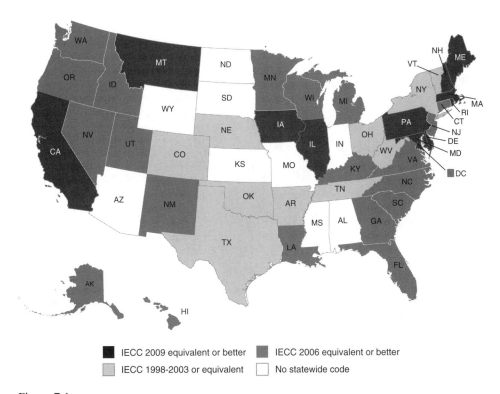

Figure 7.1
Residential building codes. Source: Database of State Incentives for Renewable Energy (http://www
.dsireusa.org).

Renewable Electricity Standards, laws that require utilities to supply a certain percentage of renewable electricity by a certain date (e.g., California must generate 33
percent of its electricity from renewable resources by 2020). Contrary to enabling a
well-lubricated national market for renewable energy, however, inconsistencies
between states over what counts as renewable energy, when it has to come on line,
how large it has to be, where it must be delivered, and how it may be traded clog
the market for renewable energy as coffee grounds clog a sink. North Carolina
set its target at 12.5 percent by 2021, whereas Maryland chose 20 percent by
2022. Similarly, each state has different notions as to what counts as renewable. In
Maine, fuel cells and high-efficiency cogeneration units count as renewable, whereas
the standard in Pennsylvania includes coal gasification and fossil-fuel distributed
generation technologies. Iowa, Minnesota, and Texas base their purchase requirements on installed capacity, whereas other states set them relative to electricity
sales. North and South Dakota have voluntary standards with no penalties, whereas

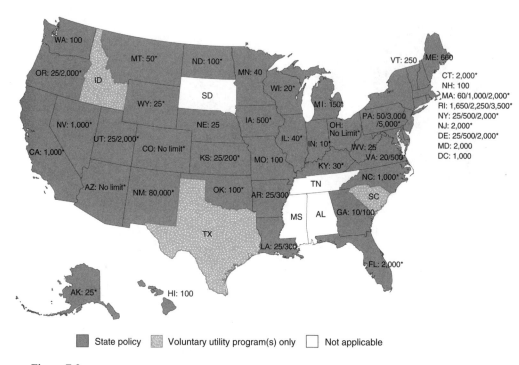

Figure 7.2
Net metering rules. Source: Database of State Incentives for Renewable Energy (http://www.dsireusa.org).
Asterisk indicates that state's policy applies only to certain types of utilities—e.g., investor-owned.

Massachusetts, Connecticut, Rhode Island, and Pennsylvania all levy different noncompliance fees. Implementing agencies and stakeholders must grapple with inconsistent state goals, and investors must interpret competing and often arbitrary statutes.

7.1.2 Economies of Scale

A second advantage to global action concerns efficiency gains from economies of scale. In the absence of international action, local actors can duplicate each other or can engage in time-consuming and complex negotiations to divide labor and resources. The drive toward the local scale generally weakens technical capacities such as data collection and research and development requiring large-scale scientific instruments. It makes little sense to have every state, city, or town measure the level, size, and type of carbon dioxide emissions, track the carbon intensity of fuels, determine their health effects, identify safe levels of emissions, and design cost-effective policy responses. Global action, furthermore, tends to enhance the identification of climate and energy problems. The initial awareness of such problems often emerges

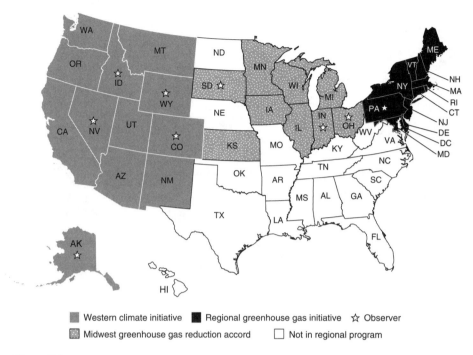

Figure 7.3
Regional carbon cap-and-trade initiatives. Source: Database of State Incentives for Renewable Energy (http://www.dsireusa.org).

from actual observation of physical change. Many "pairs of eyes" spread across the world (under a global program) are more likely to see anomalies than just those in one city.[8] Global action gets eyes looking in and into every community.

As one prominent example, consider that the US National Weather Service, a federal agency that measures rainfall with more than 10,000 gauges across the country, providing data for weather and climate forecasts, and collects snowfall data in cities and rural areas at more than 670 sites. A state-by-state or local approach to such a complicated task would incur huge transaction costs for building weather stations, improving forecasting techniques, and sharing data.[9]

Global information collection can create a unified set of indicators, making up for disparities in the competencies of local environmental programs. Without accurate information collected using standardized protocols, consumers, producers, and policy makers will find it difficult to make efficient land-use, travel, and infrastructure decisions. Localization can even lead to stagnation and inefficiency when communities become isolated from each other and lack the information necessary to promote optimal environmental policy.[10]

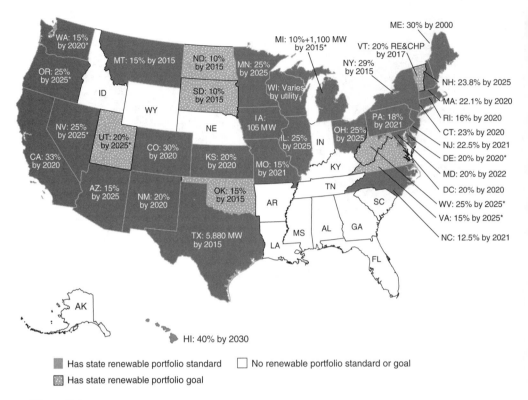

Figure 7.4
Renewable electricity standards in the United States. Source: Database of State Incentives for Renewable Energy (http://www.dsireusa.org). Asterisk indicates extra credit for solar or customer-sited renewables.

Consider acid rain, or "acid deposition," which occurs when sulfur dioxide or nitrogen oxide emissions undergo a chemical reaction in the atmosphere and become sulfuric or nitric acid. In the case of sulfur dioxide, emissions first change to sulfates—fine particles that can be transported hundreds and even thousands of miles. Sulfur dioxide can return to land in dry form, becoming sulfuric acid when inhaled or moistened, or as acid rain when washed out of the atmosphere by precipitation. Researching such a complicated form of pollution is especially difficult: the process converting pollutant emissions into acid rain occurs at high altitudes, and prevailing winds often carry the damaging precipitate far from the source. Polluting sources have little incentive to study the atmospheric science behind acid rain because the polluting regions are not the polluted regions.[11]

7.1.3 Equity and the Race to the Bottom
Global policies are able to address several equity issues that can prove difficult for localities and communities in isolation. Local environmental decision making can

create welfare losses and gaps in effective national policy. The problem with climate change is particularly acute: the rewards from restraints on GHGs will come in the politically distant future and will be distributed around the world, but the costs will be incurred in one particular location in the present. The wide distribution of expected but distant benefits in response to collective action provides an incentive for every country to encourage all to act but then shirk responsibility itself.

Global action, because of its uniformity and consistency, is able to avoid what some theorists have come to call the "race to the bottom." Richard Stewart, a scholar specializing in environmental and administrative law, defined the race to the bottom (also called the "race to laxity") as follows:

Given the mobility of industry and commerce, any individual state or community may rationally decline unilaterally to adopt high environmental standards that entail substantial costs for industry and obstacles to economic development for fear that the resulting environmental gains will be more than offset by movement of capital to other areas with lower standards. If each locality reasons in the same way, all will adopt lower standards of environmental quality than they would prefer if there were some binding mechanism that enabled them simultaneously to enact higher standards, thus eliminating the threatened loss of industry or development.[12]

Such a race to the bottom can occur for any of several reasons. Local officials may have been captured by regulatory interests, or may be unwilling to bring enforcement actions against their own agencies or local governments or companies. Community officials may also want to lure industries to relocate or construct new facilities within their jurisdiction.[13]

For example, when control over forest resources was decentralized to local communities in Cameroon, significant corporate and elite takeover of forest reserves at the village level occurred, with village elites confiscating property, embezzling funds, and facilitating deals that enhanced personal wealth at the expense of social welfare.[14] In the mining and extractive industries sector, international standards and control of resources have often proved better for local communities. In Sub-Saharan Africa, the smaller the project the dirtier it tends to be. International companies tend to have the strictest safety, health, and environmental standards, national companies tend to have mediocre standards, and local artisanal mines, tanneries, and breweries tend to have the poorest standards or no standards at all.[15]

Similarly, competition among communities has sometimes convinced them to relax environmental standards to attract industry. Companies generally consider the stringency and scope of state environmental regulation in their decisions to build new facilities or relocate, and the excessive market power held by industries over communities allows them to exert considerable influence over regulatory regimes.[16]

Other studies have found proof of a race to the bottom when looking at how individual communities in the United States have chosen to implement (or, in this case, to not implement) provisions of the Clean Air Act and the Clean Water Act. One assessment found that at least 27 states had communities operating in direct violation of the Clean Water Act. In Pennsylvania and Maryland, about one-fifth of the files reviewed in a study of day-care centers and schools showed that the drinking water at those facilities contained excessive levels of lead. One school's water supply had 4 times as much lead pollution as federal standards permit. Neither state reported such violations to the federal government. Roughly one-sixth of major industrial facilities in the United States had committed significant violations of their water discharge permits, and more than 60 percent of violations in one year involved excessive discharges.[17]

Global and national action, by contrast, has the benefit of creating a minimum standard of energy and climate security (and environmental quality) for all people. The poorer the local or state jurisdiction, the more likely it is that its regulators lack sufficient technical competence to deal with environmental problems. Global intervention therefore provides "distributive justice" and equalizes the minimum level of protection afforded to all communities. Put another way, it ensures that local actors will be limited in their ability to race to the bottom.

7.1.4 Spillovers and the Dilemmas of Collective Action

Small-scale actions by individuals, communities, or states are prone to what the University of Maryland economist Mancur Olson called the "dilemma of collective action."[18] Once some communities begin to act, other communities may decide to "free ride" on their actions. The more such communities fail to act, the more justified it is that no one else contributes. The result is collective inaction: without regulation imposed on them, communities and local actors will rarely agree to implement coordinated policies together.

Think about an analogy at a much smaller scale. If Mr. Green purchases a smaller, more fuel efficient, and less luxurious automobile than Mrs. Brown, his actions mean less consumption, less pollution, and lower gasoline prices for everyone. Yet Mr. Green's sacrifice makes it *less* likely that Mrs. Brown will see the need to change her habits, not more likely, since she enjoys the positive benefits of Mr. Green's action along with everybody else.[19] This is potentially the most serious drawback to local action meant to address climate change and strengthen energy security: as one group reduces emissions, the overall cost of producing GHG-intensive goods and services within this group will rise relative to other producing locations. A comparative advantage in the manufacturing of GHG-intensive goods will therefore shift to nonparticipating countries.

Polluters can simply move to other places that do not have restrictive policies, creating significant "leakage." A patchwork of local climate-change policies allows stakeholders to manipulate the existing market to their advantage, using regulatory loopholes to emit greenhouse gases wherever regulators are the most lax. Global action, on the other hand, prevents such spillover effects.

Three examples from the United States are quite telling. States participating in the Regional Greenhouse Gas Initiative have experienced "leakage" rates as high as 90 percent as a result of the importing of coal-generated electricity. Power plants in adjacent states have actually increased their output in order to sell into the higher-priced RGGI electricity markets.[20] Similarly, LS Power (a company based in New Jersey) proposed building a 1,200-megawatt coal plant in Early County, Georgia, to export electricity to Florida, because the plant probably would not have been approved in either New Jersey or Florida. The investing corporation is in a RGGI state with strong environmental regulations, and the intended consumers reside in a state that has a growing commitment to "green energy" and to climate-change mitigation. The plant would be located in Georgia where political leaders routinely express skepticism as to whether human actions are impacting the climate, and coal generation dominates its power supply. The proposed plant could increase the amount of carbon dioxide released by electricity generation in Georgia by 13 percent.[21] Finally, PacifiCorp, an electric utility serving customers in the Pacific Northwest, has repeatedly attempted to build coal-fired power plants in Wyoming and Utah (states without mandatory GHG-reduction targets), but not in Oregon (which has mandated stabilization of GHG emissions by 2010) or Washington (which has mandated 1990 levels by 2020).[22]

In addition to the relatively straightforward phenomenon of leakage, localized climate action in lieu of global policy sends distorted price signals. By lowering demand for carbon-intense products, local standards reduce the regional (and even the global) prices of carbon-intensive fuels. But in doing so, they provide further incentives for nearby actors without climate-change mandates to do nothing because they benefit from reduced prices. These reduced prices, in turn, encourage over-consumption in areas without carbon regulation, decrease the incentive to enact energy-efficiency and conservation measures, and discourage the adoption of renewable-energy technologies and alternative fuels for vehicles.

7.1.5 Transaction Costs and the Regulatory Commons

Addressing global problems such as climate change on a small scale or in a decentralized way leaves individual actors to negotiate solutions among themselves. Negotiation, however, may be ineffective, since parties emitting more GHGs do not fully suffer the costs of pollution and have no incentive to remedy the situation. Sources emitting less GHGs have little bargaining power, and litigation is time consuming and rife with risk and uncertainty.

For example, many local policy makers have been reluctant to pursue meaningful environmental regulation of trans-boundary or interstate pollution.[23] These local actors often disagree on questions of attribution and on whether a source state is legally responsible for pollution emanating from its facilities. They disagree on questions of causation, or on what standards of proof are appropriate to establish that their pollution injured another state. They differ on questions of liability, or on whether or not the source state causing injury has acted unreasonably or negligently. Even when these matters are agreed upon, most local actors will disagree on what remedy or course of action is appropriate. Many localities have a "litigate or settle" mentality and are therefore reluctant to engage in lawsuits in view of the uncertain outcomes and the expenses (especially attorney's fees and court costs). Few cases of pure or unidirectional transboundary pollution exist. Instead, all actors are usually complicit and responsible for pollution going somewhere; that is, pollution is reciprocal. Most local actors will therefore ignore such problems completely rather than risk admitting liability once discussions about interstate pollution commence.

Because these factors often exist together, one study noted that "the reality is that a legal system does not act like a machine, automatically churning out the prescribed response to identified problems. Instead, it represents a kind of regulatory commons, where effective action is dependent upon alliances of groups overcoming collective action barriers and pressuring administrators to respond. If structural factors act as an impediment to achieving effective regulation in the international arena, it is not unlikely that they will also frustrate effective collective action within a developed legal system."[24]

Addressing environmental and climate problems in such a decentralized system is virtually impossible. Requiring that actors negotiate solutions among themselves is time consuming and costly. Many actors will regard negotiation as ineffective, since upwind and upstream states do not suffer the social costs of pollution and thus have little political incentive to take these costs into account when setting their environmental policy. Downwind and downstream states, by contrast, have little bargaining power with upwind and upstream states. Their only leverage is the threat of a lawsuit, but such litigation is time consuming, expensive, and likely to fail on issues of causation.[25]

In this "regulatory commons," localities from which pollution or emissions originate have little incentive to stop, since they benefit from externalizing the harmful effects while enjoying the economic benefits of polluting. As the law professor Rena Steinzor notes, "For state and local officials, unrestricted devolution means that they must shoulder primary responsibility for solving difficult problems, including transboundary pollution. For industry, unrestricted devolution could mean either that chronic violators gain competitive advantage in states with weak enforcement or that a patchwork of expensive and contradictory state regulation emerges. And for the public, unrestricted devolution means not only losing ground in our steady progress to protect the natural environment, but postponing, once again, the long-overdue campaign to conquer maladies . . . that so severely diminish the enjoyment of life."[26]

In other words, devolution, or efforts to ensure that environmental policy making is always done at the smallest scale possible, when taken to its logical extreme, can cause unacceptable and irrevocable harm to social welfare.

7.2 The Benefits of Local Action

Before you begin nodding your head in agreement with what has been said so far, however, you should consider that local and decentralized actions have numerous benefits not enabled by actions at geographically larger scales. These benefits include experimentation and innovation, more flexibility, greater accountability and participation, simplicity, and positive competition among local actors that sometimes race to the top to craft better policies.

7.2.1 Experimentation and Innovation

The most basic benefit of small-scale and local actions is that they can promote experimentation and innovation. Local actions can create competition that can optimize environmental policy and create "laboratories of democracy" that experiment in crafting better policies. This experimentation can create diversity. The presence of multiple local policies can provide opportunities for trying different policies, allowing policy makers to learn from failures and replicate successes. Closely related to local actors competing to produce better policy are the idea that they can also serve as agenda setters for global policy. Local actions often lead to pressure for national or global policies that would never have arisen at larger scales.

For example, in the United States local communities, not national actors, have led the way in experimenting with and then establishing automobile emissions standards, low-emissions-vehicle programs, recycling and bottle bills, duty-to-warn measures informing consumers of exposure to harmful substances, and hazardous-waste regulation related to brownfield redevelopment programs, land transfer, and municipal solid waste.[27] Local actors were the first to implement environmental audit privilege laws, which reduce penalties for companies that voluntarily disclose and correct environmental violations, leading to greater cleanup.[28] Though the motivation for these improvements undoubtedly varies according to the differing political, economic, and environmental interests in each locale, almost every state has some communities that have passed environmental regulation before the national government did so.[29]

7.2.2 Flexibility

Local action can also promote greater flexibility. Global, top-down international regulatory systems are often unable to incorporate all the specific, detailed, temporal, and geographic information necessary to design optimal policies. This "knowledge problem" needs local and regional responses from people familiar with their own conditions. A corollary of this argument is that action at smaller scales promotes administrative efficiency, since state and local agencies are more agile and adaptive than national or global ones and thus are more able to formulate solutions well tailored to local needs and preferences. Advocates of localism have called national and international policy making an "affront to nature," because ecological systems are intrinsically variegated and diverse. Failure to take into account local environmental conditions, tastes, preferences, and economic conditions leads to a "one size fits all" prescription that is more often "one size fits nobody."[30]

Numerous examples abound. An apple orchard in upstate New York requires different pest-control strategies than one in Washington, since pest-control options vary with climate, topography, pesticide availability, and chemical resistance. The main air-pollution problem in Los Angeles is smog and particulate matter, whereas in Pima County, Arizona, it is sulfur dioxides and in the Cuyahoga Valley of Ohio it is nitrous oxides. Requiring municipalities to treat stormwater as an industrial pollution discharge or to use double liners for landfills may make sense in the Northeast, but such requirements are ill suited to arid regions with little rainfall or clay-based soils. Requiring secondary wastewater treatment makes sense in many cities but adds little value in coastal communities. One river may suffer from excessive nutrient loads, another from nutrient deficiencies. If one local water supply is required to test all contaminants for which water standards exist, even though some of those contaminants have never been found or do not exist in that particular source of drinking water, those requirements waste time and resources.[31]

Even climate change has geographically variable components. The proximity of communities to electricity generation, the prevalence of local industry, population growth, density and number of households, and existing energy-efficient technologies in use all figure in determining differences in sources and rates of GHG emissions.[32] The knowledge problem and the diversity of environmental problems, the argument goes, require local knowledge and expertise, not distant and homogenizing global action. New research in atmospheric science has also refuted the common belief that a ton of globally emitted carbon dioxide does not affect a particular location more or less than any other. Researchers at Stanford University have found that locally emitted carbon dioxide can form "domes" that trap emissions near their sources, interacting with local ozone and concentrations of particulate matter or sulfur dioxide.[33] Carbon dioxide emitted into the global atmosphere, furthermore, does not affect temperature increases uniformly, as its interactions with water vapor and other atmospheric elements will be particular to each region. The implication is that a ton of carbon dioxide can result in very different local effects connected to where it was emitted from.

7.2.3   Accountability and Participation

Local and state regulatory oversight is often more preferred and popular than national or international action. Most people find local regulators more trustworthy, capable of understanding problems and resources, and approachable. Bringing decisions about climate and energy policy closer to local citizens can therefore enhance public participation in policy making. In some ways, decentralization and localization improve accountability and self-selection. Since individuals have the mobility to choose where they will live, they can sometimes sort themselves into the jurisdictions offering the mix of policies that they prefer.[34] If a family perceives Canada's laws as too stringent, they can move to China or Chile; if Jakarta does not suit their tastes, they can head to Jerusalem or Johannesburg. Specific environmental preferences differ from place to place, and global action often depresses local self-determination by diminishing the opportunity for local participation in environmental problem solving.

7.2.4   Simplicity

Taken in aggregate, problems such as climate change and worsening energy security can be so complex that a purely global focus tends to reduce them to a set of abstractions. Such issues are often best understood when broken down and simplified into subsets of causes and problems, in essence making them simple enough to understand. In some cases a truly global picture can blur and obscure what is really going on. The British philosopher Alfred North Whitehead called this the "fallacy of misplaced concreteness," a term that describes what happens when one may mistake models of reality for reality itself. (It has also been called "eating the menu but not the meal."[35]) Reducing such problems to smaller scales, the argument runs, allows them to be better identified and then managed.

A corollary of this argument is that local action, because it is often simpler, can enable a more rapid response to changing needs and circumstances. Just as smaller ecological subsystems change faster than the larger ecosystems of which they are a part, one can argue by analogy that local policy adaptations can occur more rapidly than changes in national or global policy.[36] Most of the causes of climate change, and most of the uses of energy, occur at the local scale. Most if not all of the actions that contribute to climate change and affect energy security are local and therefore best addressed at the level of homes, neighborhoods, communities, and cities. We are all shaped in important ways by where we were born, and by the natural world we first experienced. This is always a local phenomenon, and social movements to protect the environment have usually grown from efforts to protect particular, local places (such as national parks or specific places of land), not abstract notions of "the globe."[37]

7.2.5   Positive Contagion

Though local action always has the potential to cause a race to the bottom, it can also promote healthy competition to race to the top. The race to the top acknowledges that many actions to reduce pollution (such as improving energy efficiency and releasing less toxic waste) also bring economic benefits (such as lower energy costs and fewer lawsuits). Local actors that free ride and under-protect the environment run the risk of alienating businesses that prioritize environmental quality, reducing GHG emissions, or improving energy security. As was noted above, many local actors have addressed environmental or energy problems before national and global actors to preserve environmental resources, such as clean air and wetlands.[38] Setting minimum global standards, moreover, can also prevent communities from adopting optimal standards on their own.

For example, by 1960 (10 years before the federal government implemented the 1970 amendments to the Clean Air Act), eight US states had general pollution-control laws, another nine had undertaken measures to control air pollution under their general public health laws, and eight others had authorized local air-pollution-control agencies to transcend municipal boundaries in their regulatory efforts. By 1966, ten states had adopted at least some standards for ambient air quality, which covered 14 substances as well as deposited matter. In addition, six states had emission standards covering some stationary sources before the federal government. Before 1975, when the federal government was ordered by a federal court to protect wetlands under the Clean Water Act, all 15 states with more than 10 percent of their land area in wetlands had already adopted protections.[39]

7.3   The Case for Polycentrism

Polycentric approaches—those that incorporate multiple scales and multiple stakeholder groups at once—are often able to harness the benefits of global and local action together instead of having them trade off. One important aspect of the polycentric model is its emphasis on blending of local, state, national, and global scales. Polycentric approaches imply that the sharing of power between and within these scales must be seamless.[40] Engaging multiple stakeholders can reveal diverse perspectives, build coalitions, and promote cooperation rather than competition. Having multiple regulators means that different officials with distinct perspectives review a problem; incorporating stakeholders beyond the realm of government (such as business leaders or members of environmental groups) can build coalitions and promote cooperation rather than competition. This diversity of perspective produces a broader variety of potential solutions and provides better experimentation. Overlapping jurisdiction and inclusion can encourage more appropriate levels of government to respond to climate and energy issues, and it provides more opportunities to diverse players in the policy-making process. Put another way, polycentrism captures these benefits of local action without compromising many of the benefits of global action (such as consistency and equity).

Polycentric approaches have at least four advantages over approaches that rely on local action or global action in singularity: they promote dialog, provide a regulatory safety net, enhance accountability, and maintain economies of scale.

7.3.1 Dialogue
The lack of rigidity and exclusivity encouraged by polycentrism fosters dialog and sharing of information. This can inspire a learning effect, such that others adopt better solutions and approaches. It allows "laboratories of democracy" to function while ensuring that no race to the bottom will occur.

7.3.2 Redundancy
Under a system of polycentrism, if one level of government, industry, or civil society fails to solve a problem, other layers remain available to address it. This redundancy creates a regulatory safety net and ensures that social problems are addressed through the combined application of local, regional, national, and global policies. Polycentrism can ensure that more resources are thrown toward a particular problem as approaches relying on isolated scales by themselves.

7.3.3 Accountability
Polycentrism can improve accountability and participation. Most governments do not have the resources or the personnel to implement detailed regulatory prescriptions in every community; most businesses and other organizations concentrate efforts in selected areas where their expertise is greatest, problems seem most urgent, or money can be quickly made. Forcing these stakeholders to interact with each other ensures that actual implementation is dependent on multiple layers of government and society at once. This effectively combats excessive influence of particular interest groups on elected politicians who can capture one level of government. Multiple levels of governance, the thinking goes, are harder to capture than one.

7.3.4 Economies of Scale
Polycentrism better exploits economies of scale at the national and global level by creating consistent standards. Yet polycentrism still preserves a role for local officials to experiment with policies and exceed global standards if they wish. A minimum

level of predictability and clarity is established for stakeholders no matter how far a community goes beyond a global standard, promoting equity, but local actors have the ability to experiment and innovate with how they implement it, promoting experimentation and diversity.

7.4   Challenges to Polycentrism

Though it holds many theoretical advantages, polycentrism is not without challenges. It requires a complicated amalgam of local, national, and international laws and standards, assessing similar topics and coexisting within similar jurisdictions. Divergent rules can lead to redundancy of regulation, inefficiency, and confusion as people try to figure out which laws apply to them.[41] In some cases, polycentrism can take longer to address problems. It can extend the time needed for policy resolution, because disgruntled parties can always go to the other levels of government for relief. By creating overlapping jurisdiction, regulators can blame deficiencies on other levels of governance. This can give regulators more ability to create smoke-screens, to shirk their responsibilities, and to hope that disgruntled citizens will not discern the proper target for their ire. Critical stakeholder analysis (see chapter 6) can help reveal such complex features of polycentric systems.

Notwithstanding these challenges, polycentrism clearly has the ability to ensure the best of both worlds: a degree of uniformity and consistency to force potential free riders to address inconsistencies and provide a minimum degree of action, yet without deploying the homogenizing tendencies of centralized regulation. Polycentrism exploits a middle ground between actions that rely on unitary and inflexible global standards and actions that leave local actors completely autonomous and free to do what they wish, even if it damages the climate and degrades energy security.

In other words, polycentrism combines the strengths of local and global action without adulterating policy with some of their weaknesses. Polycentrism recognizes that climate and energy problems differ substantially by region (capturing the "flexibility" benefits of local action) but also ensure that a common standard motivates all communities to act (capturing the "uniformity" and "equity" benefits associated with global action). It recognizes that policy preferences tend to be more homogeneous within smaller scales than across larger geographies. It accepts that multiple jurisdictions with overlapping duties can offer citizens more choice in setting modes of regulation (capturing the "simplicity" and improved "accountability" from local action); yet it still requires that local actors subscribe to a common set of goals and to global enforcement, minimizing "transaction costs" and the "dilemmas of collective action." Polycentrism posits that when multiple actors at a variety of scales must compete in overlapping areas, they can often promote innovation as well as cooperation and citizen involvement.[42]

7.5   Conclusion

Whatever particular mechanisms policy makers deploy to fight climate change and enhance energy security, of equal importance is the scale at which they are implemented. Local intervention and global intervention have different costs and different benefits. Local action fosters innovation and experimentation, can adapt to local circumstances and needs, and tends to reflect local interests and preferences. Such efforts are often the antidote to federally or globally imposed rigidity and "command-and-control" policy making. Global action, on the other hand, is the best way to provide uniformity and minimize transaction costs among actors, creates better economies of scale, ensures that all states bear the burdens of addressing climate change, and minimizes free-riding.

Ideal policies, however, would combine the two scales in a polycentric approach that would create multi-scalar governance. Such polycentrism would ensure that multiple layers of government and multiple stakeholders would address energy problems and climate problems. It would encourage dialog and information sharing, inspiring learning and innovation. It would stretch a regulatory safety net across a wide geography, providing a minimum level of policy action, but it would still exploit economies of scale by allowing for experimentation and flexibility within the confines of a degree of uniformity and consistency. Polycentrism would mitigate the passive bystander effect by creating important roles for multiple scales of action at once, and would reduce the risk of creating an the intellectual commons by empowering social, political, and intellectual leaders from a variety of places to contribute to improving energy security and mitigating climate change.

As the next chapter will show, polycentric approaches have already wrought much success at reversing environmental degradation associated with energy and land use, lowering greenhouse gas emissions, and improving the security and affordability of energy services in a variety of locations.

8
Case Studies

An old joke has been circulating through academic circles for some time: A professor of economics and an elected official were passing through a field one day when they suddenly stumbled into a deep hole. After they regained consciousness and inspected themselves for broken bones and physical damage, the official looked at the walls of the hole, which ran a good 10 meters to the surface. She asked the professor if he had a plan for getting them out. After pondering for a moment or two, the professor replied "Well, first assume that we have a ladder. . . ."

The joke is apt because it reminds us that academic theory does not always match reality. One scathing assessment even went so far as to argue that for the past few decades energy policy making throughout the world has consisted of professors' and theorists' assuming more and more ladders of increasing implausibility, and of politicians' climbing their way up imaginary escape routes with a "can-do, fix-'em-up" attitude.[1] The theoretical is not always practical, the potential is not always possible, and there is no substitute for empirical success.

It is with this appreciation for practical experience that we have investigated eight real-world cases in which communities, companies, and countries have taken concrete efforts to successfully reduce their emissions of greenhouse gases and to improve energy security. We began by developing an analytical framework for the selection of case studies. To be considered for this chapter, a case study had to meet the following five criteria:

• It had to address one or more of the five challenges (electricity, transport, deforestation and agriculture, waste and water, and climate change) presented in chapters 1–3.

• It had to be holistic, simultaneously addressing some combination of the technical, social, economic, environmental, and political barriers outlined in chapter 6.

• It had to be polycentric, either by mixing traditional scales (such as local/national or national/global) or by combining multiple actors (such as government regulators, business stakeholders, and civil society).

• It had to be successful, meaning that it had to meet its own goals, produce measurable benefits that tended to exceed costs, and make real and demonstrable gains.

• It had to be original. (We collected new data on known case studies or presented case studies not widely known, and we relied on original research such as interviews and field visits in addition to articles, reports, and other secondary sources for verification.)

After filtering more than 70 potential case studies according to the above criteria, we were left with the eight presented in this chapter. We chose Denmark and Germany for their efforts in the electricity sector. Brazil and Singapore for transport. Grameen Shakti in Bangladesh and China's improved cookstoves program for deforestation and agriculture. and the Oasis Project in Brazil and the Toxics Release Inventory in the United States for waste and water. Each case study begins with a short introduction before explaining the history and details of each program or policy, its costs and benefits, its challenges, and the lessons it offers. Table 8.1 presents a concise overview of our eight case studies.

To ensure that our eight case studies met all five criteria, we conducted 104 original research interviews with experts at 65 institutions in 12 countries over the course of 2 years. We also visited the sites of projects in Australia, Brazil, Denmark, Germany, India, Indonesia, Malaysia, Singapore, Spain, the United Kingdom, and the United States. More details about these interviews are presented in appendix A. When arranging our interviews and visits, we took special care to include the following:

• universities and academic institutions, such as the University of California at Berkeley, the University of Delhi, the School of Public Health at Rutgers University, Cambridge University, and the University of São Paulo

• non-governmental organizations (NGOs), such as the International Energy Foundation, the David Suzuki Foundation, the World Future Council, and the Forest Stewardship Council

• research institutes, such as the Sugarcane Technology Center in Brazil, the Risø National Laboratory in Denmark, and the Fraunhofer Institute in Germany

• regulatory agencies and government ministries, such as the Danish Energy Authority, the Singaporean Ministry of Transport, and the US Environmental Protection Agency

• Companies and manufacturers, such as Vestas, Enercon, and Solar Fabrik

• international organizations, such as the World Bank and United Nations.

Although our interviews were carefully transcribed for accuracy, to encourage candor, protect confidentiality, and meet institutional review board procedures and

Table 8.1
Programmatic summary of electricity, transport, agriculture, forestry, waste, and water case studies.

Sector	Case study	Period	Description	Polycentric component	Result
Electricity	Danish electricity policy	1970–2001	Utilizes energy and carbon taxes, R&D funds, government financing, other mechanisms to promote wind energy, energy efficiency, and combined heat and power.	Blends small-scale decentralized community control with national standards and policies.	Denmark leads the world in electricity generation from wind turbines (as a percentage of their portfolio) and wind energy manufacturing, and leads Europe in the use of combined heat and power.
Electricity	Germany's Feed-in Tariff (FIT)	1990–2009	Pays qualified renewable power providers a fixed, premium rate for their electricity over a long period of time.	Integrates residential and community producers of wind and solar energy with federal policy concerning tariffs and digression rates.	The FIT has seen renewable electricity supply jump from less than 5% in 1998 to 14.2% in 2007, created 150,000 direct jobs, displaced 79 million metric tons of carbon dioxide emissions per year, and has saved electricity customers €4.3 billion in avoided energy imports and fossil-fuel use.
Transport	Brazil's National Ethanol Program	1975–2009	Sets national targets and quotas for ethanol blending and production and promoted flex-fuel vehicles.	Includes sugarcane farmers, ethanol distillers, automobile manufacturers, and environmental groups in the formulation of national policy.	Made Brazil a world leader in ethanol production, created 3.5 million jobs, enabled ethanol distillers and sugar mills to produce electricity from bagasse, and facilitated 90% market penetration of flex-fuel vehicles.

Table 8.1
(continued)

Sector	Case study	Period	Description	Polycentric component	Result
Transport	Singapore's Urban Transport Policy	1971–2009	Restrains private automobile ownership through vehicle moratoriums and fees, levied congestion charges for roads and expressways during peak times, and vigorously promoted bus and rail mass transit.	Harnesses public-private partnerships to operate mass transit systems and works with automobile manufacturers to equip vehicles with electronic road pricing devices.	Almost two-thirds of daily trips during peak hours occur on mass transit, more than 95% of roads and expressways are congestion free, and road pricing scheme funnels $138 million in fees back into the government budget.
Deforestation and agriculture	Bangladesh's Grameen Shakti	1996–2009	Utilizes an innovative financing scheme and market-based approach to promote solar panels, biogas plants, and improved cookstoves throughout rural Bangladesh.	Enrolls communities into projects at the household and village levels but also engages district and national policy makers and international donors and lending firms.	Operates 750 offices throughout every state in Bangladesh and has installed 250,000 solar home systems, 40,000 cookstoves, and 7,000 biogas plants among 2.5 million recipients to keep biomass in the forest.

Deforestation and agriculture	China's National Improved Cookstove Program	1983–1998	Relies on a "self building, self managing, self using" policy that focused on having rural people themselves invent, distribute, and care for improved cookstoves.	Program was funded by the national, provincial, and local governments along with households themselves.	Installed 185 million improved cookstoves in 78% of all Chinese households, reduced energy use per capita in rural areas by 5.6%, and helped increase forest coverage in rural areas from 12% to 13.4%.
Waste and water	São Paulo's Oasis Project	2006–2009	Pays land owners to protect their forested land in order to preserve the forests that provide drinking water for 3,400 Brazilian municipalities.	Project is funded by an NGO, a corporate foundation, and state and local government.	Protects 700 hectares of the Atlantic Rainforest with a 100% compliance rate and minimizes the need for intensive water treatment, purification, and pumping in São Paulo.
Waste and water	United State's Toxics Release Inventory	1988–2007	Requires qualifying industrial, manufacturing, and government facilities to report on-site releases and off-site transfers of specified toxic chemicals that are then posted on a freely accessible website and database.	Project is managed by the national government but facilities compile and report their own information on their releases.	Reporting facilities decreased their releases of listed chemicals by 61% from 1988 to 2007.

ethical guidelines concerning research on human subjects we have presented such data as anonymous (though appendix A does list the names of those interviewed). Unless otherwise indicated, all the data for each of the following case studies are from these research interviews, although it is impossible to ascribe particular comments and views to specific individuals. Although those data are presented here as anonymous, we did take care to verify and cross-check all facts and numbers. For more information about the details, strengths, and weaknesses related to our methodology, see appendix B.

8.1   Denmark's Electricity Policy, 1970–2001

This case study is polycentric because Denmark combines small-scale decentralized community control of energy infrastructure with national standards and policies. It offers a paradigmatic example of how to improve energy security by researching new and innovative technologies and reducing dependence on imported fuels.

Denmark has relied on a variety of policy mechanisms, including energy taxes, research subsidies, and feed-in tariffs to promote energy efficiency, combined heat and power, and wind energy.[2] Denmark went from being almost 100 percent dependent on imported fuels such as oil and coal for their power plants in 1970 to becoming a net exporter of fuels and electricity today. The country leads the world in the exportation of wind energy technology, with a hold on roughly one-third of the world market for wind turbines. It has more wind projects integrated into its power grid than any other country in the world, and some parts of the country, such as Western Denmark, often draw more than 40 percent of their electricity from wind turbines. These achievements are all the more impressive when readers consider that Denmark has roughly the same size and population as the US state of Maryland.

8.1.1   Description

Essential to Denmark's successful approach are long-term taxes on energy fuels, electricity, and carbon dioxide, which create incentives for energy efficiency and provide the government with revenue for research on renewable energy. Higher taxes on gasoline, diesel fuel, and oil were first passed in 1974 after the oil shocks from the OPEC embargo. These measures were followed by additional taxes for coal in 1982, carbon dioxide in 1992, and natural gas and sulfur in 1996. These taxes raised nearly US$9 billion in 2005 and, cumulatively, US$25 billion from 1980 to 2005 (figure 8.1). Such taxes do mean that electricity prices are relatively high in Denmark: around 25 euro cents (32 US cents) per kilowatt-hour, versus 19 euro cents (25 US cents) per kilowatt-hour in Germany and 13 euro cents (17 US cents) per kilowatt-hour in the United Kingdom.

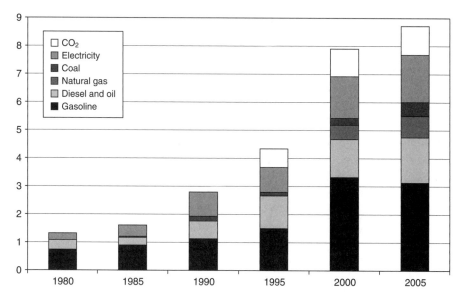

Figure 8.1
Revenue from Denmark's fuel, electricity, and carbon taxes, 1980–2005 (billions of US dollars).

The energy-related taxes were kept high after the prices of fossil fuels dropped in the 1980s and the 1990s so that the manufacturing industries could rely on stable fuel and electricity price signals. The taxes furthermore sent price signals that encouraged energy-efficiency measures in the Danish power market, and accrued government funds for R&D expenditures that were then directed at wind power, biomass, and small-scale combined heat and power units. Primary energy consumption nationally increased only 4 percent from 1980 to 2004, even though the economy grew more than 64 percent in fixed prices. The carbon tax alone added about 1.3 euro cents (1.7 US cents) per kilowatt-hour of additional income for renewable power providers.

The energy taxes also helped fund a government program prompted by researchers and local communities interested in using low-tech windmill designs to generate electricity. Danish regulators, working with manufacturers and interested citizens, took a bottom-up strategy of wind turbine development—a slow, crafts-oriented, step-by-step process including incremental learning through practical experience.[3] The Danish model adhered to learn by doing; and designers recognized the learning curve needed to perfect development and were willing to tolerate and learn from earlier setbacks. Their approach paid dividends. According to the Risø National Laboratory, from 1980 to 2005 the cost per kilowatt-hour of Danish wind turbines decreased 60–70 percent, and Danish R&D enabled wind turbines to produce 180 times as much electricity at 20 percent the cost, thanks largely to improved capacity factors, or the amount of time a turbine can generate electricity. In the same period, commercial turbine output grew by a factor of 100, from 30 kilowatts in the 1980s to 3.0 megawatts in 2006.

As Danish wind technology matured, the government supported its expansion with additional policies. Starting in 1979, the government promoted an investment subsidy that reimbursed individuals, municipalities, and farming communities for the capital costs of installing wind, solar, and biogas digesters. These subsidies initially covered 30 percent of the expense of renewable-energy systems but were scaled down periodically as the industry matured and turbine prices decreased. In 1981, the government passed a feed-in tariff (FIT) requiring utilities to buy all power produced with renewable-energy technologies at a rate above the wholesale price of electricity in a particular distribution area.[4] In 1985, an agreement was reached between the government and the electricity utilities, committing the utilities to install capacity of 100 megawatts of wind energy over a 5-year period (and fully implemented by 1992).

Also in 1985, lawmakers passed two other important policies. The Danish government established the Danish Wind Turbine Guarantee, which provided long-term financing of large wind projects that used Danish-made wind turbines, thereby reducing the risk of building larger projects and encouraging local manufacturing. The Danish Energy Authority also provided open and guaranteed access to the grid. The costs of connecting to the grid were to be shared between the owner of the wind turbine and the electricity utility. The owner of a wind turbine had to bear the cost of the low-voltage transformer and the cost of connection to the nearest connection point on the 10/20-kV distribution grid. The utilities had to cover the costs for reinforcement of the grid when needed.

As a consequence, Danish Transmission System Operators were legally obligated to finance, construct, and operate the transformer stations and T&D infrastructure for centralized wind farms and decentralized wind turbines owned by ordinary people. They were obligated to connect wind power, to extend the grid if that proved necessary, and to provide financial compensation if any of the wind power generated was curtailed. The costs of this infrastructure investment and reimbursement for curtailed power were paid for by the government, and then distributed to all customers. The distribution company had the right to reject the connection if it could prove that the costs would become excessive. In those rare cases, however, the utility was required to present alternative solutions.

To accommodate the technical difficulty of managing a highly dispersed and decentralized electricity system, the Danish T&D network is made up of mostly newer equipment with a preponderance of high-voltage transmission lines delivering power over shorter distances with low voltage levels. A significant number of transmission lines are underground, and system operators report no unaccounted loss. In other words, the grid is so "tight" that every kilowatt-hour is counted. This "tightness" is necessary so that Denmark can trade power between its Nordic neighbors and Germany. It is bolstered by the large hydroelectric and pumped hydroelectric reserves of Finland, Norway, and Sweden.

Another component of Danish electricity policy has been the aggressive support for combined heat and power facilities. After the government abolished the installation subsidy for wind energy in 1989, it quickly promoted environmentally friendly zoning in 1990 to advance electricity investments in towns and villages outside major cities. Cogeneration units were required to replace district heating units, and the use of oil, diesel fuel, and coal was prohibited and replaced by natural gas. If the local market was not large enough to cater to cogeneration, the district heating plants were required to utilize biomass. Concomitantly, all large utilities in the major cities were ordered to use biomass (especially straw), and required to obey mandatory energy-efficiency regulations. This wave of environmentally friendly conversions and improvement of efficiency drove significant investment in the combined heat and power market. Thus, cogeneration now provides roughly 60 percent of all electricity and 80 percent of heat consumed in Denmark (figure 8.2). The next closest country in Europe, the Netherlands, harnesses only 38 percent for electricity, Finland 36 percent, and every other country in the European Union less than 10 percent.[5]

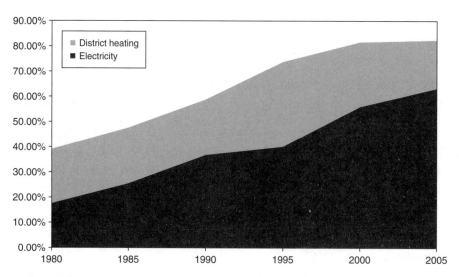

Figure 8.2
Contribution of combined heat and power to Denmark's electricity and heating supply, 1980–2005.

8.1.2   Costs and Benefits

The social and environmental benefits from Denmark's energy taxes and support for energy efficiency, wind energy, and combined heat and power have been impressive. As more renewable-energy and cogeneration units have come to replace less efficient and more polluting conventional fossil-fuel units, related CO_2 emissions have plummeted. In 1990, almost 1,000 grams of carbon dioxide were emitted per kilowatt-hour of electricity produced in Denmark. In 2005, emissions per kWh were below 600 grams. In all, the carbon dioxide emission intensity—the amount of CO_2 emitted per unit of gross domestic product—was 48 percent lower in 2004 than it was in 1980.

Denmark also used its expertise in wind turbine operation and manufacturing to create a robust export market. Danish regulators promoted the localization of wind energy manufacturing by offering a strong home market for wind technology and stable annual demand. That created opportunities through the sales of new products, jobs, and an increased tax base. Danish regulators also used their domestic turbines as a real-world laboratory to experiment with designs, lower the cost of turbine equipment, and improve capacity factors.[6]

Lastly, the greater penetration of cogeneration and wind energy has promoted the decentralization of energy supply, a transition that has brought political benefits (in the form of greater diversification) and energy-security benefits. (An electricity system with a high fraction of distributed wind also has more power plants on the grid, but also fewer individual units whose capacity is large enough to destabilize the system in the event of an accident or a terrorist attack.) Whereas in the United States and in most other countries the production and consumption aspects of energy are segregated and power plants often are "out of sight, out of mind," in Denmark energy production is close to the end users. In the 1970s, Denmark shifted from centralized generation (with fewer than 20 large-scale plants) to a decentralized model. Today there are more than 4,000 small-scale generators. Even in Copenhagen, wind turbines are fully integrated into the landscape. In public parks, near factories, and in the harbor, they are *visible*. People tend to view them as natural, whereas in the United States energy technologies are predominantly invisible. There is such strong support for renewable electricity in Denmark that when the government announced its most recent renewable-energy targets, it was promptly criticized by numerous political parties for not being progressive enough.

8.1.3   Challenges

Notwithstanding the merits of Denmark's approach to electricity policy, it has faced some challenges since 2001. The government recently switched from its time-tested method of using set feed-in tariffs to promote wind and other renewable resources to a tendering system for large-scale offshore parks. In an effort to lower prices for land-based wind mills, the tariff for renewable electricity is set at the wholesale market price plus a small subsidy. Wind developers, however, and even experts within the Danish Energy Authority, appear to concur that the current prices are too low and have stunted the growth of the wind industry. From 2004 to 2006, developers installed less than 40 megawatts of capacity, whereas they installed capacity of more than 1,000 megawatts before the changes were made. A new energy policy agreement among the political parties in the parliament is expected to adjust the tariff upward in order to meet the 2025 goals set out by the government, but this has been a long time coming.

The rapid growth of Denmark's wind-energy industry and its expansion to meet not only domestic demand but also the demand of a strong global market have created bottlenecks in manufacturing, construction, and wind farm design. The average wait for wind turbines from Vestas is now between 24 and 30 months, and LM Glasfiber, a manufacturer of turbine blades, reports a minimum two-year delay. For offshore wind turbines, delays of 12–18 months are common for cables, and delays of 18–36 months are typical for installation vessels, which oil and gas companies also use. Bottlenecks have arisen in offshore wind farm construction, due mainly to a shortage of vessels capable of mounting turbines to the sea floor as well as suitable harbors where offshore wind turbines can be assembled. Part of the explanation is the booming Chinese market; China consumes one-third of the total steel produced in the world, leads the world for wind energy installations in 2009, and its share of the world market has tripled in just 10 years. That has pushed up the world market price for raw materials, which has brought about a significant price increase for wind turbines.

A third, lingering challenge concerns price volatility. In view of the laws of supply and demand, wind power offers a low market price precisely when its output is high (because it is more likely that many turbines are generating excess power), meaning electricity is also valued the least. Conversely, wind output may be low when power is needed. This type of price volatility occurs both hourly and monthly. In January 2005, the amount of available wind capacity in Denmark dropped below 100 megawatts as turbines shut down during a hurricane, forcing system operators to switch on expensive peaking power plants and to increase imports from Germany and neighboring Nordic countries.

A final challenge connects to Denmark's overall energy policy. Though supportive of renewable energy, Denmark's energy sector is still dominated by fossil fuels. Although the government has publicly iterated a plan to be 100 percent independent from reliance on coal, natural gas, and oil, it still consumed 9 million short tons of coal in 2008.[7] Denmark also produced 15.5 million tons of oil in 2007, but renewable energy accounted for less than 3 million tons of oil equivalent. Of the total energy produced and used in Denmark, 40 percent came from oil, 23 percent from coal, 30 percent from natural gas, and only 17 percent from renewable resources.[8]

8.1.4 Lessons Learned

Still, the Danish approach to electricity policy offers a number of important lessons for other countries.

First, the Danish experience seems to prove that a carbon tax is not deleterious to the overall economy, and that, if implemented properly, it can be a useful tool for promoting wind energy, energy efficiency, and combined heat and power. The upstream carbon tax better incorporates some of the externalities associated with climate change and possesses three advantages over cap-and-trade programs: it is simpler to design, it sends price signals clearly and directly to consumers, and it helps raise government revenue that is then funneled back into R&D.

Second, as will be explored in greater detail in the case study on Germany (section 8.2), Danish taxes and research expenditures, when coupled with feed-in tariffs, could be easily replicated in other countries. Such feed-in tariffs have spurred rapid growth in domestic renewable-energy industries in both countries.

Third, guaranteed and open access to the grid helps rapidly promote renewable energy. It minimizes barriers to market entry and prevents utilities from using their power of incumbency to block renewable-energy projects on transmission-and-distribution (T&D) grounds. It increases the profitability of renewable-energy projects by shifting the costs of interconnection to the grid from a project's developer to the utility. It also helps utilities, as they can improve their debt-to-equity ratios on financed T&D infrastructure more quickly as new wind turbines and biomass facilities start generating electricity.

Fourth, the Danish model highlights the importance of participation, input, and feedback. Danish political institutions engender cooperative and egalitarian ideals that can be traced to the country's founding as an agricultural society. Small and medium-size farms occupied an overwhelming amount of the land, and the rural population was highly trained and educated. The electricity policy of Denmark has thus been characterized by strong political leadership and well-designed, consistent policy mechanisms, with a long tradition of broad political alliances and inclusive energy policies. New policies are typically negotiated with all political parties and possible stakeholders in a transparent manner.[9] As a result, the ownership of Danish energy projects is decentralized at the local level rather than concentrated in the hands of large corporations. In 2005, for example, only 12 percent of wind farms were owned by utilities; the remaining 88 percent were owned by individuals and cooperatives.[10]

8.2 Germany's Feed-in Tariff, 1990–2009

Germany's feed-in tariff is polycentric because it encourages the participation of residential and community producers of renewable energy alongside commercial and utility players in national markets. That country's feed-in tariff is an exemplary case of how to improve energy security by diversifying electricity generation and beginning to place prices on some of the negative externalities associated with energy.

Germany stands as the best example of a country that has used a feed-in tariff to promote renewable electricity supply.[11] The feed-in tariff (FIT) sets a fixed price for purchases of renewable electricity at a rate above the retail market price for each unit of electricity fed into the grid. Germany's FIT also requires power companies to purchase all electricity from eligible producers in their service area at this premium rate over a long period of time. It may not sound like much, but Germany's FIT has been a transformational tool for rapidly accelerating the diffusion of renewable electricity technologies within Germany and around the world. Since the introduction of the policy in 1990 and its modification and enhancement in 2000 and 2004, Germany's FIT has created 250,000 jobs, reduced carbon dioxide emissions by more than 79 million tons per year, saved 4.3 billion euros on energy imports, and built an industry with more than €25 billion in annual turnover for renewable power plant construction and operation. Its FIT has also seen renewable electricity supply jump from 4.8 percent of national capacity in 1998 to 14.2 percent in 2007 (table 8.2).

8.2.1 Description

German regulators formally created the country's first FIT in 1990. That particular FIT was based on the avoided costs of electricity generation. The rate of renewable energy deployment was limited by a cap on overall capacity. Wind and solar systems received the same tariff regardless of their location and size. Despite these limitations, the early FIT created significant expansion of wind and hydroelectric power facilities. In 1999, German installed wind capacity reached 4,400 megawatts. The policy helped create business networks and a community of advocates that became essential for supporting further legislation. In part because of this support, a more effective and efficient FIT, known as the Erneuerbare-Energien-Gesetz (meaning Renewable Energy Source Act), was passed in 2000. That law removed the capacity cap for most technologies; calculated tariffs above the market rate of electricity; differentiated tariffs by type, location, and size; and made permitting, transmission, and distribution of electricity more efficient. In 2004 and 2009, German legislators further modified and differentiated tariff payments, and implemented measures to

Table 8.2
Renewable energy as a share of total final energy consumption for Germany, 1998–2007.

Final energy consumption (FEC)	1998	1999	2000	2001	2002	2003	2004	2005	2006	2007	
	%										Renewable energy proportion of total electricity consumption:
Electricity generation (in relation to total gross electricity consumption)	4.8	5.5	6.3	6.7	7.8	8.1	9.5	10.4	11.7	14.2	1990: 3.4% 2000: 6.3% 2004: 9.5% 2006: 11.7%
Heat supply (in relation to total heat supply)	3.5	3.5	3.9	3.8	3.9	4.6	4.9	5.4	5.8	6.6	Composition of renewable electricity production: Wind power: 41.3% Hydroelectric power: 29.3%
Fuel consumption (in relation to total fuel consumption)	0.2	0.2	0.4	0.6	0.9	1.4	1.8	3.8	6.3	7.6	Hydroelectric power: 29.3%
Renewable as a share of total FEC	3.1	3.3	3.8	3.8	4.3	4.9	5.5	6.6	7.5	8.6	Hydroelectric power: 29.3% Biomass: 19.2% Photovoltaics 2.7% Landfill gas, sewage plant gas, biogenic waste: 7.5%

Source: German Federal Environment Ministry.

Table 8.3
Current rates offered under the German FIT scheme.

Technology	Plant size	Tariff (euro cents/kWh)	Annual degression rate
Hydro power (new)	Up to 5 MW	7.65–12.67	
Hydro power (modernized/ revitalized)	Up to 5 MW	8.65–11.67	
Hydro power (renewed)	Up to 50 MW	4.34–6.32	1%
Landfill gas, sewage gas, mine gas, and biomass	Up to 20 MW	4.16–11.67	1%–1.5%
Geothermal	Up to 10 MW	10.50–16.00	1%
Onshore wind	Location specific tariffs	5.02–9.20	1%
Offshore wind	Location specific tariffs	3.5–13.00	5% (from 2015 on)
Solar radiation (roof-mounted)	Up to 1 MW	33.00–43.01	8–10%
Solar radiation (free-standing)	All	31.94	8–10%

avoid windfall profits to power producers, such as lowering the available tariff offered to producers each year (something called "degression"). Table 8.3 depicts the FIT scheme as it exists in its most recent form. Tariffs are guaranteed for 20 years for all technologies except hydroelectric projects, which are guaranteed for 15 years.

The German FIT has been highly effective at promoting a broad assortment of renewable electricity technologies. As of late 2007, the FIT had incentivized the installation of more than 430,000 solar photovoltaic panels constituting 3,834 megawatts of installed capacity, whereas in 1991 the capacity of the installed panels was only 3 megawatts (figure 8.3). More than 90 percent of these panels, furthermore, are owned by homeowners and cooperatives, not by electric utilities or independent power providers. Growth in the wind market has been equally significant, with less than 100 megawatts installed in 1991 but 22,500 megawatts installed by the end of 2007. Every single major city in Germany has at least one wind farm.

8.2.2 Costs and Benefits
At first glance, it would make sense if Germany's FIT never had benefits that exceeded costs, since consumers are in essence paying a higher rate for renewable electricity than the existing market rate. For example, FITs cost German customers about €3.6 billion in 2004, €5.8 billion in 2006, and €8.9 billion in 2008 (the amount goes up

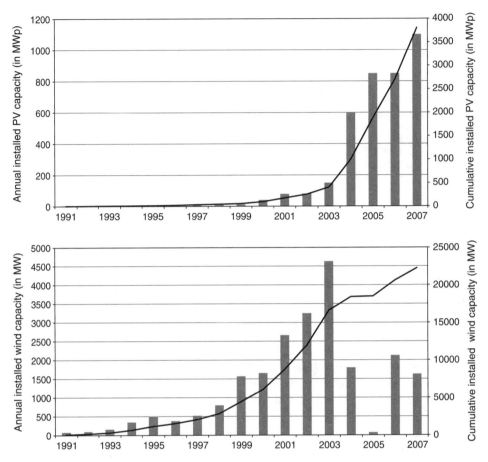

Figure 8.3
Growth of Germany's wind turbine and solar PV panel markets, 1991–2007. Black line and axis on right represent cumulative installed capacity; gray bars and axis on left represent annual installed capacity.

as more renewable-energy systems are installed to take advantage of the tariffs).[12] The cost to consumers, moreover, is not the only cost. The FIT scheme incurs additional expenses relating to enforcement and regulation (about €100 million per year) and transaction costs for grid operators (about €2 million per year).

Perhaps surprisingly, however, these costs pale in comparison to the benefits from the German FIT. These include reductions in the wholesale price of electricity resulting from less volatility, avoided pollution such as acid rain and carbon dioxide, and fewer energy imports. Because renewable electricity generators often rely on fuels that are free for the taking and non-depletable, their widespread use helps reduce and stabilize electricity prices. According to data from the German Federal Ministry of Environment, the FIT has reduced the average market price of a megawatt-hour of electricity by about €7.83, leading to overall savings of almost €5 billion in 2006. In addition, renewable electricity generators displace coal, oil, and natural-gas power plants, minimizing hazardous emissions of air pollutants. In 2005, for instance, renewable-energy technologies, mostly incentivized by the German FIT, saved the German economy €3.4 billion in pollution costs. Lastly, the FIT reduced energy dependence and the costs of importing coal and other fossil fuels, leading to a saving of €1 billion. In all the German FIT cost consumers and the government about €3.3 billion in 2006 but saved them €9.4 billion![13]

Some of these savings were in the form of displaced GHG emissions. The FIT avoided the release of 79 million metric tons of carbon dioxide equivalent in 2008. That is more than the annual emissions of Armenia, Botswana, Cambodia, Cameroon, El Salvador, Iceland, Ireland, Paraguay, and Senegal combined.[14]

The cherry on the sundae is that these calculations did not even take the numerous additional benefits of the German FIT into account. In 2008, almost 280,000 people were employed in their renewable-energy industry, 12 percent more than in 2007, and Germany led the world in the manufacturing of solar photovoltaic panels and wind energy. (The United States overtook Germany's lead in wind in 2008.) Of these jobs, at least 150,000 can be attributed to the German FIT scheme. The industry estimates that by 2020 more than 750,000 people could be working in the renewable-energy sector, which would then be larger than the German automobile industry.

8.2.3 Challenges

Though the benefits from the German FIT have been substantial, the right price for each of the tariffs had to be determined by trial and error. When prices were set too low, not enough investment occurred. When prices were set too high, they offered windfall gains to producers at the expense of consumers and an efficiency loss for the economy.[15] The early version of the FIT also did little to advance less mature technologies, such as solar photovoltaics and smaller geothermal and hydroelectric plants. As regulators gained experience with the FIT, the policy became much more complicated. Whereas the first FIT law (1990) included only five key articles of legislation, the 2000 policy had 13, the 2004 revised policy 21, and the 2009 revised policy more than 65.

The rapid growth in renewable energy, as in Denmark, has caused price inflation as a result of shortages of components, materials, and expertise. As wind farm installations increased in 2008 and 2009, rather than decreasing costs through technological learning, their rapid growth put upward pressure on manufacturing costs and delivered prices.[16] The global demand for wind turbines, driven in part by the German FIT, far exceeded the supply for these two years, contributing to a shortfall of materials. After hitting a nadir of about $700 per installed kilowatt in 2000, prices averaged about $1,420 per installed kilowatt in 2008. Since turbine costs account for 75–80 percent of the total cost of a wind project, these higher construction costs translate into higher lifetime project costs. Something similar happened with solar photovoltaic panels from 2003 to 2007. When demand for solar installations was relatively low, the high-grade silicon used to make the cells cost only $25 per kilogram. The price jumped to $400 per kilogram in 2007 as strong growth caused a global shortage of materials.

A final challenge relates to how to grow renewable energy capacity fast enough to offset expected increases in coal-fired capacity within Germany. Coal continues to be Germany's largest domestic energy resource, powering more than half of the country's electricity.[17] Because Germany remains one of the world's leading producers of lignite (brown coal), coal has strong political support in Germany, with subsidies equaling $144 per domestic ton mined in 2005—twice the similar subsidies offered in Spain, and 4 times those offered than many other European countries.[18]

Currently there are plans to construct up to 26 new coal-fired power plants utilizing domestic sources of anthracite and lignite coal.[19] Reliance on coal and other fossil fuels could increase even further in the coming years if German planners phase out their nuclear plants, which would require a substitution of 20 gigawatts of electricity capacity. As the director of the German environmental non-governmental organization Bund für Umwelt und Naturschutz Deutschland recently noted, "anyone building new coal-burning power plants cannot be seriously concerned about climate protection."[20] Yet other important stakeholders want to push renewable energy even faster, and the government has set an official target of "at least 30 percent" by 2020. In short, German policy makers appear to be divided as to whether to renew investments in coal or to switch to renewable electricity generators. The future could see Germany taking either of these paths.

8.2.4   Lessons Learned

Experience to date suggests that FITs offer policy makers the best single tool available to rapidly promote renewable energy. FITs are therefore essential for promoting a more efficient, democratic, decentralized electricity system, independent of government funding, operating with minimal degradation of ecological services, resilient to disruptions and price volatility, and highly beneficial to all income groups. Independent studies have concluded that FITs can be used effectively by any country to promote renewable energy.[21] As of early 2009, at least 46 countries had adopted FIT schemes of various kinds. (See table 8.4.)

Despite the extra initial cost to consumers to cover the expense of the tariff, FIT policies end up benefiting them in the long run by depressing electricity prices.[22] In 2007 the German FIT saved electricity customers almost 3 times what it cost them,

Table 8.4
Feed-in tariff schemes around the world. Source: M. Mendonça, D. Jacobs, and B. K. Sovacool, *Powering the Green Economy: The Feed-In Tariff Handbook* (Earthscan, 2009).

Africa	Americas	Asia	Australasia	Europe
Algeria	Argentina	India*	Australia*	Austria
Kenya	Brazil	Indonesia		Bulgaria
Mauritius	Canada*	Israel		Croatia
South Africa	Ecuador	Pakistan		Cyprus
	Nicaragua	Philippines		Czech Republic
	United States*	South Korea		Denmark
		Sri Lanka		Estonia
		Thailand		France
		Turkey		Germany
		Ukraine		Greece
				Hungary
				Ireland
				Italy
				Latvia
				Lithuania
				Luxembourg
				Macedonia
				Malta
				Netherlands
				Portugal
				Slovak Republic
				Slovenia
				Spain
				Switzerland

*Countries with states or provinces that have FITs.

and the benefits will become greater as more renewable energy capacity is brought on line. Disruptions and interruptions in supply caused by accidents, severe weather, and bottlenecks can prevent natural gas, coal, and uranium from being adequately and cost-effectively distributed to conventional power plants. Such depletable fuels are also prone to rapid price escalations and significant price volatility, and are exposed to sudden fluctuations in currency rates. Renewable fuels promoted by FITs, in contrast, are sometimes free, are less prone to speculation, do not have to be transported (with some exceptions), and insulate the power sector from dependence on foreign suppliers. This is how they save consumers money.

Because FITs create consistency and predictability for renewable energy financiers, investors, manufacturers, and producers, they bolster domestic renewable-energy industries by creating hundreds of thousands of high-paying jobs. Renewable electricity generators prompted by FITs involve a highly skilled workforce and modernize the local industry base. Using renewable energy makes local businesses less dependent on imports from other regions, frees capital for investments outside the energy sector, and serves as an important financial hedge against future spikes in energy prices. Whereas investments in conventional fossil-fuel and nuclear power plants send money *out* of the economy, investments in renewable electricity generators keep money *in* the economy. About half of every dollar expended on conventional electricity leaves many economies (and in some areas 80–95 percent of the cost of energy leaves local economies), whereas every dollar invested in renewable electricity can produce $1.40 of gross economic gain.[23]

Rather than pass a single FIT policy and then passively wait for it to work, German regulators constantly monitored and adjusted FIT rules and rates, differentiated tariffs by technology, and started degression to keep prices low. The German FIT, initially implemented in 1990, was amended and augmented in 2000, 2004, and 2009. To minimize uncertainty in the market for renewable energy, these changes were implemented slowly and with input and feedback from manufacturers, power companies, environmental groups, and other stakeholders. The tariffs were differentiated by technology type, size, and location to give regulators more control over the sorts of renewable-energy technologies they wanted to support, and to ensure that even the less mature systems (such as solar panels) received incentives. Degression of FIT rates also ensured that investors were rewarded for earlier investments in renewable energy so that their social and political benefits begin accruing sooner.

8.3   Brazil's Proálcool Program and Promotion of Flex-Fuel Vehicles, 1975–2009

Our Brazil case study is polycentric because it shows how, in the formulation of national energy policy, integration of sugarcane farmers, ethanol distillers, automobile manufacturers, and environmental groups can encourage the rapid adoption of an alternative transport fuel and of vehicles that do not rely on petroleum. It also shows how energy security can be improved through the displacement of gasoline and oil in the transport sector.

Brazil created its Proálcool program in November 1975 to increase ethanol production and substitute ethanol for petroleum in conventional vehicles. Brazil not only surpassed the initial goals of their program within its first 3 years, but emerged to become the second-largest producer of ethanol in the world, making 26.9 billion liters of ethanol in the 2008–2009 season (figure 8.4). Annual ethanol production for Brazil now involves 7 million hectares of planted land, more than 72,000 farmers and sugarcane producers, 390 sugarcane mills, 240 distillery plants, 3.5 million direct and indirect jobs, a harvest of 569 million tons of sugarcane, 106 million tons of bagasse (a by-product used to make electricity), 30 million tons of sugar, and revenues of $20 billion. The program is so successful that it is no longer necessary for the government to subsidize ethanol directly. As figure 8.4 shows, a liter of Brazilian ethanol produces more than 30 times as many jobs as an equivalent liter of oil, coal, or hydroelectricity and has 3 times as much energy as any other type of ethanol.

During the period of the Proálcool program, as table 8.5 shows, fermentation times for ethanol production have been cut in half, yields have almost tripled, and efficiencies for sugarcane extraction, fermentation, distillation, and electricity generation have improved steadily. Faced with the prospects of uncertain ethanol prices in the early 2000s, the government began heavily promoting flex-fuel vehicles (FFVs) that can run on ethanol and gasoline, a strategy that has proved so successful that 90 percent of the new cars sold in Brazil in 2009 were FFVs.

8.3.1 Description
Although Brazilian researchers experimented with different blends of ethanol in the 1920s, an aggressive program did not really emerge until 1975, when the Brazilian

Table 8.5
Notable technology improvements in Brazilian ethanol production, 1975–2005.

	1975	2005
Milling capacity (tons crushed/day)	5,500	14,000
Fermentation time (hours)	16	8
Extraction efficiency	93%	97%
Fermentation efficiency	82%	91%
Distillation efficiency	98%	99.5%
Distillery overall efficiency	66%	86%
Boiler efficiency	66%	88%

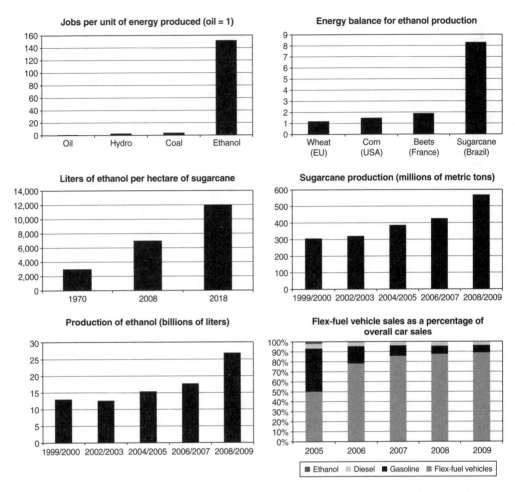

Figure 8.4
Employment and energy aspects of Brazilian ethanol production and flex-fuel vehicle use. Sources: J. Goldemberg, "The Brazilian biofuels industry," *Biotechnology for Biofuels* 1, 2008, no. 6: 1–7; H. Machado-Filho, "Climate change and the international trade of biofuels," *Carbon and Climate Law Review* 2, 2008, no. 1: 67–77; Ministry of Agriculture, Livestock, and Food Supply, *Brazil and Agribusiness* (DF, 2007).

government initiated one in response to the 1973 oil crisis. The Programa Nacional do Álcool, abbreviated to Proálcool, was created after high gasoline prices resulted from the OPEC oil embargo. Brazilian regulators hoped that ethanol development would insulate the economy from the vagaries of the global market. Proálcool had two primary components: a compulsory mandate that all gasoline had to have at least a 10 percent blend of ethanol, and a voluntary mandate that automobile manufacturers design vehicles to run on 100 percent ethanol.[24] The program has four distinct phases.

In the first phase (1975–1979), annexed distilleries (that is, distilleries attached to sugarcane mills) produced anhydrous ethanol from molasses. The ethanol was then blended with gasoline, in mixes ranging from 5 percent to 25 percent for use in cars. The program was focused almost exclusively on growing sugarcane for ethanol production and adapting existing plantations and factories to produce ethanol. The blending of ethanol started slowly, beginning in São Paulo and the northeast before expanding to the rest of the country. At first, researchers focused on developing an engine that could operate entirely on ethanol (called a "neat" ethanol engine), but after a while they partnered with automobile manufacturers to design an entire vehicle. During this first phase of the program, ethanol production increased from 220 million liters (58.1 million gallons) in 1975 to 2.8 billion liters (739.7 million gallons) in 1979.

The second phase of the program (1979–1984) focused on increasing ethanol production and increasing the efficiency of distillation processes. Independent sugarcane distilleries began to emerge. No longer coupled to sugar mills, these newly autonomous distilleries produced anhydrous ethanol (for blending) and hydrous ethanol (pure) from both sugar and molasses. To use hydrous ethanol, regulators also shifted the program from gasohol to promoting neat-ethanol vehicles, which began to enter the commercial market in 1980. The government regulated the pump price of hydrous ethanol to be equivalent to 64.5 percent of the price of gasoline and used higher gasoline taxes to pay for any true differences in price between the two fuels. As a result, pure ethanol was always cheaper than its alternatives, purchases of gasoline helped fund the ethanol program, and drivers were incentivized to purchase neat-ethanol vehicles. In 1981, 90 percent of all new vehicles sold in Brazil were neat-ethanol vehicles. That same year, public mistrust and suspicion about high ethanol prices and the poor performance of ethanol-fueled vehicles arose. In response, the government worked with auto manufacturers to improve ethanol-powered engines, capped the price of ethanol to stabilize prices, and ran an educational program to inform consumers. By 1983, consumer confidence in ethanol resumed and the sale of ethanol-fuel vehicles reached 84 percent of total vehicle sales in Brazil. By the end of the second phase of the program, in 1984, there were 5 million neat-ethanol vehicles and most service stations sold exclusively ethanol or ethanol and gasohol.[25]

Various stakeholders became involved in the program. The government protected sugarcane workers by giving them bonuses related to the value of sugar and ethanol produced each year. Though working conditions on sugar plantations and ethanol distilleries were indeed strenuous, workers in São Paulo received wages 80 percent higher than the agricultural-sector average.[26] Environmental groups were invited to solicit feedback, and as a result the government mandated that rainforest land could not be used for new sugarcane plantations, processing facilities, or ethanol plants. Brazilian automakers were enrolled early on, and in 1979 they reached an agreement with the government to produce cars that ran both on blends of ethanol and gasoline as well as purely on ethanol. This agreement with all of Brazil's major domestic manufacturers resulted in the production of 250,000 ethanol-only vehicles by 1980 and 350,000 by 1982.[27] Partnerships were later made with international automobile companies such as Ford, Fiat, and Volkswagen. The government sought the involvement of gasoline distributors, providing them with low-cost pumps and tanks for ethanol.[28] As ethanol production increased, partnerships were cemented between ethanol producers and electricity suppliers, since bagasse—a residue of the process of making ethanol from sugarcane—can be used as a fuel for power plants.[29]

During its third phase (1985–1990), the program survived drastic changes in the political and economic structure of Brazil. In 1986, 76.1 percent of all automobiles on the road in Brazil, and 96 percent of all new vehicles sold there, burned ethanol. However, the military dictatorship governing Brazil came to an end in 1985, and the newly democratic government began to scale the program back. Low oil prices in 1987 and 1988 convinced regulators to reduce ethanol subsidies, prohibit federal funds from being used to purchase sugar, and restrict government investment in ethanol infrastructure. The relaxation of government involvement also provoked what one interview respondent called a "purification of the market" by consolidating ethanol producers and driving "uncompetitive and inefficient" firms out of the industry. By the end of 1988, a shortage of ethanol (with stable demand) precipitated drastic price increases, which caused people with neat-ethanol cars to become frustrated and switch to gasoline and diesel vehicles. From mid 1989 to mid 1990, sales of ethanol cars dropped to only 11 percent of all sales.

The fourth and current phase, lasting from about 1991to 2009, has seen moderated policy of support for ethanol and an almost complete replacement of neat-ethanol vehicles with flexible-fuel vehicles. In 1993, in response to political concern about increasing dependence on imported oil and to intense lobbying from the sugar industry, a federal decree was passed requiring that all gasoline be blended with 22 percent anhydrous ethanol. In 2003, bounded blending limits were set: a minimum of 20 percent ethanol and a maximum of 25 percent. In 2003, as drivers continued to abandon ethanol-only cars for their conventional counterparts, the government began incentivizing FFVs through reduced tax rates and fuel taxes. With sensors in the fuel system that recognized the ethanol level of the fuel and managed the engine accordingly, Brazilian FFVs were capable of running on any blend of ethanol with gasoline, on gasoline alone, or entirely on ethanol. In 2004, the year FFVs entered the market, they accounted for 17 percent of new car sales (versus 3 percent of ethanol-only cars). Their market share has since risen to 90 percent. Able to run on either pure ethanol or gasohol, taking advantage of whichever is cheaper, these cars have proved popular. Thus, the only currently remaining government incentives related to ethanol are the blending mandate and reductions in fuel and motor vehicle taxes for the purchase of FFVs.

8.3.2 Costs and Benefits

The most significant benefit of Brazil's ethanol program is that it has driven the price of a gallon of ethanol well below the price of a gallon of gasoline or diesel fuel. Much of the explanation has to do with the physical properties of sugarcane ethanol. As a motor fuel, sugarcane ethanol has a much higher octane (usually above 98) than gasoline (usually an octane rating of 80), so it can be used in engines with a higher compression ratio (12:1 instead of the 8:1 common in gasoline engines) that therefore burn fuel about 15 percent more efficiently.[30] Relative to other sources of ethanol, such as corn and sugar beets, sugarcane has a higher energy content and can be grown on less land (at less cost) to produce the same amount of energy per a gallon. One can get 150 tons of biomass per hectare of sugarcane on first cut, versus only 8–9 tons per year for grains and 15–20 tons per year for corn. Whereas the latter two crops have to be replanted, sugarcane sprouts again by itself and can be harvested every 12–18 months. The energy inputs for sugarcane are therefore about 8:1 (for every one unit of fossil energy put into the ethanol lifecycle, one gets eight units out of it). Beets, corn, and wheat all have ratios less than 2:1.[31] In monetary terms, it cost about 20 US cents to produce a liter of ethanol using Brazilian sugarcane, 47 cents to produce a liter of ethanol from corn, and 97 cents to produce a liter from cereals in 2007. For Brazil, these benefits translate into real savings at the pump. In 2008, a liter of ethanol at a typical gasoline station in Brazil cost less than *half* a liter of gasoline.

The ethanol program has also greatly reduced Brazil's dependence on imported oil, with substantial macroeconomic savings. In 1970, before Proálcool started, Brazil was dependent on imports for 80 percent of its oil supply, yet in 2009 more than 60 percent of demand for motor fuel was met with ethanol. Although the net cost to the central government in terms of subsidies for Proálcool was about $11 billion from 1975 to 1985, the country saved $55 billion in avoided oil imports. From 1975 to 2000, the entire ethanol program cost the government about $30

Figure 8.5
A Petrosul station in Piracicaba. Note prices of ethanol (álcool) and gasoline. Source: Benjamin K. Sovacool.

billion but saved $30 billion *every 2 years* at an average oil price of $50 per barrel. Moreover, these economic benefits are only the direct ones. By some accounts, every dollar invested in ethanol produces another three dollars in indirect economic benefits. Confirming that the benefits from ethanol go beyond merely displacing oil imports, two independent economists not affiliated with the Brazilian program estimated that the country's external debt was $100 billion less than it would have been in the absence of ethanol production.[32]

Another important benefit from Brazil's ethanol program has been electricity generation from bagasse. Sugarcane cultivation produces as a residue bagasse, the fibrous part of the sugarcane plant that remains after its stalks have been crushed and their juice extracted. Bagasse is such a versatile material that it has even been used to make disposable diapers and greeting cards.[33] One major component of the Brazilian program is that virtually all of the energy needs for the production process—the electricity required to crush, ferment, distill, and refine sugarcane into ethanol—are met by on-site bagasse-fired electricity generators. In 2008, the ethanol industry produced so much bagasse that they met all of their energy needs *and* exported additional electricity to the grid at a low cost of 6.5 US cents per kilowatt-hour. Though this price was still higher than that of the cheap electricity coming from Brazil's vast network of hydroelectric dams, it was competitive enough so that bagasse-fired electricity displaced the need to operate 3 gigawatts of installed fossil-fuel capacity. Making electricity from bagasse is becoming so profitable that although the typical sugar mill today receives 56 percent of its income from sugar, 43 percent from ethanol, and 1 percent from electricity, the industry expects ethanol to attain a share of 54 percent and electricity a 16 percent share by 2025.

Sugar, ethanol, and bagasse are not the only beneficial products of the ethanol industry. The sugarcane and ethanol production process brings with it a variety of other co-products and industries. In the state of São Paulo, for example, the ethanol industry created 50,000 new jobs, 80 percent of which were expected to be retained in the next 10 years. Apart from employing almost 4 million workers in 2009, sugar mills are presently responsible for managing 600 schools, 200 day-care centers, and 300 hospitals throughout Brazil. A sample of 47 mills in São Paulo found that more than 90 percent offered health care and dental care, life insurance, and transportation. Eighty percent offered meals and pharmaceutical care, and 84 percent had profit-sharing programs.[34] Indirect benefits can be just as significant as those that are direct and measurable. Filter cake and vinasse, two residual substances left after sugarcane alcohol distillation, can be used as fertilizer. Four substances from the ethanol fermentation process—whole yeast, protein concentrate, yeast extract, and the cellular wall—have great value as secondary products. Every liter of ethanol produced creates an excess of 30 grams (1.1 ounce) of dry yeast. When the annual production of ethanol was 15 billion tons, the industry created 450,000 kilograms (992.1 pounds) of yeast, proteins, and extracts as by-products.[35] These can be used to enhance the protein content of pastas, biscuits, and breads, can be manufactured into vitamins, or can be put into salad dressings, soups, and snacks. Resins from sugarcane ethanol have also been used to make biodegradable foam packaging. One company currently sells a biodegradable sugarcane Styrofoam substitute that can be formed into trays for fruits and vegetables, packaging for electronics, mulch for the planting of new tees, and carpets that absorb chemicals.[36]

As a perhaps odd benefit (in view of the recent controversy over connection between biofuels and deforestation), sugarcane cultivation can improve soil fertility and restore degraded land. Whereas growing wheat and soybeans usually requires irrigated water and fertilizer, sugarcane is a grass that can grow naturally and unassisted in Brazil without external inputs. When the right variety is planted in the right place, sugarcane planting can maintain soil fertility and improve it in formerly degraded land areas, create habitats for species and insects, and prevent runoff and sedimentation. Sugarcane also acts as a carbon sink since carbon dioxide is sequestered in its roots (which are not uprooted during harvesting). Each hectare of sugarcane removes 10–28 tons of carbon dioxide per year, whereas the same space planted with corn removes only 2–4 tons. Planters in Brazil also rely on a rich variation of sugarcane types to minimize monocultures. Each plantation has only 5–7 percent of its land area covered with one variety of sugarcane, a practice that promotes diversity and traces back to indigenous farming practices in the sixteenth and seventeenth centuries.

Because of these attributes—greater energy content, by-products such as bagasse and vinasse, local production and farming—Brazilian ethanol has a much lower carbon and pollution footprint than gasoline, diesel fuel, and other biofuels. Though sugarcane burning can release particle pollution into the air, regulators have made a concerted effort to transition away from burning sugarcane to the use of mechanized tilling. These efforts reduced sugarcane-associated particle pollution by 3,900 tons in the period 2000–2007, and the use of ethanol resulted in a 6,500-ton reduction in volatile hydrocarbon emissions in the same period.[37] Though some studies have shown that the carbon footprint for ethanol can be high, these erroneously presumed that forests were cleared for sugarcane production when in fact a majority of plantations operate on converted coffee plantations and degraded pastures.[38] Although the combustion of ethanol in automobile engines is not benign (ethanol is a significant source of aldehyde emissions and peroxyacetyl nitrate pollution), every kilometer fueled by ethanol releases less particulate matter, carbon monoxide, volatile organic compounds, lead, benzene (a carcinogen), 1-3 butadiene, sulfur oxide, and carbon monoxide than gasoline.[39] Some GHGs, including nitrogen oxide, methane, and carbon dioxide, are emitted from nitrification and de-nitrification through the use of fertilizer, soil transformations, poorly drained soils, and motorized equipment; however, lifecycle GHG emissions are much lower for sugarcane ethanol than for gasoline—ethanol releases 0.6 kilogram of carbon dioxide per liter, gasoline 1 kilogram.[40]

In Brazil in 2008, about 27.5 million metric tons of GHG emissions were avoided as a result of ethanol's replacing gasoline, and an additional 5.7 million tons were avoided through the use of bagasse-fired electricity.[41] From 1970 to 2008, about 806 million metric tons of carbon dioxide equivalent have been avoided from the use of ethanol all together, an amount greater than the emissions from Australia, Fiji, New Zealand, Papua New Guinea, and the Solomon Islands for one year.[42]

8.3.3 Challenges

At the top of the list of the challenges present by the ethanol program and by reliance on FFVs is the fact that sugarcane is a weather-dependent crop whose production can surge during the right conditions (creating an excess of supply) or fall dangerously low during a drought (leading to severe shortages). Good Brazilian weather in the 1990s brought ethanol prices as low as 10 US cents per liter; and an unexpected surplus in late 2008 (due to more plentiful harvests than anticipated and increased demand for ethanol because of higher oil prices) sent prices plummeting again, leaving 30 million tons of sugarcane unprocessed. Furthermore, sugarcane cannot grow everywhere; although 100 countries grow it, the best climates are limited to tropical areas. Climatic conditions explain why the top 10 countries produce 80 percent of the world's sugarcane.

Second, the ethanol industry in Brazil is becoming more mechanized, consolidated, and capital-intensive. As a result, its energy and water inputs may increase and its ability to offer jobs may decrease. An average Brazilian sugar mill in 2008 processed 1.4 million tons of sugarcane and managed a plantation of about 80,000 hectares (197,685 acres), a yield facilitated by mechanized planting and harvesting. Such mechanization, however, can severely compact soil and damage the bottom halves of sugarcane stalks, creating losses of 5–10 percent. Mechanization also makes it easier to centralize production in the hands of corporate conglomerates instead of local farming cooperatives, and reduces the amount of labor needed to harvest each hectare of sugarcane. By 2017, it is expected that mechanization will have displaced 180,000 sugarcane cutting jobs, although as many as 100,000 of these jobs probably will be absorbed by the industry performing other duties such as operating tractor equipment, monitoring satellites, and managing computers.[43] Part of these job losses will also be offset by the longer growing season that mechanization offers. In the past, sugarcane planting began in May and ended in November, but now, owing to mechanized planting of precocious cultures and avoided burning, planting begins in April and ends in January.

A third challenge relates to a growing worry that Brazil may become too powerful in the global ethanol industry. Brazil provided about 40 percent of the world's sugar and more than 50 percent of its ethanol in 2006 (though the percentage dropped to 35 in 2009).[44] That same year Brazil exported 6.1 billion US dollars' worth of sugar to 113 countries and 1.6 billion US dollars' worth of ethanol to 40 countries.[45] Policy makers in Asia, Europe, and North America have expressed concern about becoming dependent on Brazilian suppliers and about the possibility for Brazil to become a cartel similar to OPEC. If such a cartel does begin to form, they believe that shifting from gasoline to ethanol would merely be replacing one type of dependence with another.

A fourth challenge concerns the thermodynamic efficiency of ethanol production. Though sugarcane ethanol is more efficient than ethanol and biofuels from other sources, and has many advantages over fossil fuels, current production is far from optimal. One ton of sugarcane has about 7,400 megajoules of energy but produces only 1,950 MJ (85 liters) of ethanol and 216 megajoules (60 kilowatt-hours) of electricity, resulting in a conversion efficiency of 29.3 percent (and similar to the poor efficiencies of thermoelectric power plants). The most efficient ethanol-production techniques that recycle trash and produce ethanol close to its point of consumption can get this number as high as 33 percent, but substantial room for improvement remains.

Fifth, there are limits to how much electricity generation from bagasse can be increased. Brazil already relies on a network of hydroelectric dams to produce power at very low prices, often below 3 US cents per kilowatt-hour unless a rare drought occurs. Shortfalls in electricity demand have classically been met by importing cheap hydroelectric power from Argentina and Paraguay. Though bagasse-fired electricity could theoretically provide about 60 percent of Brazil's electricity using existing amounts of waste and residue, Brazil currently has no need for electricity that is more expensive than these cheap sources of hydropower. Furthermore, most of the bagasse is currently used on site to power ethanol fermentation, distillation, and production.

Sixth, some aspects of the Brazilian program are unique and would be difficult to replicate elsewhere. In the 1970s, when the program was set up in response to the oil shocks, Brazil was a military dictatorship run by General Ernesto Geisel, a former president of the state-owned oil company Petrobras. Because of Geisel's experience with energy, convincing him of the need for Proálcool was relatively easy, and the dictatorship enabled the program to be centralized and enforced. Similarly, many countries lack the political support of a strong agribusiness lobby and sugar industry. The support of these stakeholders for the ethanol program made political consensus less difficult and ensured that the program was well protected in the parliament.

Some popular commentators have argued that ethanol production in Brazil directly contributes to illegal logging and the destruction of the Amazonian and Atlantic rainforests.[46] A closely related argument is that, though sugarcane cultivation may not be occurring directly in the rainforests of Brazil, the land occupied by sugarcane plantations has pushed cattle grazing and soybeans farming activates into previously pristine grasslands and forests.

These criticisms harbor an intuitive logic. However, they do not readily apply to Brazil. The primary drivers of deforestation in Brazil are unregulated cattle ranching and soybean cultivation, not ethanol production. The Amazon is not well suited for growing sugarcane—it is full of slopes greater than 12° (not ideal for sugarcane production), it is too wet, it allows for only a three-month harvest, and it has only two sugar mills, one of which is dedicated entirely to supplying sugar for soft drinks. More than two-thirds of sugarcane production occurs in the state of São Paulo, hundreds of kilometers south of the Amazon, and almost all sugarcane production nationwide occurs on traditional farmland, not grassland or newly converted forest-land. From 1992 to 2003, for example, 94 percent of sugarcane production expansion and harvesting occurred on existing farms, with no new land taken.[47] Farming trends in São Paulo also show that sugarcane cultivation has caused the density

Figure 8.6
Map of sugarcane crops in Brazil. Source: J. Goldemberg, "The Brazilian biofuels industry," *Biotechnology for Biofuels* 1, 2008, no. 6: 1–7.

of cattle to increase on existing land rather than provoking ranchers to claim new land.[48]

Interestingly, one recent assessment found that ethanol production in Brazil could be increased to 102 billion liters (an increase of about a factor of 5) and still require no new forestland.[49] Further, this assessment took special care to exclude all environmentally sensitive and protected areas, existing forests and reserves, areas with slopes greater than 12°, and all land currently occupied by permanent and temporary existing crops (such as soybeans, corn, wheat, banana, and cassava).

Finally, there is a belief that biofuels production in Brazil has been responsible for rising food prices.[50] Factual evidence seems to disprove the "food versus fuel" claim. Because sugarcane is hearty and dense in its biomass, less than 1 percent of the arable land in Brazil is dedicated to ethanol production, whereas 4 percent is used for soybeans, 3 percent for corn, and a much larger proportion for cattle. In 2008, Brazil produced its largest grain harvest ever—142 million tons of corn, soybeans, rice, beans, cotton, wheat, sorghum, peanuts, castor-oil plants, sunflowers, and other products—but it also boasted the largest sugarcane yield ever.[51] From 1990 to 2007, grain production in Brazil rose to a record volume of 135 million tons, an increase of 140 percent, yet in the same period ethanol production rose 56 percent.[52] In 2006, when the Agricultural Ministry sent 47 researchers to visit 353 production facilities in 19 states to map the expansion of sugarcane plantations, they found that volume of national grain and sugarcane production (combined) increased 217 percent from 1996 to 2007 thanks to advances in productivity and higher yields, while planted area increased only 28 percent (11 million hectares). In the same period, the area of sugarcane cultivation for alcohol production increased by only 3 million hectares but resulted in a 116 percent increase in productivity.[53]

Moreover, sugarcane cultivation usually is integrated with other food crops. One agronomic characteristic of sugarcane fields is that they have to be rotated every 6 or 7 years to maximize harvests and replenish soils. This means that every year in Brazil 12–16 percent of the space occupied by sugarcane crops is sown with other cultures, such as peanuts and soybeans. Clearly, Brazil is producing record levels of food and biofuel at the same time. The more likely culprits behind rising food prices are agricultural subsidies, financial speculation, and surges in the cost of oil that then increase the costs of mechanized agriculture.

8.3.4   Lessons Learned

Though not all countries can grow sugarcane, the Brazilian experience with ethanol and flexible-fuel vehicles offers at least four lessons.

First, flexibility is important. The Proálcool program was versatile in its production requirements. The government did not impose technology requirements on producers, and instead allowed firms to select their own production processes to meet targets. Some producers built new facilities; others modified existing plants and retrofitted. Producers had the option of switching back and forth between ethanol and sugar depending on market prices. If the price of sugar rose in comparison to the price of ethanol, many plants would temporarily switch, and vice versa. This type of flexibility married the ethanol program to the sugar industry; it also ensured that the risks associated with producing ethanol were relatively short-lived, cheap, and reversible.

Second, the Brazilian program provided only limited direct subsidies to ethanol producers and is a testament to the utility of phasing out subsidies once they have accomplished their goals. Although Brazilian regulators relied on a mix of direct subsidies, tax breaks, preferential low-interest loans, and mandates to create a well-defined and stable market at the beginning of the program, these efforts were never intended to cover all or even most of the costs associated with ethanol production. When the government mandated in 1980 that ethanol be sold at a rate cheaper than gasoline, the market difference between the two fuels was paid for by consumers themselves rather than taxpayers or the government. In 1985, the US ethanol industry filed a formal complaint with the US International Trade Court alleging that Brazil's subsidies for ethanol constituted dumping and illegitimate trade practices used to enhance the competitiveness of Brazilian exports. The US International Trade Court's own independent investigation found that Brazil's subsidies amounted to less than 3 percent of total value of ethanol production.[54] Brazilian regulators then scaled back even these subsidies once the ethanol industry had matured. The late 1980s and the early 1990s saw the government deregulate and privatize sugarcane and ethanol producers while removing and then eliminating government expenditures on the program. The responsibility for production shifted to the industry, which had little difficulty continuing development because the program was very much industry-driven and focused on applied research from the start.

Third, Proálcool and the promotion of FFVs emphasize the utility of continuous monitoring and adaptation to external events. Brazilian regulators set an ever-changing mandate of blending gradually increasing percentages of ethanol with gasoline that gave distributors and users time to adapt. The percentages increased slowly (from 4.5 percent in 1977 to 15 percent in 1980 and 25 percent in 2004), giving producers and users the ability to adjust, and also enabled targets to be raised or lowered depending on market conditions. Similarly, when sales of ethanol-only vehicles began to plummet and consumers began purchasing gasoline-fueled vehicles to take advantage of falling oil prices, the government responded by encouraging the purchase of FFVs that could run on both ethanol and gasoline.

Fourth, Proálcool demonstrates the necessity of focusing simultaneously on supply of a new technology and demand for it. Program managers worked on separate problems in teams, so challenges relating to ethanol storage and use were not segregated from production or distribution. These teams labored simultaneously on new planting, harvesting, and production techniques for sugarcane at agricultural institutes, on the development of engines that could run on ethanol (at the Aeronautical Technological Center), and on new varieties of sugarcane (at biotechnology institutes). The Brazilian style was to focus on both "supply" and "demand" aspects of ethanol, providing soft loans to sugarcane growers and ethanol distillers but also providing incentives and tax credits to purchasers of ethanol-fueled vehicles. The Brazilian government used Petrobras, the state-owned oil company, to ensure favorable prices for ethanol by inflating gasoline prices. Taxi drivers received tax breaks for converting their vehicles to run on pure ethanol, and the government converted its vehicles to run on it.

8.4   Singapore's Urban Transport Policy, 1971–2009

This case study is polycentric because Singapore harnesses public-private partnerships to operate mass transit systems and works with automobile manufacturers to equip vehicles with devices that make electronic road pricing practicable. Singapore has pursued a synergetic approach to urban transport policy that involves both "supply-side" and "demand-side" elements as well as "carrots" and "sticks."[55] Aspects have included restraint of vehicle ownership and vehicle moratoriums, steady improvement of public mass transit, road pricing schemes, and the provision of real-time information to drivers. "Supply-side" components have invested in train and bus infrastructure and constructed electronic road pricing schemes, whereas "demand-side" components attempt to alter behavior in favor of public mass transit by restricting the number of private vehicles through quota systems and higher vehicle fees. The Singaporean policies rely on a mix of incentives and disincentives; "sticks" raise the costs of driving a private automobile through purchase taxes and usage fees, whereas "carrots" encourage public transport and more efficient driving practices.[56] The Singaporean Ministry of Transport estimates that almost 5 million trips per day (about 60 percent) are made on mass rapid transit, light rail transit, and buses—impressive figures for a country that has a population of fewer than 5 million.

8.4.1   Description

After Singapore achieved independence from British rule (on June 3, 1959), Singaporean government planners focused on housing, on jobs, and on minimizing corruption. The resulting industrialization caused proliferation of factories and homes away from the city center. As a result, the number of private vehicles in Singapore went from 70,100 in 1960 to 142,500 vehicles in 1970. Among the factors in the growth in automobile ownership were greater household incomes, suburbanization of housing away from the city center, and public transport that was notoriously slow and unreliable.[57] (The only modes of public transport were buses and taxis with infrequent service and uncoordinated schedules.)

In 1967, concerned that the traffic situation would quickly become unmanageable, government leaders embarked on a State and City Planning Project with the assistance of the United Nations and other international experts. Shaped by Singapore's geography and its limited resources of land, the project resulted, in 1971, in a concept plan that provided a framework for spatial and urban development oriented to creating a city center and road corridors; the plan was updated in 1991 and 2001.[58] The project highlighted the fact that patterns of driving and vehicle ownership were rapidly becoming unsustainable within Singapore, and brought the matter of transport to the attention of senior government officials.[59] Officials responded by implementing a number of mechanisms to restrict the supply of, and curb the demand for, private vehicles. The most influential of these mechanisms are presented in table 8.6.

The cornerstone of Singaporean transport policy in the 1970s was purchase and usage fees for private automobiles.[60] In 1972, the government raised the import duty on automobiles from 30 percent to 45 percent and introduced an Additional Registration Fee (ARF) to further inflate the cost of purchasing a new vehicle. In 1974, the ARF was raised to 55 percent of the open-market value of a car; by 1983, it had been increased to 185 percent, which means that a car having a market value of 15,000 US dollars had an ARF of nearly $28,000.

In 1975, to reduce congestion in Singapore's central business district, the government supplemented the aforementioned taxes and fees with an Area Licensing Scheme (ALS), the first congestion tax in the world. The ALS charged drivers for entering a "restricted zone" of congested traffic areas during peak hours and required them to purchase daily or monthly licenses available at designated booths and gasoline stations. At the beginning of the program, the daily fee to enter the restricted zone was about US$2. Police personnel were stationed at various entry points into the business district to record the plate numbers of vehicles lacking proper ALS permits, whose owners were then required to pay a sobering fine equivalent to US$50. The scheme reduced traffic into the restricted zone by 45 percent, decreased accidents by 25 percent, and increased traffic speed by 20 percent.[61]

In 1990, the government initiated a Vehicle Quota Scheme (VQS) to further restrict the number of vehicles in Singapore and avoid the volatility seen under the ARF scheme. (Vehicle sales tended to slow immediately after an increase in the ARF and to escalate sharply when people speculated that a decrease in the ARF was imminent.) Under the VQS, Certificates of Entitlement (COE) were auctioned and the growth rate of automobiles was limited to 3 percent per year. The VQS required potential automobile purchasers to pre-bid for a COE at quarterly (and then

Table 8.6

Major measures introduced to curb road congestion in Singapore, 1972–2009. Sources: M. Goh, "Congestion management and electronic road pricing in Singapore," *Journal of Transport Geography* 10, 2002: 29–38; *Land Transport Master Plan* (Singapore Ministry of Transport, 2008).

	Measure	Concise description	Success rate
1972	Additional Registration Fee (ARF)	Extra levy imposed on new vehicle, priced at 5–140% of the vehicle's cost.	Only initially. Scheme was revised in 1974 and 1975.
1975	Area Licensing Scheme (ALS)	Restrict access to CBD* from 7:30 a.m. to 2 p.m. on Saturdays through purchase of supplementary licenses.	Initial decrease in traffic into the CBD was 45%. By 1988, decrease was not sustained, owing to increase in employment in the CBD.
1987	Mass Rapid Transit (MRT)	Serves heavy passenger transit corridors.	Ridership rose from 346 million in 1998 to 360 million in 1999, an increase of 14 million.
1990	Vehicle Quota System (VQS)	COE is introduced, i.e., new car population allowed to increase at 3% in tandem with road capacity growth. Motorists now must bid for the right to own a car.	With VQS, 41,000 fewer vehicles were registered between 1990 and 1993.
1994	Off Peak Car (OPC) Scheme	Offer new and existing car owners the option to save on car registration and taxes in return for lower car use.	Not very successful as most motorists preferred ready use of car for convenience.
1995	Road Pricing Scheme (RPS)	Manual road pricing scheme introduced for linear passage vehicle flow, i.e., remove bottlenecks at congested expressways or arterials outside CBD.	Traffic volume along RPS monitored expressways decreased by 41% from 12,400 to 7,300 vehicles while public transportation travel speed increased by 16%.
1998	Electronic Road Pricing (ERP)	Automated road pricing to reduce the 147 enforcement personnel needed for RPS and replace ALS, OPC and RPS.	Traffic volume on ERP monitored roads decreased by 17%.
1999	Light Rail Transit (LRT)	Serve as passenger feeder to existing MRT network.	Currently carrying payload of 39,000 passengers daily.
2009	VQS Revised	VQS modified to limit the growth of new cars to 1.5% per year	Expected to further reduce car ownership.

*central business district.

monthly) auctions, and, once purchased, each COE was valid for only 10 years. COE bids reached significant levels during times of high demand, with some luxury automobile owners paying an additional amount equal to US$58,000 for certificates. In 1994, the government augmented the VQS with an "off-peak" scheme that allowed drivers who were willing to agree to drive only during non-rush-hour periods and to put a conspicuous red number plate on a car to pay lower usage and ownership fees. In 2009, the VQS system was revised to reduce the growth rate of automobiles to 1.5 percent per year in response to concerns about inadequate road space and increased traffic congestion.

In the 1990s, convinced that rates of automobile ownership and traffic congestion were still at undesirable levels, the government started a road pricing scheme. This scheme was practically identical to the Area Licensing Scheme, but it was designed to cover major expressways—those wishing to use expressways during peak times had to purchase a special permit. Charging similar rates to the ALS, the RPS spread traffic to other routes and times. After it was extended to include the East Coast Parkway, the Central Expressway, and the Pan Island Expressway, 16 percent of motorists stopped using expressways during RPS operational hours, and about 8 percent switched to other roads. About three-fourths of all drivers switched the timing of their journeys to before 7:30 a.m. or after 9:30 a.m., relied on public transport, carpooled, or abandoned their trip.[62]

In 1998, the ALS and the RPS were consolidated into a single electronically collected toll system that charges drivers each time they use the most congested roads at peak times. This system, known as Electronic Road Pricing (ERP), utilizes 80 gantries around the city to collect money from dashboard-mounted devices.[63]

The ERP system has many advantages over the earlier ALS and RPS programs because it is able to differentiate charges according to level of congestion, time of travel, and actual road use. ERP prices change throughout the day, varying for different half-hours from US$2 for a car during rush hour to 30 cents during less congested periods, and also varying for different types of vehicles, with buses and large trucks getting charged as much as US$6. The ERP allows for "shoulder pricing," in which charges can increase before the peak and then decrease after it.[64] The ERP can also tag vehicles moving as fast as 100 kilometers an hour (62 miles per hour) instead of having them slow down at traditional toll booths, and it rations vehicle flow efficiently as it charges drivers directly, immediately, and automatically.[65]

These innovative mechanisms would not have been nearly as successful if they had not been coupled to rigorous investments in public transit. Efforts began in 1973 with improvements of buses, including the forced mergers of bus companies, the imposition of a professional unified bus company, the reorganization and streamlining of bus routes, the banning of independent taxis, and the creation of bus lanes in major corridors.[66] A US$10 billion Mass Rapid Transit (MRT) rail system opened

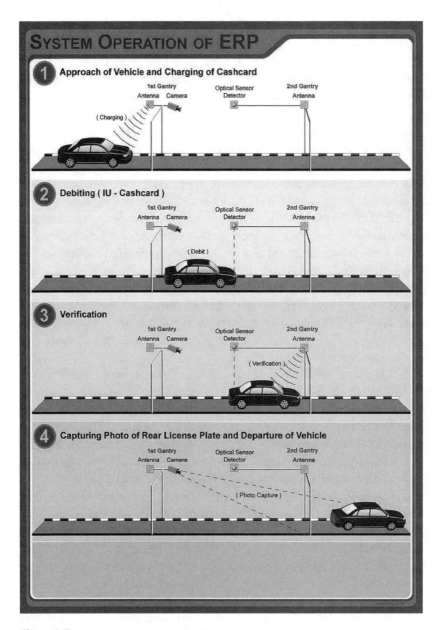

Figure 8.7
Singapore's Electronic Road Pricing system. Source: L. T. W. Hin and R. Subramaniam, "Congestion control of heavy vehicles using electronic road pricing: The Singapore experience," *International Journal of Heavy Vehicle Systems* 13, 2006, no. 1/2: 37–55.

in 1987 to support the 12,600 buses and 20,000 taxis in the city, and in 1999 light rapid transit—a fully automated rail system equivalent to the "People Movers" found in the US—was added.

The government provided the initial capital to build the MRT and maintains investments in rolling stock, but private companies recoup operating costs from passenger fares. Some of the money used to build and maintain this system comes from the revenues generated from the VQS and from ERP. The system is so heavily used that it has not received the social stigma of being "low class" often associated with mass transit in Bangkok, Detroit, Kuala Lumpur, or Los Angeles. At this writing, the government is expending another $15 billion to extend existing MRT routes (the Tuas extension to the East-West Line and a North-South extension to the Marine Bay Area) and to build three new lines (the North Shore Line, the Downtown Line, and the Circle Line).

A final component of Singapore's urban transport policy is exemplified not by any single policy but by a collection of mechanisms that offer real-time signals and information to drivers and mass-transit commuters. To better control urban traffic, the Green Link Determining System (GLIDE) allows traffic signals to be coordinated through centralized control and monitoring. An Expressway Monitoring and Advisory System (EMAS) compiles data on traffic conditions on expressways and provides drivers with up-to-date information. A TrafficScan system equips taxis (and other vehicles) with Global Positioning Satellite receivers and disseminates information about traffic conditions to drivers with GPS receivers, and to Internet users via a website.[67]

8.4.2 Costs and Benefits

The collective benefits from Singapore's urban transport policies can be divided into four areas.

First, Singapore has lower private vehicle ownership than Western cities of similar size, economic activity, and income. Though the fleet of vehicles in aggregate has more than doubled since the 1970s, its expansion has barely outpaced population growth, with about 100 cars per 1,000 people in 1970 and only 130 cars per 1,000 people in 2008, not counting foreigners.

Second, Singapore's efforts have improved traffic flow and reduced congestion. The Area Licensing Scheme led almost immediately to a 50 percent reduction in traffic, and travel speeds increased from an average of 11 miles per hour (17.7 kilometers per hour) to 21 miles per hour (33.8 kilometers per hour) during the first year. Electronic Road Pricing has also reduced traffic volume into the central business district by about 10–15 percent during peak operation hours relative to the ALS,[68] and it has shifted driving habits so that morning peak traffic is down 7.2 percent, mid-day traffic is down 7.6 percent, and off-peak traffic is up 28 percent.[69]

More than 95 percent of expressways, roads in the central business district, and arterial roads were "congestion free" during peak periods from 2006 to 2008.[70] Congestion pricing and information schemes are estimated to have increased traffic throughput at intersections and expressways, resulting in a net cost saving of US$30 million per year (attributable to shorter delays and less time spent in traffic jams). Less traffic congestion also serves to lubricate commerce and increase productivity. This means that the entire Singaporean transport network is more efficient, with motorized passenger travel in Singapore consuming about 12,000 megajoules per capita (MJ/cap), versus 30,000 MJ/cap for a typical Australian city.[71]

Third, taxes and fees and programs such as the Vehicle Quota Scheme and Electronic Road Pricing provide the government with hundreds of millions of dollars of revenue. Fixed vehicle taxes, purchase and ownership fees, and Certificates of Entitlement account for about one-fifth of all government revenue, creating funds that are then invested back into transportation infrastructure and into housing, education, health care, and other socially desirable programs. Though implementation of the ERP system cost about US$1,400 million, annual revenue from the program was initially US$350 million and operating costs only US$7 million, so the system paid for itself in 5 years.[72] Currently the ERP costs US$20–25 million a year to operate but produces annual revenues of US$100 million.

Fourth, many components of Singapore's transport system are cheaper than their counterparts in other major metropolitan areas. The average price of a Mass Rapid Transit ride in Singapore is US$0.91. The comparable prices are US$1.15 in New York and Tokyo, US$1.40 in Hong Kong, and US$2.45 in London; the average price of a bus ride in Singapore is US$0.67, versus US$0.87 in New York, US$0.96 in London, US$1.11 in Hong Kong, and US$1.38 in Tokyo. Mostly because of improved traffic flow, the average peak taxi fare for a 9-kilometer trip in Singapore is US$9.39, versus US$12.35 in Hong Kong, US$14.62 in New York, and US$24.56 in London.[73]

8.4.3 Challenges

Singapore continues to face a host of transport-related challenges. First, demand for travel and aspirations to ownership of private automobiles are increasing. The Singaporean Land Transport Authority expects private travel demand to increase from 8.9 million journeys a day in 2008 to about 14.3 million journeys a day by 2020.[74] However, between 1997 and 2004 the share of public transport during morning peak hours decreased from 67 percent to 63 percent. Less than 15 percent of people commuted to work by car in 1980, but about 25 percent did so in 2000,[75] and the total vehicle population grew from 670,000 in 1996 to 850,000 in 2007.[76] The number of vehicular trips has more than tripled, from 2.7 million trips in 1981 to 7.8 million trips in 2005.[77] At the same time, Singapore is running out of space for cars, with 12 percent of its land area already occupied by roads and little physical space for growth. Even the Vehicle Quota Scheme, which now restricts car ownership to a growth rate of 1.5 percent per year, still permits ownership to increase substantially in the long term.

With rising levels of affluence and changing demographics, the expectations of many Singaporeans have been realigned to value the increased mobility and luxury offered by private automobiles. In absolute terms, the number of vehicles (and the GHG emissions associated with operating them) continues to increase. The decreasing prices of cars (due to increased competition from overseas suppliers and the availability of cheaper models) adds to this challenge. A new Toyota Corolla cost about the equivalent of US$66,000 (inclusive of all taxes and fees) in the 1990s but costs only the equivalent of $40,000 today, and some cars from China and Korea sell for less than the equivalent of $30,000. These falling costs may motivate people to purchase the maximum number of cars permitted under the VQS.

Furthermore, many of Singapore's policies have been technology intensive and/or capital intensive. Electronic Road Pricing, for example, employs a combination of complex radio frequencies, imaging and smart card technologies, optical detection, cameras, and computers working simultaneously. The ERP necessitated 10 years of planning, testing, and preparation, including a pilot program that involved fitting 250 vehicles with transponders and monitoring 4.8 million transactions.[78] Before the ERP began operation, the government managed a free program to outfit 97 percent of all vehicles in Singapore with transponders. Foreign visitors who bring cars to Singapore must rent a battery-powered device so as to be able to travel on ERP roads. Each of these policies required substantial time, human resources, and money to implement.

Because the combined effect of Singapore's policies is to increase the expense of owning a private vehicle (notwithstanding the falling prices mentioned above), those policies have contributed to class division between people who can afford them and those that cannot. For example, a person who pays the equivalent of US$174,000 for a luxury sedan or a sports car may not be influenced by an average daily ERP charge equivalent to US$2.40. Making vehicles more expensive, moreover, may also enhance their status as luxury items that middle-class workers yearn to own.[79]

Shifting traffic away from urban and congested areas means that more traffic is diverted to arterial streets full of homes, schools, and other activity centers. This intrusion of automobiles into residential areas raises concerns about noise, air pollution, and safety, especially when one considers that a high proportion of traffic-related deaths in Singapore are among pedestrians.[80]

Additionally, the urban transport policies of Singapore, though they have helped curb traffic and congestion, still heavily favor motorized forms of transport. That is, they give high priority to maintaining traffic flow for automobiles, buses, and taxis, and to incentivizing commuters to use the MRT. This trend may run counter to attempts to improve the safety of non-motorized forms of transport (walking, jogging, running, skating, bicycling), and improving the traffic flow of private automobiles also removes the speed advantages for mass transit.[81] Plans for the future also reflect this bias: there is much discussion of community car owning schemes, an expansion of the off-peak program, and a transition to electric vehicles, but less effort to encourage walking and bicycling.

Restraint of private vehicle ownership and government control of mass transit can also engender a tendency for providers of public transport to presume that they have captive customers who cannot afford private vehicles. As a result, these providers can come to view the demand for their services as inelastic, eroding effort to improve services or increase choices for their customers.[82] According to one assessment, the revenues collected by the government from road transport have, ever since the 1960s, systematically exceeded government expenditure on road transport factors ranging from 3 to 6, which implies that they were not invested back into improving the transport system.[83] Also, many buses and MRT cars are already completely full during peak times and very uncomfortable for those needing seats, such as the elderly, the physically disabled, and the pregnant, but the companies that operate them have not yet invested in better service.

Finally, although Singapore has relatively few cars per capita, it has a large number of cars per area of land. In 1997, Singapore had about 350,000 cars (95 percent of them privately owned) out of 650,000 registered motor vehicles traveling the country's 3,060 kilometers (1,900 miles) of roads, resulting in an automobile density of more than 210 motor vehicles per kilometer (130 vehicles per mile). This very high density compares with 70 vehicles per kilometer (43 vehicles per mile) for London, 45 vehicles per kilometer (28 vehicles per mile) for Tokyo, and 33 vehicles per kilometer (20 vehicles per mile) for the United States (although this last comparison may be unfair in view of the size of the US).[84] Another assessment noted that if all the vehicles currently in Singapore were to occupy the available roads at the same time, it would give rise to nearly bumper-to-bumper traffic, with 1.1 vehicles per 5 meters, or 16.4 feet, of road.[85] Granted, this statistic partly reflects the success of Singapore's transportation policy (that is, people choose to keep their cars off the road during peak hours, and don't all drive at the same time), but it does underscore the fact that the city has a prodigious number of cars.

8.4.4   Lessons Learned

By far the most distinctive element of Singapore's transportation policy is the restrictions on automobile ownership. Many other cities and governments have used some combination of tolls, investments in mass transit, road pricing, and other factors such as fuel economy standards or fuel blending mandates, but Singapore actively limits the growth of motor vehicles to no greater than 1.5 percent through the current VQS. The Area Licensing Scheme and Electronic Road Pricing help provide efficient flow on the traffic system.

Although the Vehicle Quota System is certainly distinctive, a combination of mechanisms and incentives was needed to reduce car ownership, improve traffic management, raise the cost of roads, and encourage mass transit. Buses, trains, and light rail service are pillars of Singapore's transport strategy, but the system is complemented by the integration of transport and land-use planning, by restrictions on car ownership, and by fees for congestion. These elements were done as part of a synergistic strategy, and they underscore the importance of a holistic approach to transport planning.[86]

Singapore was quick to create strict penalties for noncompliance. Existing penalties for having insufficient balances on ERP cash cards or for not having an ERP card to begin with are well above the cost of ERP rates, minimizing the chance that people will "risk it" and try to scam the system. ERP fines are also automated using number-plate recognition. According to records for 2004, about 260,000 ERP transactions occurred daily but less than 0.5 percent involved violations of cash cards with insufficient balances. One major element here is that people still have the choice and freedom to own cars and drive; these are just influenced by policies designed to incorporate the full cost of driving into vehicle purchases and use. Singapore's mechanisms send signals backed by penalties; one chooses whether or not to follow them.

Singaporean planners continually seek feedback from motorists and citizens, and they have continually adjusted their programs to make them more flexible and adaptive. Extensive public communications and education efforts were conducted long before the launch of any major efforts. Before the ERP was in place, for instance, brochures on how the system operated were sent to every motorist, and the system was switched on informally at zero charge months before operation in order to make it more familiar to drivers. A hotline was set up to handle queries about the scheme, and their feedback was incorporated into the first round of pricing. ERP charges are also reviewed and adjusted every 3 months to take traffic conditions into account, with charges often increasing or decreasing depending on average vehicle speeds in congestion areas. After finding that roads in some parts of Singapore were not congested on Saturdays, regulators eliminated tolls for those roads. ERP rates differ not only by location, but also for vehicle type and over time. Similarly, the COE charged under the VQS has varied with demand for vehicles, from as high as over US$100,000 to as low as US$1.

Finally, when the scholastic literature in the 1970s centered on how urban transport planners should use price mechanisms to reduce congestion, create bus lanes, and use other "thrifty" ways of addressing problems, the Singaporean government listened. They were not afraid to implement policies unpopular with drivers that ended up being good for the public at large. Many drivers in Singapore, noted one interview respondent, "hate the VQS, find the ERP painful, and are overly critical of Singapore's transport policies." Tremendous opposition to major components of Singaporean policy exists, yet the government continues to restrict car ownership because it recognizes that some policies may be unpopular but necessary. Government officials attempt to mitigate this resistance through public education and the incorporation of feedback. This general philosophy, independent of the actual policies, may also help explain Singapore's success.

8.5   Bangladesh's Grameen Shakti, 1996–2009

This case study is polycentric because Grameen Shakti (GS) enrolls communities into renewable-energy projects at the household and village levels but also engages district and national policy makers, international donors, and lending firms. It provides an outstanding example of how to utilize grass-roots, off-grid energy sources to fight energy poverty and improve the availability of energy services, an important dimension of energy security.

Grameen Shakti is a nonprofit company that provides finance and technical assistance for renewable-energy projects to the rural population of Bangladesh. On the

Table 8.7
Programmatic summary of Grameen Shakti's activities and achievements, July 2009.

Total offices	750
Number of districts covered	64 out of 64 districts
Number of villages covered	40,000
Number of island covered	16
Total beneficiaries	2.5 million
Total employees	5,000
Solar Home Systems (SHS) Installed	250,000
Improved Cook Stoves (ICS) Installed	40,000
Biogas Plants Installed	7,000
Installed solar capacity	12.5 MW
Daily solar output	50 MWh
Number of trained technicians	2,575
Number of trained customers	97,996
Expansion plan for SHS units, 2009–2012	1 million
Expansion plan for biogas units, 2009–2012	500,000
Expansion plan for ICS units, 2009–2012	10 million

basis of experiences from the Grameen Bank's microcredit program, GS promotes solar home systems, small biogas plants, and improved cookstoves to reduce deforestation, fight poverty, and provide energy services. Other novel factors in the approach taken from GS include a focus on matching energy supply with income-generating activities, relying on local knowledge and entrepreneurship, utilizing community awareness campaigns, and innovative payment methods including fertilizer, livestock, and cash.[87] As of July 2009, GS operated 750 offices throughout every district of Bangladesh and had installed 250,000 solar home systems, 40,000 cookstoves, and 7,000 biogas plants. It plans to accelerate its expansion so that by 2012 a million solar home systems, half a million biogas plants, and 10 million improved cookstoves are in place in Bangladesh.

8.5.1 Description

Inspired by the success of the Grameen Bank (a specialized bank that gives small loans, referred to as "microcredit," to poor women), Grameen Shakti was established in June 1996 to distribute renewable energy systems to the rural population. The organization was created to promote knowledge and awareness about renewable energy, to provide technical training on solar energy to the rural workforce, and to overcome the high initial cost of installing solar and biogas systems. Taken from the Sanskrit word for "energy" or "empowerment," Grameen Shakti (which literally means "village energy" or "village empowerment") was kicked off with a US$750,000 (€580,300) loan from the International Finance Corporation and a number of small grants; it quickly repaid these loans and became a self-sustainable nonprofit company.[88] As of March 2009, the organization had 12 divisional offices, 94 regional offices, and 5,000 employees, mostly engineers. It received a small amount of funding from the Infrastructure Development Company Limited (a government organization in Bangladesh that finances rural-based renewable-energy companies), but all the rest from sales. Its programs promoting solar home systems, biogas, and improved cookstoves have been the most significant.

The solar home systems (SHS) program draws on the decentralized nature of solar photovoltaic panels to provide electricity to off-grid and inaccessible areas in rural Bangladesh. The SHS program targets those areas that have little to no access to electricity and limited opportunities to become connected to centralized electricity supply within the next 5–10 years. The ease of operating solar panels, their long lifespan, avoidance of combustible fuel, and lack of pollution make them an ideal choice for remote areas, and the SHS offers microcredit schemes to enable homeowners and businesses to acquire the necessary capital they need to finance installation. A single 50-watt panel (imported from Japan), a battery, and associated equipment cost about $400 and can power four lamps and one black-and-white television.

Under the SHS program, an interested person makes a down payment to cover 15–25 percent of the system's cost and then repays GS by means of a low-interest loan. In view of the expense of kerosene and diesel fuel in rural Bangladesh, solar systems typically pay for themselves in 3–4 years. A purchaser then owns a system that lasts 20 years without fuel costs. Customers can also elect to share in the cost of larger systems under a solar micro-utility scheme that allows shopkeepers or villages to share in making a 10 percent down payment with a repayment schedule of 42 months. GS offers free maintenance while a loan is being repaid, and trains interested clients in maintenance and operation at no additional cost. As of 2009, GS has installed more than 250,000 solar home systems with a peak production capacity of 50 megawatts. About 10,000 clients have availed themselves of the micro-utility financing option, and the program is growing at a rate of 10,000 new clients per month.

GS has established 45 Grameen Technology Centers (GTCs) that have trained more than 40,000 female heads of households in the proper use and maintenance of solar panels. GS engineers pay monthly visits to households during their financing

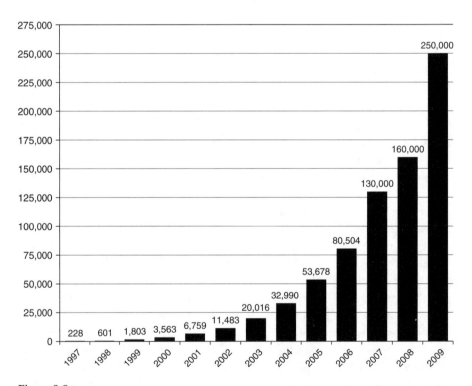

Figure 8.8
Growth of the Grameen Shakti Solar Home System Program, 1997–2009. Source: research interviews.

period, and they offer their services for a small fee, afterwards, if a client signs an annual maintenance agreement with GS. The GTCs have trained more than 5,000 women in advanced solar maintenance, enabling them to become full-time specialists who travel around the country to service GS clients for this small fee. Apart from providing repair and maintenance services, GTC-trained female technicians also assemble solar accessories such as lampshades, charge controllers, and inverters, and GTCs run exposure programs for rural school children to increase their awareness about renewable energy.

To further lower costs, GS funneled money back into R&D on solar devices, and GS engineers designed and patented original low-cost ballasts and charge regulators that are more efficient and less expensive (often by more than 50 percent) than the ones previously available on the local market.[89] Furthermore, GS offers an inclusive warranty that includes a buy-back component: if the area in which a buyer lives gets connected to the national grid, the buyer may return the system to the organization.

Grameen Shakti's biogas program promotes 3-cubic-meter biogas plants suitable for use in homes. One of these can provide enough gas and heat for cooking three meals a day for an average-size family. The biogas program also can be subscribed to on a commercial scale, with larger systems offering enough gas to meet the energy needs of restaurants, tea stalls, and bakeries. By relying on biogas, these units minimize reliance on traditional forms of biomass, animal dung, and charcoal (with their negative environmental and social impacts), and also protect communities from disease by enhancing sanitation. The plants harness gas obtained from livestock and even human excrement. Researchers estimated in 2006 that Bangladesh had more than 30 billion cubic meters of potential biogas from livestock and another 10 billion cubic meters from humans.[90] The biogas program quite literally has people using their own waste to meet their energy needs.

The financing scheme supporting the biogas program is similar to the SHS. A purchaser pays 25 percent of the total cost of each system as a down payment, then repays the rest in 24 monthly installments with a 6 percent service charge. Buyers are also encouraged to construct their own plants under the supervision of GS engineers. Biogas plants at the community scale have proved quite effective, as many people in rural Bangladesh live as joint families or in joined households with dozens of people close to each other, so they can easily share a biogas system. GS manages a special program in Chilmari (the northern part of the country) that provides farmers and communities with the livestock in addition to biogas plants so that they have adequate "fuel." By June 2009, more than 7,000 biogas units had been installed. Follow-up evaluations have found that 90 percent of the plants installed under the project are still in operation and that more than 90 percent of the households that use them meet their fuel demand exclusively from these plants.

One interesting offshoot of the biogas program has been the production of high-quality organic fertilizer as a by-product of the biogas plant. The use of this fertilizer has reduced the need for chemical fertilizers by 30–40 percent at many farms and fisheries, and those not wishing to use the fertilizer have sold it commercially. A 3-cubic-meter cow-dung-based biogas plant can produce 8 metric tons of slurry, equivalent to 224 kilograms (494 pounds) of urea and more than 1,200 kilograms (2,646 pounds) of fertilizer. Mrs. Mohammad Abdur Razzak, for example, owns a large chicken farm of 2,000 egg-laying hens and hoses the coop's waste into an underground chamber, where it ferments and releases biogas into a pipe that is connected to her cooking stove. Her animals produce so much biogas that she makes an extra US$71 (€55) per month by renting 10 cookstoves and the excess gas to her neighbors, and the leftover slurry that isn't converted into gas is sold to local farmers as fertilizer.[91]

The Improved Cook Stoves (ICS) program distributes one-, two-, and three-mouthed clay cookstoves. These cut fuel use by half and have chimneys that create a smoke-free cooking environment, improving air quality within the home. These efficient cookstoves not only result in less fuel consumption (typically reducing fuel needs by 40–50 percent); they also facilitate shorter cooking times and generate more heat. GS has recently started an ambitious program to install 10 million of these systems by 2012, and it has trained more than 2,000 local youths and women to manufacture sell, and repair improved cookstoves. A special effort has been made to promote improved cookstoves to operators of restaurants.

8.5.2   Costs and Benefits

By far the most significant benefit from Grameen Shakti's programs has been the provision of modern energy services to segments of the population that would not otherwise enjoy them. Only 30–35 percent of the population of Bangladesh has access to the national electricity grid, and the solar panels provided by GS not only light homes and offices but also schools, mosques, and fishing boats; they also provide electricity to operate radios, mobile telephones, electric fans, computers, soldering irons, drill machines, water pumps, and battery chargers. Solar electricity also avoids the use of kupi lamps fueled by kerosene, a fire hazard.

A second important benefit has been a significant reduction in deforestation. Bangladesh is a biomass-centered energy system, with trees and bamboo meeting about 48 percent of all domestic energy requirements, followed by agricultural residues (36 percent), dung (13 percent), and fossil fuels.[92] Widespread destruction of forests has occurred to satisfy energy needs, with homestead forest cover reduced to 8 percent of its original area,[93] and half of Bangladesh's natural forests have been destroyed in a single generation by people collecting fuel-wood.[94] Even when fuel collectors avoid cutting down trees and take only fodder and kindling, they can still devastate forests by removing important nutrients from the ecosystem.[95] The programs operated by GS keep trunks, branches, shrubs, roots, twigs, leaves, and trees in the forest by minimizing the need for woody biomass and increasing the efficiency of cookstoves.[96]

A third benefit is higher income and employment. Direct employment comes from GS itself, which has 5,000 full-time employees and whose training programs provide education to entrepreneurs wishing to form businesses related to solar panels, biogas plants, farming, and cooking. Indirect benefits come from the use of modern energy services. Improved lighting enhances educational activities such as reading and writing, extends the working hours in markets to beyond dusk, and powers mobile telephones that enable communication and enhance skills. The use of solar panels and biogas plants also minimizes the exposure of communities to volatile and escalating prices for conventional fuels and biomass. As deforestation has accelerated in Bangladesh, the demand for wood has outpaced the supply, causing the price of wood to increase from 25 takas (US$.35, €.27) in 1980 to 90 takas (US$1.27, €.98) in 1991 and 120 takas (US$1.69, €1.30) in 2007. When put into the context of the typical household budget, about half of the annual income of rural households can be spent on fuel.[97] As the calorific value of these fuels is low, they also require large volumes and therefore force people to spend up to 5 hours a day collecting wood that they could otherwise spend making money or gaining a formal education. Furthermore, a study of rural fuel use in the Noakhali district of Bangladesh documented that 40 percent of families collected wood from their own homesteads and 13 percent from the market, but 47 percent depended on them both, meaning they spent both time and money meeting their fuel needs.[98]

A fourth benefit relates to public health. Reliance on biomass combustion for cooking and lighting produces a significant quantity of hazardous pollutants, including fine particulates, nitrogen oxide, and carbon monoxide, which are typically emitted within the home. One interview respondent, who compared this type of exposure to "living within a giant cigarette," noted that homes reliant on biomass tended to have higher rates of acute respiratory infections, eye problems, low birth weights, and lung cancer. In one survey of rural Bangladesh, more than half of the women reported headaches, lung disease, asthma, *and* cardiovascular disease related to cooking with biomass fuels.[99] Fuel collection also presents a health hazard, with many women carrying more than their weight in fuel hundreds of kilometers per month. These adverse health effects can lock families into poverty, as they increase expenditures for medical care while also diminishing productivity.[100]

8.5.3   Challenges

From a standpoint of sustainability, the biggest challenge to the work being done by Grameen Shakti relates to higher living standards. The energy services and employment generated by GS create higher incomes that then lead to greater material aspirations and rates of consumption. It is one thing to use solar and biogas power to light schools and cook meals, but quite another to use them to power televisions and to support hoards of commercial livestock. GS provides technology and enables people to escape energy poverty, but it does not ensure that they become sustainable stewards of our planet. It empowers them to create their own lifestyle, which can be environmentally sustainable or not. (To be fair, GS rigorously defends its position by noting that livestock can improve food security and that televisions can be used to spread knowledge and to warn populations before and during natural disasters.)

In a macroeconomic sense, the programs being done by GS have yet to make a significant dent in the national energy policy of Bangladesh. Less than one-third of the country's 150 million people still have access to electricity, and so far the hundreds of thousands of solar panels and biogas units promoted by GS have made only a small contribution to national energy supply. The utilization of renewable energy remains insignificant when compared with the combustion of fossil fuels by electric utilities,[101] and the government lacks the resources and the political will to comprehensively promote renewable energy resources on a national scale.[102]

Some of the systems promoted by GS, such as solar panels, have risen in cost and/ or face technical challenges. Although solar panels are expected to reach grid parity—i.e., produce electricity cheaper than large-scale power plants— in Bangladesh within the next decade, better batteries, warranties on imported solar panels, recycling and disposal of panels, and their operation during longer periods of reduced sun during the monsoon season are all of concern. GS programs also require a substantial down payment that are still beyond the means of the poorest members of many communities.

Although solar panels are applicable practically everywhere in Bangladesh, biogas plants are not. Biogas plants are comparatively capital intensive in Bangladesh: they require at least two large animals constantly producing waste (humans alone do not produce enough). And because they rely on digestion (a biological process) instead of combustion (a thermochemical process) to produce heat and energy, their performance can vary. One respondent commented that biogas units were "renowned for being temperamental." Such units require lots of water and are usually most attractive only for a middle class of Bangladeshi villagers who are wealthy enough to afford livestock but cannot afford liquefied petroleum gas or electricity. Furthermore, because biogas units run on methane derived from waste, even small leaks can drastically overwhelm the carbon benefits of displacing biomass, since methane is a much more potent greenhouse gas than carbon dioxide, although GS planners attempt to address this problem through the standardization of biogas pipes.

The greater use of solar panels and biogas plants reduces the need to connect (or at least the profitability of connecting) rural Bangladeshis to the national electricity grid. Because of this, GS employees have reported resistance from the Bangladesh Rural Electrification Board, which sees many of GS's programs as obstructing its plans to extend the centralized grid into rural areas. (Fortunately, other parts of the Bangladeshi government appear to recognize the importance of GS programs and support them through tax exemptions, special funds, and a renewable portfolio standard aiming for 10 percent of national renewable-energy supply by 2020.)

Finally, natural disasters continue to cause widespread damage among the rural Bangladeshi population. Floods, landslides, and tsunamis are among only a few of the recent events that have either directly damaged GS solar panels, stoves, and biogas units or indirectly destroyed cultivated and arable land, thereby reducing the capital villagers have available to make the down payment for GS technologies.[103]

8.5.4 Lessons Learned

The experience with Grameen Shakti suggests multiple lessons for other communities and countries, especially those with emerging economies. Its success underscores that capacity building can go hand in hand with distributing renewable energy systems (and eventually improving health and reducing deforestation). GS focuses on training to develop local expertise relating to energy projects, including expertise which ensures that systems are better maintained and knowledge of the kind that often leads to entrepreneurship. The effectiveness of GS is based on a simple formula of meeting local needs for energy with simple technology constructed and maintained by a community workforce and used to make villages and communities self-sufficient. These systems work best when owned, operated, and repaired by the people themselves. And everything used in the SHS program except the solar panels and the batteries now is produced within Bangladesh.

A closely related lesson is that big and expensive technologies are not always needed to provide large segments of the population with electricity. One of the "most interesting lessons gleaned from the SHS program is that small remains beautiful," noted one former GS employee. If solar home systems and biogas plants can electrify remote homes and provide heat for cooking, relatively small things such as better meals, radios, light bulbs, and telephones can make a huge difference in individual lifestyle and enjoyment.

Grameen Shakti has also done an excellent job of linking its products and services to other local businesses, and integrating its technologies with other programs. For example, GS connects the use of biogas units in homes and shops with the livestock, poultry, agriculture, and fishery industries. Clients who wish to own their own biogas unit can also purchase livestock, and clients who do not wish to use the fertilizer created as a by-product from biogas units can sell it to local farmers, aquaculturists, and poultry ranchers. Similar linkages have been made in the promotion of GS's solar panels, mobile telephones, compact fluorescent light bulbs, and LED devices.

The success of GS's programs serves as a warning against subsidies and against giving items away. GS highlights the importance of having local communities pay for renewable-energy projects themselves—so they become invested in how they perform—but using financing to overcome the first-cost hurdle by means of low-interest loans and installments. The focus on economic sustainability, ensuring that GS operates as a not-for-profit company rather than a full-fledged NGO, also means any excess revenues generated by sales can put back into the company. "Any profits, we recycle," noted current managing director of Grameen Shakti, Dipal Barua.[104]

Grameen Shakti has taken a progressive stance toward community acceptance, promotion, and advocacy. Its Grameen Technology Centers often conduct large demonstrations of solar and biogas devices, and GS employees sometimes go from door to door to familiarize communities with technology. GS engineers consistently work with village leaders to distribute brochures, hold science fairs at local elementary schools, and host workshops for policy makers. GS recruits women and youths and trains them as technicians. It runs a scholarship competition for the children of SHS owners. It allows clients to return their system at a reduced price to the organization, and it gives free maintenance and training to all current clients of the SHS program so that they can care for and maintain their systems by themselves. The basic idea behind these elements is that people must come to trust the energy technology they use as well as the organization that provides it.

8.6   China's National Improved Stove Program, 1983–1998

This case study is polycentric because China synergized local, provincial, and national attempts to disseminate cleaner-burning cookstoves to rural populations. It illustrates how energy security can be improved by increasing the efficiency of fuel used for one of the most essential energy services around the world: cooking.

Despite a trend toward urbanization in recent years, about 60 percent of the population in China is still rural, and 90 percent of all households in China still combust solid fuels (wood, biomass, coal) in their homes.[105] To prevent shortages of these fuels among the rural population and to minimize the degradation of forest lands and agricultural soils, the Chinese Ministry of Agriculture implemented a National Improved Stove Program (NISP) from 1983 to 1998. During this period, the NISP was responsible for the installation of 185 million improved cookstoves and facilitated the penetration of improved stoves from less than 1 percent of the Chinese market in 1982 to more than 80 percent by 1998. The cookstoves being installed in China in 1994 were equivalent to 90 percent of all the improved stoves installed in the world. As a result, China's energy use per capita declined in rural areas at an annual rate of 5.6 percent from 1983 to 1990.[106]

8.6.1 Description

Rural China has grappled with population pressures, deforestation, and fuel shortages for more than a millennium. During the late 1970s and the early 1980s, millions of households experienced shortages of wood and fuel. Anecdotal stories flourished about peasants burning their furniture to keep warm, combusting uprooted plants and grasses to cook food, and chopping down bridges for firewood. The Ministry of Agriculture started the National Improved Stove Program in 1983 to respond to these challenges under the direction of the State Development Planning Commission and the Ministry of Finance. Though the characterization of phases is a bit rough (owing to the overlap between national and provincial programs), the first phase of the NISP ran from about 1983 to 1990 and entailed the use of subsidies to incentivize county and provisional government officials and technical institutions to distribute safer and cleaner cookstoves to rural residents.

The justification for embarking on a major cookstove program was related to the fact that most stoves in use at the time had efficiencies averaging only 10–12 percent, so as much as 90 percent of the energy content of the wood or charcoal burned in them was wasted. In some cases, existing cookstoves could be markedly improved by something as simple as adding a chimney or putting more insulation around the stove; in other cases, replacing older stoves with new ones yielded drastic increases in efficiency.

The newer stoves had a grate and an improved combustion chamber, almost always had a chimney, and often required a switch from charcoal or polluted wood to a "healthier" fuel (soft biomass, crop residues, firewood). They utilized higher-temperature ceramics, fire-resistant materials, and longer-lasting metals. They had more insulation, and a better frame that guided hot gases closer to cooking pots. They could cook more food at once, and many had coils around the combustion chamber to heat water while cooking was in process. Some improved stoves were connected to radiators or space heaters so that heat could be recycled and/or vented to other rooms, and some sent heat through pipes directly into a *kang* (a brick platform for a bed). Some improved stoves were "fuel flexible" and could combust coal and biomass, although doing so required a homeowner to insert a different combustion chamber for each fuel. Improved stoves were also often aesthetically pleasing, with beautifully designed tile and artwork, and thus were something to be proud of and handed down to children.

Aside from conserving energy, the major motivations behind the NISP were to reduce the physical hardship of collecting fuel (a task mostly done by women), to increase the earning capacity of rural communities, to reduce pressure on rural forests, and to consolidate rural energy efforts into a single agency rather than having them divided among four ministries. The Ministry of Agriculture set up a Bureau of Environmental Protection and Energy (BEPE) in 1983 to manage the program. The

BEPE adopted a "self-building, self-managing, self-using" policy focused on having rural people themselves invent, distribute, and care for cookstoves, and it set up pilot programs in hundreds of rural provinces.[107]

The BEPE's management skipped many administrative levels of the Chinese government. It disbursed subsidies directly from the central government to the counties, so that provincial officials had no say in how funds were spent, but it then enrolled provincial governments in monitoring and evaluating. Under the first phase, the BEPE signed 100 contracts with the most needy rural counties; these counties then received funding for 3 years through seven rounds of county selection. Counties were chosen on criteria including shortages of fuel wood, high reliance on woody biomass, and a local willingness to devote resources to the program. At the local level, the key implementing agencies were County Rural Energy Offices (CREOs).

Despite this somewhat complicated structure, the role of the national government in the NISP was surprisingly limited. One primary aim of the NISP was to train local stove builders and installers so that they could carry out most of their work autonomously. Combustion chambers and ceramic tiles were often manufactured at the scale of villages (100 batches a day) instead of in huge industrial plants. Some builders specialized in stove construction; others retained unique knowledge about how to incorporate stoves into homes. CREOs often hired these local builders to design and install stoves, then sent demonstration teams to visit households and show villagers how to operate and maintain the new stoves.

CREOs also exhibited a great variation in how they chose to spend the money given by the BEPE to promote improved stoves. Officials in Wuhua County, for example, decided that only households with improved cookstove would be allowed to cut wood in the mountains, purchase wood at a preferential price, or access the municipal woodpile. The officials also authorized the poorest households to obtain parts for their stoves without charge, awarded well-performing townships with cash prizes and bonuses, penalized village leaders who chose not to participate, and fined craftspersons for making traditional stoves. Yet in the more industrial Jiangjin County, officials gave special priority to local enterprises that manufactured improved stoves and focused their efforts on incentivizing installations of cookstoves in new houses.[108]

The NISP proved so successful in its early years that it became a centerpiece of the Seventh Five-Year Plan for 1986–1990, which called on using improved stoves (along with firewood forest plantations, small hydropower units, and biogas stoves) to deliver energy services to the rural Chinese. During this period, the total number of stoves installed jumped from 63.6 million to 120 million.

In the second phase of the program (1990–1995), with shortages of biomass and woody fuels alleviated in many counties and rural incomes rising, the government began scaling back both national and provincial subsidies for the program. Officials made a major push for efforts related to commercialization, certification, and standardization. Regulators used tax breaks and low-interest loans to incentivize rural energy companies to become the primary distributors of stoves, although the government retained the responsibility for training and the provision of technical support. This phase also saw a shift in the development of stove models from those manufactured on site to more industrial units that could be mass produced and pre-assembled to lower costs and improve quality control.

The third and final phase of the program (1995–1998) witnessed a shift to extension efforts only, the government leaving development and dissemination mostly to private actors (with some oversight from local governments). The national government also began a transition to other efforts, such as providing energy-efficiency labels for appliances and promoting wind turbines and solar panels. Still, by the close of the program in 1998, nearly 27,000 members of the Ministry of Agriculture staff were engaged in rural energy promotion at the county and township levels, and there were nearly as many county-level rural government agencies (2,024) as there were counties (2,126). That same year, the Ministry of Agriculture claimed that more than a billion people, living in 185 million of China's 236 million households, had received improved biomass or coal stoves.[109]

8.6.2 Costs and Benefits

The NISP brought at least five distinct benefits.

First, there was a measurable decrease in the rate and extent of deforestation in rural China. Although these statistics are a bit dated, about 4 percent of China's

Table 8.8
Households adopting improved stoves under the NISP and provincial programs.

	NISP households (millions)	Households under provincial programs (millions)	Total households/ year (millions)
1983	2.6	4.0	6.6
1984	11	9.7	20.7
1985	8.4	9.5	17.9
1986	9.9	8.5	18.4
1987	8.9	9.1	18
1988	10	7.5	17.5
1989	4.5	5	9.5
1990	3.6	7.8	11.4
1991–1998	7.8	57.2	65
Total	66.7	118.3	185

standing forests are used as fuel wood and roughly 13 percent of cultivated land in China is used to grow fuel wood.[110] Improved cookstoves, by doubling and sometimes tripling the energy efficiency of cooking, therefore, need one-third to one-half as much fuel. In combination with the results of promoting biogas stoves and diffuse firewood plantations along with improved cookstoves, China's efforts protected about 3 million hectares (7.4 million acres) of young forests from being damaged each year and increased the forest coverage around rural homes by 12–13.4 percent from 1989 to 1999.

Second, improved cookstoves vent pollutants out of homes through chimneys, and those that burn biomass avoid the use of coal and the resulting arsenic and fluorine poisoning.[111] Traditional cookstoves release numerous hazardous pollutants directly into the home, including carbon monoxide, particulate matter, nitrous oxides, sulfur oxides, and a range of volatile organic compounds. These pollutants are leading causes of chronic obstructive pulmonary disease, lung cancer, acute respiratory infections, and carbon monoxide poisoning, especially among children and the women doing the cooking.[112] When roughly quantified, the pollution exposure from indoor combustion of biomass and coal in China is responsible for about 380,000 premature deaths and 3.2 million disability-adjusted life years. Put another way, the indoor smoke released from solid fuels is more dangerous than obesity, road traffic accidents, urban air pollution, and unsafe sex in China, combined (figure 8.9).[113] Pollutants released from an unvented stove inside a home are about 1,000 times as likely to go down someone's throat (a concept known as the "intake fraction" of pollution) rather than as pollutants released from a vented stove. Because of close proximity to the source of pollution, a one-ton-per-year reduction in particulate emissions from a household stove yields 40 times the health benefits of a similar reduction at a coal-fired power station.[114]

Third, the NISP helps to socially empower those who must otherwise spend hours each day collecting fuel wood and cooking. In China, much as in Bangladesh, many individuals walk thousands of kilometers carrying large burdens of fuel each year, which leaves them less time to do other more productive and enjoyable activities and which increases the risks of broken bones, arthritis, exhaustion, and malnutrition. One recent assessment calculated that improved cookstoves in China had saved 1.4 million households 4 million working days per year that otherwise would have been spent on collecting fuel.[115]

Fourth, because the use of improved stoves means that a portion of biomass (usually straw or rice husks) is now returned to the soil as fertilizer, the NISP has helped increase the organic material content for previously marginalized soils, controlled water loss, and prevented erosion. Over the course of the NISP, the utilization of organic fertilizer in rural farmland increased by 74 percent for some provinces and the use of chemical fertilizer declined 57 percent.

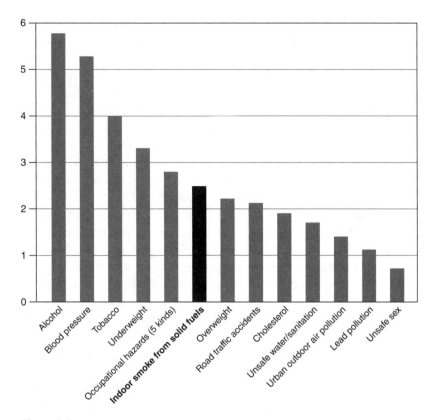

Figure 8.9
Percent of disability-adjusted life years for China in 2000. Source: J. Zhang and K. Smith, "Household air pollution from coal and biomass fuels in China," *Environmental Health Perspectives* 115, 2007, no. 6: 848–855.

Fifth, because forests act as carbon sinks, and because the burning of wood and of biomass can release GHGs, the use of improved stoves reduces the carbon footprint of rural communities. One study calculated that the use of the improved stoves promoted by NISP has reduced carbon dioxide emissions by 82 million tons per year,[116] roughly equivalent to the reported annual carbon dioxide emissions of Austria.[117]

8.6.3 Challenges

Two challenges encountered by the NISP were the scale and complexity of China's rural areas and the question of what counts as an "improved" stove. "One person's improved stove is another one's traditional junk," commented one of the interview respondents, "and one decade's improved stove is another one's obsolete model. We now realize that the term 'improved' is not a very good one, and instead [should] name the stove or define it by its physical properties." When the NISP began, however, managers did not really define the term "improved." The multiplicity of "improved" stove and fuel types can quickly confound attempts to evaluate impacts and design general rules. For example, a random subsample of 396 households selected from a 3,500 household survey about rural fuel use in China found a large variation in types of fuel and stoves used, with multiple types of fuel being burned within the same household in different rooms for different purposes and during different seasons.[118] Some villages reported 34 fuel-stove combinations in the summer and 28 in the winter, and the average household uses 2.6 types of fuel. Such complexity makes detailed assessment of stove performance difficult.

Promoting improved cookstoves is not always successful at reducing indoor air pollution and GHG emissions, especially if coal stoves are used for space heating and water heating. Widespread use of coal stoves without flues in rural China has caused fluorine poisoning, and coal stoves are substantially more carbon-intensive than biomass stoves. The NISP never stipulated reducing air pollution or fighting climate change as explicit goals, but rather placed higher priorities on fuel efficiency and rural economic growth. This meant that not all new stoves had chimneys, nor were they meant to burn only clean fuels.[119] And as coal has become cheaper and more available in China than other fuels, substitution of coal for biomass has occurred.

Many of the NISP cookstoves no longer meet the needs of Chinese households. The stoves installed during the 1980s and the 1990s are not as efficient as the newest models, and still emit carbon dioxide (from burning wood) and black carbon (a potent GHG). Ensuring that these stoves had chimneys produced significant public health benefits, but a chimney merely relocates emissions outside; it does not stop those emissions from damaging the climate. Indeed, about one-quarter of the world's emissions of black carbon come from household biomass stoves. Consequently, a large proportion of the Chinese population is still exposed to dangerous levels of indoor air pollution. As recently as 2007, a study of rural Chinese households monitored for exposure to particulate matter in kitchens and living rooms found that nearly all household stove and fuel groupings had unacceptably high pollution levels, with some homes experiencing pollution levels twice as high as health regulations allowed.[120] Clearly, a large fraction of rural population remains chronically exposed to levels of pollution far higher than those determined to be safe.

Although the NISP has lessened the rate of deforestation in China, it has not proved sufficient to abate it entirely. The most recent national forest inventory for China (released in 2005) documented that although overall forest cover in China increased slightly from 1999 to 2003, the increase masked the fact that timber harvesting had continued to devastate mature and old-growth forests and the fact that almost all of the increase was attributable to the planting of young saplings.[121] Attempts to stop deforestation are tainted by corruption and profit motive. One study, which collected data from 28 provinces over five census periods, revealed that most forest managers themselves harvested more than government quotas allowed in order to raise their income.[122] Perversely, the study found that the average state forest manager harvested beyond the quota by at least 50,000 cubic meters per year, and that higher quotas often brought a faster decline in forest growth. In China's southwestern Yunnan Province, interviews with actual villagers and illegal loggers revealed that deforestation was driven by complex social, political, and economic factors that would not be remedied by improved cookstoves.[123] Resentment of the government, increasing inequalities among peasants, financial hardship, and falling prices for agricultural commodities have all incentivized peasants to destroy forests in exchange for revenue.

8.6.4   Lessons Learned

One lesson from the NISP that is easily replicable for other countries is the value of continual independent monitoring and evaluation. Each year of the NISP, pilot counties were selected on a basis of need, and only 100 new counties (those with the most acute fuel shortages, with the requisite managerial and financial resources, and/or with plentiful raw materials) were chosen per year. (This contrasts sharply with the Indian cookstove program, in which stoves were distributed everywhere all at once. That program struggled in places with inadequate distribution and training channels, and quickly developed a poor reputation among villagers.) Cookstove designs were consistently upgraded and tested through national cookstove competitions, and the stoves themselves were certified and standardized. Officials conducted on-site inspections and checks to monitor progress and to ensure villagers were using cookstoves properly.

Another lesson is that local enforcement and involvement can be a prerequisite for a program's success. Under the NISP, the BEPE gave counties the authority to implement components of the program themselves and retained only the role of monitoring and evaluation. County officials focused on town-scale production and often required local technicians and manufacturers to participate, sometimes through positive incentives (such as bonuses or subsidies) and sometimes through sanctions (such as fines for producing outdated stoves or restrictions on what types of stoves could be installed in new buildings). The BEPE sponsored national stove competitions in 1982 and 1986 to highlight improvements in stove design and to encourage further innovation. Program officials sought feedback from users and from local artisans to ensure that stoves were built to suit their preferences, instead of predetermining what households wanted without asking them or imposing standardized stoves upon them. Therefore, stove makers utilized locally available materials to keep costs low rather than relying on special or imported materials; stoves were easily mass produced rather than individually built; and they burned wood of any size and at times additional fuels such as agricultural residue, wheat, maize, and rice stalks rather than burning only small pieces of wood.[124] The philosophy behind such inclusion was that long-term behavioral change is best encouraged when those who directly use a technology have a stake in improving it.

Yet another lesson is that, from the start, the NISP was designed so that government would have only a limited role. During the final phase of the program, the national government was responsible only for certification and monitoring; the remaining aspects of cookstove design, installation, and commercialization were left to the marketplace. When overall government expenditures on the NISP are put in context, they show that the national government supported 860 counties with a total budget equivalent to only US$1.4 million (€1.1 million) per year, and local governments contributed only an additional US$17.9 million (€13.9 million) per year. This amount corresponds to less than US$25,000 (€19,300) spent per village. Households paid for most of the materials and installation themselves, and over the entire course of the program government expenditures covered only about $1.28 (€.99) per stove, less than one-tenth of the total cost of slightly more than $13 (€10) per stove.[125] In the first 7 years of the program the government spent a combined US$200 million (€154.8 million), but private actors spent more than US$1 billion (€773.9 million). The philosophy behind this component of the NISP was that people will care more about things they have to work for, and that people will take more pride and responsibility in a cookstove's performance if they have more of their own money at stake.[126] Ultimately, the national government's role was limited to dissemination of stoves, technical advice, quality control, and a small subsidy, nothing more.

Closely connected to how much money was spent was *how* it was spent. Rather than create an interagency task force or spread responsibility for the NISP among multiple ministries, one ministry (the Ministry of Agriculture) was placed in charge. That ministry quickly created a new agency (the BEPE) to manage the program. The BEPE's approach was simple and direct and relied upon a network of service and implementing organizations spread across 38 provinces and municipal regions, with more than 1,500 regional energy offices in 2,300 counties. The program was not directed from the top; rather, responsibility was quickly shared and devolved. The NISP underscores the essence of polycentric action, for its funding was the joint responsibility of multiple scales: the national government provided about 3 percent of the funds over the course of the program, provincial and local governments provided 14 percent, and households and businesses provided 83 percent. Relying on the national government to contract with local governments, which then contracted with individual companies and villages, avoided delays and reduced costs; it also allowed flexibility and experimentation with how the program was implemented in different regions. The program did not set upper financial limits on the types of stoves that could be installed, so many regions chose high-quality stoves that owners cherished and were proud to display. This, again, contrasts with the Indian program, which set a limit of $5 (€3.87) per stove in an attempt to keep costs low and then saw a majority of the 35 million stoves installed over the course of the program fail in the first 6 months because they were too cheap.

The NISP demonstrates that, in contrast with a policy of providing rural areas of China with huge chunks of energy from centralized power plants, relatively small changes at the household and village levels can add up quickly and can accumulate. Energy policy too often focuses on influencing urban patterns of consumption, owing to cities' high population density and the relative ease of affecting large numbers of people with a single policy. Yet the NISP reminds us that much of the world's energy is consumed in rural areas where population density is low and resources are dispersed. In these areas, low-level technologies done at a small scale with a short-term investment period can be just as important as capital-intensive technologies done at a larger scale.[127]

Finally, the order in which the elements of the NISP were carried out implies that a particular sequence is most effective.[128] The program began with county surveys to determine the regions that most needed cookstoves. It then targeted areas where fuel was scarce or costly, indoor air pollution was understood to be an important problem, and local actors expressed a desire to embrace new technology. Once optimal regions were identified, stoves were designed with feedback from the local users themselves so that they best met their preferences. In some regions, stoves that could utilize multiple fuels were deemed best; in other regions, stoves that decreased the time needed to gather and collect biomass; in still other regions, stoves that could be quickly mass produced by local craftspersons. Local actors were therefore viewed as active participants in the program. Once mass production of stoves began and commercialization got underway, the government scaled back its role in the program. Subsidies were carefully considered and calculated so that a majority of expenditures came from households themselves, and networks for commercial distribution were established on a basis of market demand rather than government largesse. The last phase of the program involved demonstration visits, monitoring, and follow-up surveys to ensure that the NISP continually adapted and responded to challenges.

8.7   The Oasis Project in the Atlantic Rainforest of Brazil, 2006–Present

This case study is polycentric because the Oasis Project enrolled civil-society actors, a corporate foundation, and state and local governments in a scheme to improve water supply and stop deforestation. It demonstrates how energy security can be improved by preserving biodiversity and reducing the need for water treatment and pumping. (Water is energy intensive, and energy is water intensive.)

The Oasis Project is a payment scheme that protects parts of the remaining Atlantic Rainforest in Brazil in order to preserve drinking water and biodiversity. About 93 percent of the Atlantic Rainforest has been destroyed; the remaining 7 percent contains crucial watersheds and drainage basins that provide water to metropolitan areas in Brazil. The Oasis Project pays land owners to protect their forested land in order to preserve parts of the forests that provide drinking water for 3,400 Brazilian municipalities.[129]

8.7.1   Description

The Oasis Project is managed by the Fundação O Boticário de Proteção à Natureza, a non-governmental organization founded in 1990 by O Boticário, one of the world's largest manufacturers of perfumes and cosmetics. The O Boticário Group pledges about 0.8–1.1 percent of its net income per year to the Fundação O Boticário, which is composed mostly of environmental scientists, biologists, zoologists, and forest engineers. Fundação O Boticário (headquartered in Curitiba, Brazil) manages the Salto Morato Nature Preserve in Guaraqueçaba and the Serra do Tombador Nature Preserve in Cavalcante, Goiás, operates public education exhibits relating to nature in urban centers, and makes small grants related to ecology, conservation, and effective nature protection all over Brazil. To date, the Foundation has invested 12.8 million reals (equivalent to 7.5 million US dollars or 5.8 million euros) in 1,100 environmental projects, has a full-time staff of 35, and spends about 7.7 million reals ($4.5 million, €3.5 million) a year.

The Oasis Project started in 2006 with the basic idea of getting the business community involved in sponsoring ecosystem services and protecting the Guarapiranga Watershed Basin, the Billings Watershed, and the Capivari-Monos and Bororé Colônia Environmental Protection Areas surrounding the city of São Paulo (figure 8.10). These areas, which together encompass about 82,000 hectares (202,600 acres), supply water to the São Paulo municipal district. They need urgent protection. The rate of population growth is strongly positive in the forested areas that supply the water to these basins (and negative in denser urban areas), and the net loss of forest cover was a rate of 5.1 percent from 1991 to 2000.[130] From 1989 to 1996, Guarapiranga alone lost 15 percent of its vegetation cover and suffered an urban growth rate of 50 percent, precipitating local water shortages and exorbitant costs associated with water treatment and purification.[131] Clearing forests, furthermore, means that water pulses down and contributes to more erosion, but soil can act as natural storage and absorption when forests are properly maintained.

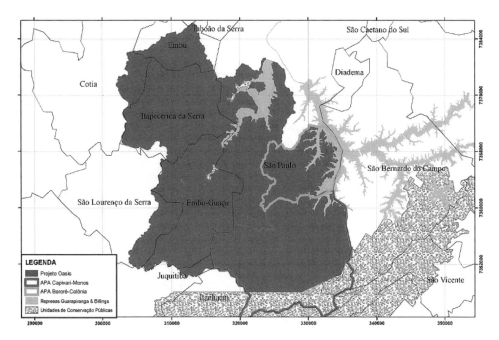

Figure 8.10
Map of Oasis Project, Guarapiranga Watershed Basin, Billings Watershed, and Capivari-Monos and Bororé Colônia Environmental Protection Areas. Source: research interviews.

The Oasis Project involves three primary stakeholders: the Fundação O Boticário itself (which selects areas for protection, collects donations, distributes them to land owners, and manages the program), the sponsors (most of them large corporations that elect to donate money in order to promote community welfare and enhance their corporate social responsibility programs), and the land owners or landholders (who receive financial resources to maintain their land). The Mitsubishi International Corporation Foundation was the first sponsor to join the program. In November 2006, it donated 681,800 reals ($400,000, €310,000) spread across 5 years with the intention of protecting 2,500 hectares (6,178 acres) of private land from 2007 to 2011.[132]

By far the most unique feature of the Oasis Project is its progressive and multi-phase process of selecting land for preservation. "Forest engineers" identify the most important areas to be protected, conduct environmental valuation of individual properties, establish "rewards for services" contracts between the Foundation and land owners, and then continually monitor contracted areas. Researchers began by identifying the most critical aspects of the Billings and Guarapiranga watersheds, which provide water for 4 million residents in São Paulo. About 49.5 million hectares (122 million acres) were identified as Priority 1 areas, meaning that they help filter and transport clean water *and* are at the greatest risk of deforestation; 17.5 million hectares (43.2 million acres) were given Priority 2 status, meaning that they help filter water but are at less risk of deforestation; and 15.2 million hectares (37.6 million acres) were given Priority 3 status, meaning that they are primarily at risk of deforestation but have less significant water benefits.

Perhaps the most novel component of this process was the creation of a watershed-valuation index to be used in the selection of parcels of land after Priority 1 and Priority 2 areas were designated. This index was designed to assign values to forested land on the basis of three variables: maintenance of water quality (avoided water treatment), erosion control (avoided reservoirs and dams), and production and storage of water (avoided pumping). Researchers determined that a typical hectare of ideal and perfectly forested land was worth about 324 reals ($19, €147) in total, or 85 reals ($50, €39) a year in terms of production and storage, 68 reals ($40, €31) a year in erosion control, and 170 reals ($100, €77) a year in terms of providing and maintaining water quality.

Researchers then approached land owners living in high-priority areas about agreeing to keep their land forested in exchange for ecosystem payments. Since no area of land was perfect, most land owners were offered rates equal to 66–94 percent of the ideal rate, at an average rate per hectare of 68 reals ($146, €113) per year in total or roughly $8,000 per year per property (although some received substantially more or less because of unique properties of their forested land). Payments are disbursed from the foundation every 6 months. As of early 2009, thirteen large land owners had subscribed to protect a total of 657 hectares (1,620 acres). Foundation representatives visit land owners twice a year to monitor changes in land use and to verify payments, and also offer informal free technical assistance to help with planning and capacity building related to building fences to keep illegal poachers at bay, avoiding the contamination of springs, creating watershed communities, and making repairs and restoring land without damaging the forest.

If the Fundação O Boticário had the funding, it could immediately triple the size of the Oasis Project to add 1,280 hectares (3,160 acres) of Priority 1 areas. Land owners have already indicated their intent to join the program, but the Foundation lacks the resources to establish land tenure and fund the contracts. Managers of the Oasis Project expect that it could be rapidly expanded to protect 80,000 hectares (197,684 acres) if the state and municipal governments of São Paulo adopted and replicated Oasis Project. The state of São Paulo is in the process of creating a water fund, a small water charge of .017 real (less than one US cent, or €.008) per liter of water to fund a scaling up of the Oasis Project. The fund is on schedule to begin collecting money at the end of 2009, and project managers expect that this next phase of the project will produce at least 170 million reals ($100 million, €77 million) in revenues per year.

8.7.2   Costs and Benefits

The biggest benefit of the Oasis Project is that it protects watersheds and forests that are already protected under law but are still being deforested and destroyed. Brazil has 256 conservation parks, some even protecting the Atlantic rainforest, but most of these are "paper parks" created by executive order without additional funding or provision for implementation. Many of these conservation parks were also created without consultation with or compensation to land owners, and state and local governments do not have the staff or the financial resources to protect them. The greater São Paulo region also has 29 state parks, 24 ecological stations, 13 preserved forests, two state reserves, and two forest nurseries, yet deforestation and water pollution continue to outpace their positive influence. Many of these local laws and regulations, including watershed-protection agreements, are simply not enforced.

One unfortunate aspect of these types of protection is that they make it appear that conservation is being achieved even though destruction continues. Moreover, international agreements such as the Clean Development Mechanism of the Kyoto Protocol contribute to the dilemma because they also exclude lands already "protected" by national law as being eligible for tradable carbon credits. Simply put, the Oasis Project protects forests that international agreements already presume are protected and therefore ignore.

A second benefit unique to the Oasis Project is that it explicitly protects water according to a scientific index of watershed valuation. Many types of conservation areas or forest reserves aim to protect water but also have a number of other competing goals that can conflict with water-management strategies. If the only goal of forest managers was to sequester carbon, for instance, it could make sense to replace natural forests with rows of eucalyptus trees, which have more carbon-storage potential. The Oasis Project prioritizes water quality, erosion prevention, and water conveyance. The properties participating in the Project in 2007 had an average of more than 70 percent of their land still covered by natural vegetation.

A third benefit of the Oasis Project is that it generates income and forms attachments between communities and the forest. Many of the land owners interested in the project currently receive nothing for their forested land, as that land is not cultivating crops or supporting livestock. Most of the families currently living in the Atlantic Rainforest make very little money selling vegetables or making crafts and have annual incomes of less than 8,523 reals ($5,000, €3,869). These land owners often end up doubling or even tripling their annual income by becoming part of the Oasis Project. This creates a strong incentive to preserve the land. As of June 2009, the Oasis Project had 100 percent compliance, and not a single hectare of protected land had been deforested.

8.7.3   Challenges

In its current form, the Oasis Project is not sufficient to ensure water quality and availability for all or even a significant part of São Paulo, which continues to be degraded and contaminated by industrial activity. Part of the problem is that until recently water has been free within Brazil and widely considered to be inexhaustible. People had to pay for its treatment and distribution, but not for the water itself. Overcoming this notion that water is and should be free—without value—is likely to require a tremendous educational effort.

Second, the Oasis Project works only so long as the ecosystem payments it offers exceed what land owners could make from deforesting their land. Put another way, the price signals in favor of preserving land must outweigh those that incentivize converting forest land to agriculture and other uses. In the local culture of most parts of the Atlantic Rainforest (and elsewhere in Brazil), it is acceptable to "clean" the forest in order to open the "front door" to agricultural and cattle-raising activities.[133] Forests have great value as potential sources of income and real estate, and many Brazilians believe that in their natural state they have little to offer the local population. Thus, the Oasis Project must continually ensure that its payments provide the proper incentives to preserve the land.

A third challenge, closely connected to the second, is funding. The Oasis Project works only if it can find corporate foundations and companies to "sponsor" parts of the Atlantic Rainforest. So far, the Project is running mostly on money from the Fundação O Boticário and an initial contribution from the Mitsubishi Corporation International Foundation. If it fails to find other corporate sponsors, or if its plan to expand the program through funding from local and regional governments falls through, the Project will collapse.

A fourth challenge faced by Oasis Project managers is establishing land rights and tenure. The attorney responsible for creating legal contracts between land owners and the Fundação O Boticário commented that in general the constitution and the transmission of property rights can be "very informal" in Brazil. In some cases "contracts" or "deeds" for land are verbal and passed down generation to generation, in other cases they are notes handwritten by people on their deathbeds, and in some cases they do not exist at all. Attempting to ensure that these land owners enter into binding and legal agreements can be very difficult.

Finally, protecting the Atlantic Rainforest has proved difficult in part because of its size. Parts extend across multiple states of Brazil and even into Argentina and Paraguay, which makes jurisdiction over policy and coordination of conservation activities tricky. The actual watersheds feeding the municipalities of São Paulo fall partially under the cities' jurisdiction and partially under that of São Paulo state and that of the neighboring state of Parana.

8.7.4 Lessons Learned

The simplest lesson from the Oasis Project is that nature is better than technology as a provider of clean water. The central premise behind the Oasis Project is that, because naturally forested land has great value in terms of preventing erosion and cleaning and conveying fresh drinkable water, people ought to be paid to keep that land forested. Approaches such as investing in water-purification plants and in water-treatment facilities are often more expensive than these simple efforts to preserve forests. The Oasis Project takes the path of least resistance and offers direct biannual ecosystem payments to the owners of this land if they promise to avoid altering it.

The Oasis Project also demonstrates the importance of flexibility in assessing land and disbursing payments. Instead of giving land owners a flat rate for their land, the Oasis Project utilizes a watershed-valuation index to determine which land is the most valuable. The funding for the payments now comes from an environmental NGO (Fundação O Boticário) and a multinational corporation (Mitsubishi), and soon municipal and state governments will contribute to it. And the land owners receive more than merely cash; they also get capacity building and assistance to encourage them to manage their land optimally and to form watershed communities.

Finally, the Oasis Project is eminently scalable. It can be implemented at virtually any level of spatial organization, from a few hectares to the entire forest. This replicability probably explains why Oasis Project managers have been approached by water planners, environmental groups, and forestry stakeholders about applying similar programs to the Brazilian part of the Amazonian rainforest and in other countries.

8.8   The Toxics Release Inventory in the United States, 1988–2007

This case study is polycentric because the Toxics Release Inventory is managed by the federal government but asks companies and industries to report information about their pollution. It addresses the waste and water dimensions of energy production and energy security, with electric utilities one of the predominant groups participating in the program.

The Toxics Release Inventory (TRI) is a freely accessible database that provides information on about 650 chemicals released into the environment in the United States. In its latest incarnation, the TRI contains data on how nearly 22,000 industrial and government facilities used, managed, disposed, and recycled their hazardous wastes. The goal of the TRI is to offer information that then empowers citizens and non-governmental agencies to hold polluters accountable for how they manage toxic chemicals, especially those that spill or seep into the water, air, and land. From 1988 (when the program began) to 2001, total on-site emissions of chemicals into the air were reduced 70 percent, from 2.2 billion pounds (998 million kilograms) to 650 million pounds (295 million kilograms); discharges into surface waters were reduced 65 percent, from 42 million pounds (19 million kilograms) to 15 million pounds (6.8 million kilograms); and releases into the land were reduced about 40 percent from 3.2 billion pounds (1.5 billion kilograms) to 1.9 billion pounds (862 million kilograms).[134]

8.8.1 Description

The driving force behind the creation of the TRI was the monumental increase in the use and discharge of chemicals during the twentieth century. From the 1940s to the early 1990s, the US chemical industry grew twice as fast as the US economy as a whole.[135] The increasing exposure of people to accidental and chronic releases of chemical pollutants, a string of industrial accidents in the mid 1980s, and the Bhopal disaster in India prompted New Jersey to pass its Toxic Catastrophe Prevention Act. That act, passed in 1986, created the first chemical release inventory program in the US.

Regulators quickly decided that a national inventory was needed after an internal report circulated within the US Environmental Protection Agency (EPA) warned that the agency's "multifaceted environmental control programs" were suffering from "a 'Tower of Babel' syndrome, with many different regulatory tongues speaking in the mutually incomprehensible languages of toxic chemicals."[136] The federal TRI was proposed as a way of providing national and state regulators with a common reference point for the collection of data relating to pollution. It was hoped that this database would create a more comprehensive picture of releases and would increase public knowledge about contamination.

The federal government followed through on the EPA's advice and passed the TRI, and on October 17, 1986, President Ronald Reagan signed the Emergency Planning and Community Right-to-Know Act (EPCRA).[137] Section 313 of the EPCRA created the TRI, and the TRI explicitly required that the owners and operators of selected industrial facilities that manufactured, used, processed, or imported particular toxic chemicals to report annual releases and emissions of those chemicals to the EPA. The TRI was structured to provide regulators and the community-at-large information about chemical accidents, hazardous substances stored or held by companies, chemical health effects, and routine releases of toxic chemicals into the environment.

The TRI initially applied to only 20 chemical categories out of hundreds, and to only a limited number of facilities. A facility was required to report releases only if it was classified as a manufacturer, had 10 or more fulltime employees, and handled more than specified amounts of TRI chemicals.[138] Once a year, beginning in 1988, a facility had to report to the EPA details of its operations; the estimated amount of TRI chemicals released into the air, into specific water bodies, or to areas of land on site, and also transfers of chemicals to off-site facilities; methods used to treat waste streams; and activities that managers were taking to reduce the production of waste.[139] Facilities reporting to the TRI had to use "existing and readily available data," but were not required to monitor actual quantities and concentrations and could instead rely on "reasonable estimates." The EPA was empowered to fine any person failing to comply with TRI requirements $25,000 per day per violation, and each failure was considered a separate violation.

The TRI accomplished four initial tasks: First, it created a broadly accessible, objective, open-ended, cross-media metric of chemical pollution (a proxy for environmental performance) that was not tied to any particular regulatory standard. It stipulated uniform rules about how information concerning those releases had to be reported, compiled, and released to the public. Second, it enabled people to monitor performance, benchmark, and identify best practices. Third, it compelled and also enabled polluting facilities to track their own performance and then compare, rank, and assess their performance relative to other firms (since the information was public, it could be utilized to assess partners, peers and competitors) and relative to other production processes. Fourth, it enhanced the transparency and accountability of chemical pollution to investors, regulators, and the public.[140]

By distributing more accurate information about pollution, the TRI created strong mechanisms for monitoring and assessment. Firms could self-monitor by producing a series of periodic reports on the releases of pollutants at each of their facilities. Competitors could peer monitor to highlight their comparative performance and to augment accountability to standards of best practice. Legislators could regulatory monitor polluters by establishing baselines, profiles, and trends in pollution at many levels (facilities, firms, communities, states) to make comparisons between them and determine the efficacy of environmental standards. Civil-society groups and individuals could monitor exposure within a specified area and thus could organize boycotts, arrange for pickets, file lawsuits, place political pressure on elected officials, and socially ostracize employees.[141] Investors, insurers, lenders, and even employees could monitor the environmental performance of firms to assess potential liabilities and analyze the environmental impact of certain suppliers and products. Such monitoring created unprecedented access to information about toxic releases, helped illuminate understanding of the risks associated with pollution, and made it practicable to set reduction targets and to measure progress.

Although it has gone through some important changes since 1988,[142] in its current form the TRI tracks releases of about 650 toxic chemicals from 21,996 facilities in the US. The EPA makes this information available at http://www.epa.gov/tri and also in printed reports.

8.8.2 Costs and Benefits

The simplest and most direct benefit of the TRI has been substantially reduced releases of listed chemicals. Of course, the reduction of these chemicals over time cannot be attributed entirely to the TRI; other factors probably included more stringent environmental regulations, substitution, recycling, improved technology, higher energy prices, and moving manufacturing facilities out of the country. Still, from 1988 to 2007 facilities decreased their on-site and off-site releases of TRI listed chemicals by 61 percent, a reduction amounting to almost 2 billion pounds of chemical pollution in 2007.[143] That is equivalent to 18 times the weight of the *Titanic*.[144]

Releases of TRI chemicals have also decreased dramatically in recent years for the most polluting sectors—metals mining, electric utilities, primary metal production, chemical manufacturing, hazardous waste recovery, paper and pulp production, and food and beverage manufacturing—with an overall reduction in on-site and off-site releases of more than 50 percent from 1998 to 2007 (from about 3 billion pounds to less than 1.2 billion). Figure 8.11 shows the specific decrease in releases from the four largest of these sectors. Note that for some reason electric utilities have seen only a slow reduction of toxic releases.

One reason why listed companies have curtailed toxic releases could be that negative TRI listings have been shown to affect stock prices. One study analyzed the top 40 companies listed on the TRI from 1989 to 1992 and found that a listing often prompted investors to update their expectations of future pollution-related expenditures and liabilities such as the probability of accidents and likelihood of litigation. As a result, those companies had their largest stock-price declines on the day new TRI information became available to the public, and the companies thus became subsequently more concerned about reducing pollution than their industry peers.[145] Another study found that higher pollution figures from the TRI resulted in negative publicity toward the firm, and that therefore TRI information had a significant impact on stock, with an average loss of $4.1 million in stock value for TRI firms on the day pollution figures were released.[146]

The TRI has also been associated with improved benchmarking and environmental performance. One meta-survey of the TRI found that many companies began reducing releases voluntarily to improve their images. Further, a TRI listing was strongly correlated with companies' beginning to produce annual reports of their environmental performance.[147] Another study of the US chemical industry (responsible for 53 percent of TRI releases that year) utilized TRI data to benchmark their performance against competitors and convinced firms to reduce their pollution in order to improve employee and investor relations.[148]

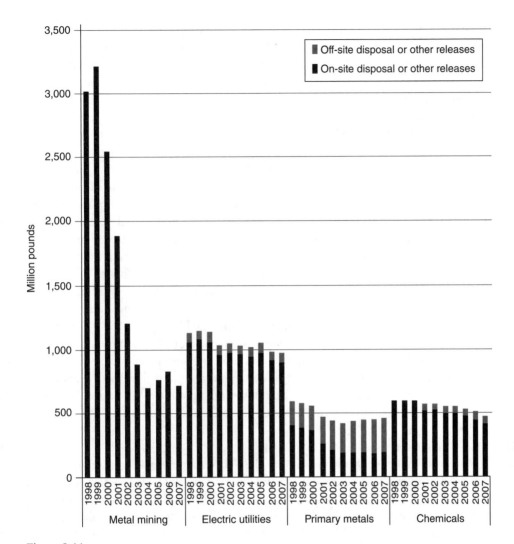

Figure 8.11
Total disposal and releases of chemicals listed in TRI, 1998–2007. Source: US Environmental Protection Agency 2009.

One ancillary benefit has been a reduction in the need for public expenditures for the purpose of managing releases of TRI chemicals. Treating pollution once it has been released into the environment is very expensive. One interview respondent suggested that it can cost more than $100 million to remove a single kilogram of mercury or lead from drinking or surface water. Lowering the amount of pollution released therefore lessens the need for capital-intensive water treatment, treatment that is already expected to cost $75.1 billion in the next 20 years merely to filtrate and disinfect water supplies.[149]

A final benefit has been the creation of additional environmental protections. Information listed in the TRI has founded broader environmental and health legislation at the state, national, and international levels. By developing a "Green Index" with TRI data, it became possible to compare the environmental performance of states, and to rank communities according to indicators relating to the purity of air, water, and land.[150] TRI information was instrumental in spotlighting concerns over the unequal distribution of releases near minority communities ("environmental racism"), and was used by various communities to create targeted pollution taxes and fees.[151] The publication of TRI information enabled environmental advocacy groups to verify compliance with other environmental laws.[152] In 1990, the sponsors of the Clean Air Act Amendments made extensive use of TRI data on air emissions to convince legislators to begin capping and reducing sulfur dioxide emissions, and dozens of states have used TRI data to determine pollution thresholds, enact their own legislation, and target facilities that have high amounts of pollution.[153] And at least 30 countries have adopted the use of databases modeled on the TRI.[154]

8.8.3   Challenges

The greatest challenge to the creation of the TRI was fierce and continuing opposition from industry. During the debates in Congress in 1986, some industries argued extensively for alternatives, ranging from an inventory with fewer listed chemicals to outright rejection of any type of list whatsoever.[155] In 1994, when the EPA decided to nearly double the number of chemicals to be reported in the TRI, the Chemical Manufacturers Association initiated litigation. The Troy Corporation, the National Oilseed Producers Association, and the N-Methylpyrrolidone Producers Group quickly sued. Each of these cases was decided in favor of the EPA.[156] In the late 1990s, when the EPA broadened the scope of the TRI yet again, some industries responded by mounting a public-relations campaign to claim that release of new data would enable "industrial espionage" through "reverse engineering," as competitors could use TRI data to determine how a facility worked, even though these claims were untrue.[157]

Another challenge relates to the accuracy of the TRI database. The TRI does not have a fully accurate list of toxic releases, because many companies rely on approximations and estimates instead of direct measurements. Some companies may be unable to obtain complete or current information from their suppliers, and that makes it difficult to satisfy reporting requirements.[158] The particular releases for a specific company can also vary greatly from year to year as operators close and/or relocate facilities, adopt new manufacturing processes, rely on new suppliers, or find safer or cheaper substitutes. Adding to the complexity, the list of TRI chemicals and reporting requirements has changed over time. For example, in 2006 the EPA (ostensibly under pressure from the George W. Bush administration) changed TRI reporting requirements to raise the threshold at which a company was required to report releases from 500 pounds (227 kilograms) to 5,000 pounds with an additional requirement that only 2,000 pounds (907 kilograms) of that pollution had to be released directly into the environment.[159] Yet in April 2009,[160] the EPA revoked the changes (ostensibly under pressure from the Barack Obama administration) and restored the initial requirements, creating a discrepancy within the TRI over the levels of pollution historically reported.

In addition, the TRI's coverage of pollution sources and of polluters is incomplete. The hundreds of substances compiled in the TRI are only a fraction of the 80,000 or so chemicals commercially used in the US. This broader list of uncovered substances includes carcinogens (such as pesticides and emissions from coke ovens) and many greenhouses gases. At the start of the program, in 1988, the TRI covered only 5 percent of all toxic chemical releases, and it excluded facilities with fewer than 10 employees and facilities that did not manufacture substances. These latter two categories of excluded facilities encompassed most of the dry cleaners, electric utilities, mining and farming operations, and chemical storage facilities in the country, as well as a number of federal agencies and facilities that have since been included but were not originally listed.[161]

The TRI does not even cover all potential sources of the chemicals that it does list. Nonpoint sources (that is, unfixed or mobile sources) such as automobiles, boats, and airplanes, along with point sources too small to track, such as individual farms or personal septic systems, commonly release millions of tons of fertilizers, herbicides, insecticides, oil, grease, bacteria, and nutrients each year that are unreported and untracked by the TRI. Listed toxic chemicals found in products such as televisions and phones, moreover, are not included, since they do not constitute "waste."[162]

A third challenge involves the scope of the TRI and its exclusion of risk data. The TRI compiles data on chemical releases only; it does not discuss the risk that those chemicals have on human health and the environment. TRI chemicals differ greatly in toxicity and in potential for exposure. Some chemicals are very serious threats in extremely low doses; others will have different impacts depending on sunlight, heat, and the surrounding environment.[163] In addition, the TRI reports on the releases disclosed by a particular firm or within a certain area once a year, but it does not reveal the nature of how these chemicals were released (i.e., whether they were dumped in large quantities in one spill or released gradually). This type of information relating to the rate and duration of exposure is central to estimating the adverse health effects of pollutants, but is not included.[164]

A fourth challenge relates to the accessibility and user-friendliness of the TRI database. Surveys have shown that citizens are interested in learning about toxic pollution but tend not to know that the TRI exists or how to use it. EPCRA did not consider public education and outreach to be within its scope, and many people who have tried to make use the TRI have complained that it is cumbersome. Some critics have argued that the TRI "overwhelms the public with unnecessary information that is difficult to understand."[165]

A fifth challenge is the strategic manipulation of what some facilities elect to disclose. Data for the TRI are self-reported and only checked by the EPA to be "fairly accurate."[166] Some studies have demonstrated a mismatch between monitored releases of actual chemicals and the reported releases within the TRI, especially for concentrations of lead, nitric acid, and polycyclic aromatic hydrocarbons.[167] Other studies have noted that firms can intentionally under-report or over-report emissions so that their competitors will mis-estimate their performance.[168]

A sixth and possibly more disturbing challenge is that facilities can strategically move wastes to other countries to lower the releases reported to the TRI. In the mid 1990s, a few companies were able to achieve "phantom reductions" by transferring materials to non-manufacturing subsidiaries.[169] Perversely, some companies have been accused of continuing to release pollution in order to lower property values, which in turn makes nearby land cheaper to purchase for future expansion.[170] Other companies have relocated manufacturing facilities or certain chemicals to facilities overseas in order to escape TRI reporting requirements. One study noted that some companies were able to show TRI reductions of 50 percent for some chemicals by shifting production to areas with cheaper sources of labor and energy in Asia, South America, and Africa.[171] The truly dangerous aspect of this overseas relocation is that from an environmental standpoint it doesn't matter where some pollutants are released—all that matters is that they are released. Releases of mercury, for example, are long-lived, often spread thousands of miles away from their source, and bioaccumulate. The trans-boundary nature of mercury emissions is so great that about 70 percent of the mercury falling on the US is from sources in other countries.[172]

Seventh, the TRI may have diminishing returns in its ability to compel firms to reduce pollution. One insightful study of the TRI's impact on corporate behavior suggested that the magnitude of pressure to reduce pollution (both externally from the press and internally from employees and shareholders) depends more on the extent to which TRI information was unexpected. Negative publicity can lead to significant declines in stock prices, but such declines tend not to be directed toward the largest polluters, only those that were not "known" as polluters to investors and the public. Similarly, public pressure to reduce pollution occurs only insofar as communities have the necessary education and income to use the TRI database and comprehend pollution patterns; firms may not feel as much pressure in areas where people are disempowered or less politically active. The implication is that the provision of information about known polluters will have diminishing effects, and that over time the TRI will have less effect as people either come to know and accept rates of pollution, or are incapable of understanding the dynamics of that pollution in the first place.[173]

8.8.4 Lessons Learned

Notwithstanding the challenges cited in the preceding subsection, the TRI clearly underscores the power of information and measurement to alter behavior and reduce pollution. The TRI essentially changed the property rights concerning information about chemical use and pollution; before its creation, data on toxic releases were not in the public domain. Putting such data in the public domain focused attention on companies, generated press stories, sparked some dialog with local communities, and gave investors and regulators ways to measure relative performance. Simply providing information on toxic chemical releases through a centrally located, widely disseminated, freely available database such as the TRI has induced meaningful reductions in pollution without the need for permits, environmental standards, timetables, and mandates. The TRI gives communities a chance to understand and then address pollution, to monitor it, and to then make informed decisions about what rates of exposure are acceptable to them.[174] The TRI distributes that information to shareholders, investors, and employees of reporting facilities as well, enabling them to monitor and track the quantities of hazardous waste their companies discharge. The reporting procedures required by the TRI make communities and companies aware, often for the first time, of the substances they are routinely releasing into the environment.

A second lesson is that information-disclosure and community right-to-know programs such as the TRI, once in place, tend to facilitate important mechanisms that then further curb pollution. TRI data have been used for supporting environmental improvement by setting priorities, as well as encouraging voluntary reductions; utilized to track the progress in pollution reduction, and to support regulatory initiatives; employed to support citizen action and participation, and also used by civil-society groups to raise awareness about pollution and polluters.[175] The TRI emphasizes that information on pollution itself has multifarious and far-reaching potential to strengthen environmental protection, stigmatize polluters, and embolden the public all at once.

8.9   Conclusion

These eight case studies cover a variety of different sectors from electricity and transport to waste and forestry, in both emerging and developed economies, at various moments in time. Yet despite these differences in history and context, each illustrates the effectiveness of polycentrism, of blending actors and scales in tackling energy and climate-related problems. They demonstrate the socio-technical nature of energy and climate challenges, and the value of designing policies that reflect the types of barriers causing the socio-technical gap. They also illustrate the importance of designing and implementing policies reflecting the lifestyles and needs of the citizens they assist, while also improving energy security and reducing damage to the Earth's climatic system.

The stability and commitment of public entities to their policy approaches and programs is a common denominator of our cases, as is an emphasis on speed, scope, and scale—on programs and practices that can be quickly scaled up and implemented almost anywhere, tailored to the unique needs of different political and social environments.

In the end, our cases also affirm the value of the case-study method—that is, of investigating a phenomenon within its real-life context to explore causation in order to find underlying principles. The intention is to bring a more accurate understanding to complex issues and to provide students and policy makers with detailed, contextual information.

9

Conclusions

At 10 o'clock on a cold winter night, figures emerge from a dark seawall and walk in the direction of a brightly lit power plant. They climb two 10-foot razor-wired electrified security fences, cross a parking lot, and continue toward the gentle hum of boilers and generators. They enter an unlocked part of the generator house and cause a giant turbine to crash. Before leaving, they put up a poster, handmade from an old bed sheet, that says "No New Coal!" They walk out the way they walked in, hop back over the fence, and recede into the darkness.

Though such an event may sound like a fantasy, it happened on November 28, 2008 at the fossil-fuel Kingsnorth power plant. The most heavily guarded power station in the United Kingdom, with £12 million ($18.5 million; €14.4 million) worth of security, was unable to prevent intruders, monitored by closed-circuit television cameras, from breaching its defenses, sabotaging one of the plant's 500-megawatt turbines, and leaving a homemade poster protesting coal. The act forced the coal- and oil-fired plant to suspend electricity generation for 4 hours and caused a temporary 2 percent reduction in emissions of greenhouse gases over the entire United Kingdom.[1]

A few months later, on June 23, 2009, James Hansen, an eminent climatologist and the director of the National Aeronautic Space Administration's Goddard Institute for Space Studies, was arrested, with 30 other demonstrators, for obstruction and impeding traffic after crossing onto the property of Massey Energy Company in Coal River Valley, West Virginia, as part of a protest. Massey Energy is the biggest company conducting mountaintop removal in West Virginia. Mountaintop removal is a technique that blasts the caps of mountains off in order to more easily retrieve their coal. The protest was part of a string of increasingly dramatic actions in the United States objecting to the Obama administration's announcement that the US Environmental Protection Administration will reform, but not abolish, mountaintop removal.[2]

Two months later, and halfway around the world, in August 2009, the Hay Point Export Terminal near Mackay, Australia, which can load 45 million tons of coal per year, was closed for 2 days after more than a dozen youths scaled a 50-meter wall and chained themselves to a loading port. The incident prevented the export of 180,000 tons of coal and caused a loss of $14 million in revenues and royalties. Ten of the activists were jailed.[3]

Finally, jump ahead to the United Nations Climate Change Conference in Copenhagen, Denmark, in December of 2009. Although the activists from the United Kingdom, from the United States, and from Australia risked their lives and reputations to bring attention to climate change, many of the delegates to the conference exercised unsustainable consumption habits. More than 1,200 limousines were ordered. One delegate demanded 42 vehicles for his delegation. Almost no delegates used electric cars, bicycles, or hybrid vehicles. The Copenhagen airport saw 140 extra private jets, and most planes had to fly into Copenhagen to drop people off and then fly to Sweden to park. Menus at the conference featured fish, scallops, caviar, and paté de foie gras. The eleven-day conference generated 41,000 tons of CO_2 equivalent, equal to the amount of CO_2 produced globally by all commercial air travel in a single day.[4] That the Copenhagen conference (similar to the following year's Cancun Conference) reached no binding agreement on climate change was all the more unfortunate. It was more a "triumph of spin over substance"[5] and "short on detail and devoid of obligation."[6]

The trespassers in the United Kingdom, the United States, and Australia ostensibly believed that the risks associated with their crimes and acts of civil disobedience were less important than the threats posed by the continued damage from coal-fired electricity and its GHG emissions. Yet their leaders accomplished very little, if anything, in the global climate negotiations, while emitting prodigious amounts of carbon dioxide. How can we make sense of these conflicting trends? Virtually everything we do in our daily lives somehow contributes to climate change, and we don't always have intruders or demonstrators reminding us that the electricity we buy, the appliances we use, and the cars we drive have important impacts on our atmospheric and ecological footprint.

How serious are the current threats to the global climate and to energy security—defined as equitably providing available, affordable, reliable, efficient, environmentally benign, properly governed and socially acceptable energy services to citizens? All humanity has to do to ruin the planet for future generations (by melting the Arctic glaciers and Greenland ice sheet, flooding low-lying islands and coastal cities worldwide, reducing biodiversity, polluting the air, and acidifying the oceans) is continue along the current trajectory. We humans, and our energy and industrial systems, are now the most voracious predators in the oceans and the most successful terrestrial carnivores. We dominate all major ecosystems, from grasslands and forests to fisheries and rivers. The only exception is deep marine vent communities, and we are already beginning to explore commercial applications there. In this way human beings more resemble a ravaging bloom of algae or a mold enveloping fruit than protective or even prudent stewards of our planet.[7]

The influence and impact of human actions on the climate has become so pronounced that the planet has entered a new geological epoch, moving from the 12,000-year-old Holocene to the fittingly titled Anthropocene, which means "human" and "new."[8] If the entire 4.5-billion-year history of the Earth is condensed into a single day, human beings do not even appear until 77 seconds before midnight, and the entire industrial age lasts less than a second. Think about this simple fact: human beings have managed to pollute the atmosphere and disrupt the global climatic system in less than one second of the Earth's 24-hour history. According to the population ecologist William Rees, "human activity is the most powerful geological force altering the face of the planet, and the erosive pace is accelerating."[9] If we could speed up time and view things from outer space, the global economy and its energy infrastructure would appear to be literally crashing into the Earth like an asteroid.[10] This philosophy that our age-old system of matter and energy is incompatible with our money-based culture, expressed by Marion King Hubbert as quoted in the introduction to our book, is reflected in what we present as the book's five most salient conclusions.

9.1   The Socio-Technical Nature of Climate and Energy Challenges

The energy-security and climate-change challenges confronting the world are neither entirely technical nor entirely social; they are socio-technical. That is, they involve not only technologies (including physical devices, objects, infrastructures, systems, and tools), but also people (who are motivated by human values, habits and routines, cognitive limitations, and cultural beliefs). As the Dutch professor of technology and society Wiebe Bijker cleverly put it,

Purely social relations are to be found only in the imaginations of sociologists, among baboons, or possibly on nudist beaches; and purely technical relations are to be found only in the sophisticated reaches of science fiction. The technical is socially constructed, and the social is technically constructed—all stable ensembles are bound together as much by the technical as by the social. Where there was purity, there is now heterogeneity. Social classes, occupational groups, firms, professions, machines—all are held in place by intimately linked social and technical means. . . . Society is not determined by technology, nor is technology determined by society. Both emerge as two sides of the sociotechnical coin.[11]

This simple conclusion has profound implications for energy and climate research. It means that attempts to improve energy security, to mitigate emissions of greenhouses gases, or to adapt to climate change merely by building new technologies will be wholly insufficient.

This is not to say that such efforts are unimportant—indeed, many of the key GHG-reducing and adapting technologies presented in chapters 3 and 4 need further research to improve performance and lower costs. Some of these technologies have been around for decades, but others are quite novel and in early stages of commercialization. Table 9.1 lists 48 technologies and practices currently available for deployment and 44 that have to be developed further. Such commercialization efforts, however, must be coupled with attempts to eliminate market and policy failures that contribute to persistent socio-technical gaps—the shortfall between technically possible solutions to energy security and climate change, and actual social choices. They also must be strengthened by efforts to educate and inform consumers, to overcome biases and apathy, to shift cultural values and behavior, and to incentivize people to use new technologies along with old and familiar ones that already work. Relatively simple changes in individuals' lifestyles, such as consuming less energy at home, bicycling instead of driving to work, eating less meat, and purchasing second-hand or used items, can add up to significant climatic benefits. In short, the socio-technical dimension of energy and climate change necessitates holistic and complementary solutions that avoid looking at only one face of the socio-technical coin. Individuals' behaviors can be just as important as the development of new technologies.

Climate change and energy security are thus "wicked problems" that involve sets of interdependent systems and infrastructure, meaning actions targeted at one part of the system will affect other components in unplanned ways, producing unintended or unpredictable effects. Fixing the problem is not like building a better engine or flying an airplane, but about the more complex and ambiguous difficulties associated with engaging stakeholders, altering social values, and aligning political regulations.[12]

9.2   Justification for Government Intervention

The complex socio-technical nature of climate and energy challenges offers a robust justification for government intervention. Chapter 5 explored numerous market failures and barriers that exist on both social and technical planes, including technical obstacles (externalities, high costs, infrastructural limitations) and social, regulatory, and institutional obstacles (policy failures, utility monopolies, energy-price volatility, lack of knowledge, lack of training, lack of information).

Governments can do much to overcome these impediments, from putting a price on carbon to introducing a range of innovative and effective complementary policies, some regulatory and others voluntary. If targeted to overcome behavioral barriers (loss aversion, asymmetric information, habits, heuristics to deal with overwhelming deliberation costs), these policies can transform markets. Options include

Table 9.1
Some technologies for climate-change mitigation, geo-engineering, and adaptation.

	Commercially available today	Technically feasible but not yet commercially available
Energy end use and infrastructure		
High-efficiency transportation	Hybrid electric vehicles (HEVs) and plug-in HEVs	Long-range all-electric cars
	Flex-fuel vehicles	
	Clean diesel engines	Vehicle-to-grid systems
	Adaptive traffic-control systems	
High-performance buildings	Low-emissivity windows	Zero-energy homes
	Compact fluorescent light bulbs	Solid-state lighting
	Exterior insulation finish systems	Smart roofs and dynamic wall insulation
	Heat pump water heater	Integrated heat pump systems
Industry	Advanced heat exchangers	Microwave processing of metals
	Cogeneration	Nano-ceramic coatings
	Fiber optics for combustion measurement and control	Pressure swing adsorption
	Resource recovery and utilization	Isothermal melting
	Super boilers	Cokeless ironmaking
		Oxy fuel firing
Electricity grid and delivery infrastructure	High-voltage DC transmission	DC appliances
	High-temperature superconductor transmission lines, generators, motors, and transformers	
	Composite-core, low sag transmission lines	
	Flywheels	
	Smart meters	

Table 9.1
(continued)

	Commercially available today	Technically feasible but not yet commercially available
Energy supply		
High-efficiency fossil power	Distributed generation Integrated gasification combined cycle Direct and indirect cycle stationary fuel cells Fischer-Tropsch reactors for solid-to-liquid fuel conversion Oxy-fuel combustion	Advanced power systems Coproduction of hydrogen and nuclear power
Hydrogen power and fuels	Forklifts Hydrogen production from natural gas and biomass Backup power High pressure hydrogen storage tanks	Fuel cells for vehicular applications
Renewable electricity and fuels	Wind power Solar thermal power Concentrating solar power Silicon and thin film superconductors for photovoltaic solar power Geothermal energy Tidal turbines Pumped hydroelectric storage Reservoir (large) and run-of-river (small) hydroelectric power Biochemical reactors for conversion of sugar to ethanol Gasification and pyrolysis systems to produce biofuels Biodiesel	Turbines for low-wind regimes Wave power Enhanced geothermal energy Ocean power Algal fuels Cellulosic and advanced ethanol

Table 9.1
(continued)

	Commercially available today	Technically feasible but not yet commercially available
Nuclear fission	Advanced light-water reactor designs Pressurized heavy-water reactors	Advanced boiling-water reactors European pressurized-water reactors Adaptation of nuclear power plants to "dry" closed-loop cooling systems
Capturing and sequestering carbon		
Carbon capture	Chilled ammonia capture process Amine scrubbing	Post-combustion capture Oxyfuel combustion Pre-combustion capture
Geologic storage	CO_2 injection with oil or methane recovery Geological monitoring and modeling methods for CO_2 fate and transport	Saline formations Deep-seam coal beds
Terrestrial sequestration	Cropland, forestland, and grazing management with advanced information technologies	Genetic engineering to enhance biological carbon uptake
Reducing other greenhouse gases		
Methane from energy and waste	Landfill gas collection and use Aerobic and anaerobic bioreactor treatment	Coal-mine methane
Methane and nitrous oxide from agriculture	Advanced agricultural sensors and controls Centralized digester technologies for manure management Controlled release fertilizers Nitrogen transformation inhibitors	
High-global-warming-potential gases	Substitution of SF_6 use with fluorinated ketones in the magnesium industry Distributed and secondary loop refrigeration in supermarkets	Magnetic refrigeration CO_2 refrigerant

Table 9.1
(continued)

	Commercially available today	Technically feasible but not yet commercially available
Nitrous oxide from combustion and industry	Abatement technologies for nitric acid production such as non-selective catalytic reduction	Advanced catalysts
Geo-engineering		
Solar radiation management	White roofs and pavements	Sulfate particles in the stratosphere Cloud brightening from salt spray
Carbon-cycle engineering	Carbon capture via biochar burial Direct capture of CO_2 from air	Ocean iron fertilization Increased ocean alkalinity
Adaptation		
Sea-level rise	Planting mangroves Coastal land acquisition Building storm surge barriers Building bridges Erecting seawalls	
Drought	More harvesting of rainwater Adjusting planting dates and crop variety	Developing drought-resistant crops
Extreme temperatures	Opening of cooling centers	Reflective roadway systems
Permafrost melt	Building dams Insulating railway lines	
Ecosystem adaptation		Species transplantation Maintaining biodiversity gene banks
Informational infrastructure	Environmental monitoring Weather forecasting Early warning systems	
Institutional infrastructure	Land use planning	
Total	70	44

increasing research expenditures for low-carbon technologies, sponsoring neighborhood workshops to personalize information about clean-energy choices, reforming subsidies and designing incentives to overcome social impediments, implementing payments for ecosystem services, and requiring extended producer responsibility.

However, achieving the necessary market transformations will require a much deeper understanding of the barriers facing particular policies and the ways to overcome them. Individuals and institutions will have to change their behaviors so that marketable and effective energy and climate technologies and policies become widely adopted.

9.3   Speed, Scope, and Scale

Although intervention by governments is important, it is often much more effective when implemented with a diversity of institutions, and with the speed, scope, and scale required to repair the planet. Individuals, cities, and corporations must act alongside regulators and government officials. As the German sociologist and philosopher Jürgen Habermas once wrote, "in the process of enlightenment there can only be participants." The same holds true for climate change and energy security: we must *all* participate.

Individual behavior, for example, is a discrete and often overlooked source of GHG emissions. Individuals are often not listed in government reports as a source of emissions, yet some studies have noted that individual behavior (actions under the direct, substantial control of a person but not undertaken in the scope of their employment) accounts for 32 percent of annual CO_2 emissions in the United States.[13] Individual behavior, such as driving automobiles, eating food, taking vacations, and using electricity in the home, was responsible for 4.4 trillion pounds (2 trillion kilograms) of CO_2 emissions in the US in 2000, while the entire industrial sector emitted only 3.9 trillion pounds (1.8 trillion kilograms). Emissions from such individual behavior in the US accounted for about 8 percent of the world's carbon dioxide emissions—more than the total emissions of individual countries such as Canada, South Korea, and the United Kingdom, and more than emissions from the global chemical manufacturing and petroleum refining industries.[14] (See figure 9.1.)

Individuals, however, can alter many of their daily practices to reduce emissions substantially. They can, for instance, use less energy-intensive goods and services, drive more efficient cars, purchase better electric appliances, eat less meat, and conserve water. They should be viewed not as passive recipients loosely connected to climate change, but as active participants whose lifestyles centrally contribute to energy and climate problems. The situation brings to mind Rachel Carson, nature writer and advocate of environmental ethics, who wrote in 1962 that "the human

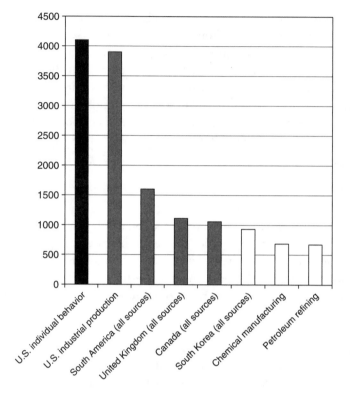

Figure 9.1
Carbon dioxide emissions from individual behavior in the United States compared with other sources (billions of pounds). Individual emissions include those associated with household electricity use, personal motor vehicle use, personal air travel, and mass transport, but exclude emissions attributed to work, industrial production, and travel undertaken in the course of employment. Source: Michael P. Vandenbergh and Anne C. Steinemann, "The carbon-neutral individual," *New York University Law Review* 82, 2007: 1673–1745.

race is challenged more than ever to demonstrate our mastery—not over nature, but of ourselves."[15]

The business community is also mobilizing, and many multi-national corporations are becoming forces for environmental progress and energy security. Companies are beginning to appreciate that it is in their mutual best interest to keep the climate stable. This is often because many of the best things companies and manufacturers can do to reduce emissions also improve efficiency and create profits. Since environmental standards are becoming stricter, building clean facilities today prevents the need for expensive retrofits in the future. Energy-security-strengthening and climate-stabilizing practices also help in recruiting and retaining employees and in improving their morale.[16]

9.4   The Power of Polycentrism

The approaches that have been most successful at mitigating GHG emissions and improving energy security have been holistic ones taken by governments, individuals, and a broad array of stakeholders in a polycentric manner. Eight striking examples of polycentric programs and policies that have worked in the real world were presented in chapter 8. These case studies show that, with the right mix of technologies and public policies, communities, countries, and major emitters can reduce their emissions drastically, accelerate deployment of new energy systems, restore large tracts of degraded forestland, reduce pollution greatly, and diversify their energy resources. The policies illustrated in figure 9.2, organized by scale from local to global, demonstrate the frequency of different scales of engagement. There are no truly global climate-change or energy-security policies that a majority of the world's GHG emitters all operate within. Rather, there are many local, small-scale, and regional initiatives, most of which have emerged in the past 10 years. These efforts reveal the ability to intervene quickly, but this virtue of speed is not always matched by proper scope or scale. Although many of the effective policies discussed in the case studies could be replicated all over the world, near-term success remains tenuous without a global framework to prevent carbon leakage, address equity issues, and ensure economies of scale.

9.5   Coordinated, Progressive, and Consistent Policies

Although each case study offers a set of different lessons, there are some common themes. The most successful efforts have been coordinated, progressive, and consistent—coordinated in the sense that policies were implemented in concert with important stakeholders from industry and civil society, progressive in that they all had mechanisms to adjust and respond to emerging challenges, and consistent in that constant changes to program structure were kept to a minimum and were transparent when they did occur. The eight successful efforts that were the subjects of our case studies all relied on diversification of a variety of mechanisms to promote more sustainable policy. They depended on a suite of technologies promoted by a host of policies that influence both technical dimensions (such as performance and price), and social measures (such as knowledge and behavior). They also show the inadequacy of attempting to address energy and climate issues from only the "supply" side (building more technologies) or the "demand" side (altering the behavior of consumers). The best programs focused on aspects of both by lowering costs and improving technologies in addition to shaping values and demand for products.

We are aware that many skeptics will respond that the costs are too great, that technology isn't ready, that polycentric approaches are too complicated to work,

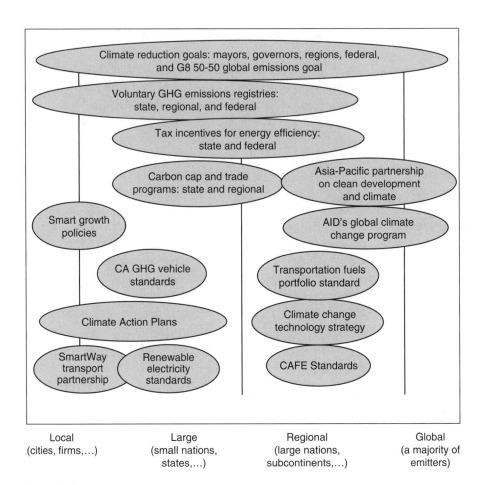

Figure 9.2
Scales of climate policy.

that too much time is required to shift to new energy systems, or that government shouldn't intervene. Let us remind those readers that countries are already paying billions of dollars (and possibly trillions, according to some estimates) in transfers of wealth to sustain their autocentric lifestyles, and to address environmental damages, climatic changes, power outages, energy price spikes, and deteriorating public health. Compensating for these massive damages does not "cost" society more. It merely shifts previously external costs to the market, where everyone can see them, and forces the polluters (and consumers complicit in such pollution) to pay for such external costs.[17]

Those who express dismay about the urgency of a transition to new energy and climate technologies should consider the history of automobiles, military aircraft, cell phones, jet engines, computers, the Cold War, and the case studies presented in chapter 8. Only a few thousand automobiles were on the road in the United States by 1900, but by 1910 there were more than a million. In 1939, it was thought that the US was able to build no more than 2,000 airplanes a year, but by 1946 there were 257,000 in service. Use of the Internet and the number of personal computers has, on average, doubled every 3–4 months since their invention. The first commercial mobile telephone service, launched in 1977, had only a few thousand subscribers, but 20 years later the number had surpassed 3.3 billion. In 1950 there were no jet airliners. In 1960, 7 percent of airliners were jet powered. By 1990, 80 percent were. Today there are more than 18,000 commercial jets.[18] The United States and Soviet Union spent the equivalent of about 10 trillion US dollars (7.7 trillion euros) on the Cold War, enough money to replace the entire infrastructure of the world at the time. Every school, hospital, roadway, building, and farm could have been purchased for the cost of a political movement, one based on *possible* threats and *potential* destruction instead of the *real* risks already posed from climate change and unsustainable sources of energy.[19]

Skeptics should also read chapter 8 closely. Denmark was able to install more than 6,200 commercial wind turbines in less than 10 years. Germany installed 430,000 residential solar panels in 9 years. Brazil became the world's leader in ethanol production in less than 5 years. Singapore effectively manages an average daily traffic volume of 270,000 vehicles and 970,000 overall vehicles with electronic road pricing and other demand-side tools. Grameen Shakti in Bangladesh came to operate hundreds of offices in every state and has installed 250,000 solar home systems, 40,000 cookstoves, and 7,000 biogas plants in 15 years. The Chinese installed 185 million cleaner-burning cookstoves in about 10 years. The Oasis Project in São Paulo protects watersheds and forests with a 100 percent compliance rate. In the United States, the Toxics Release Inventory enables 300 million citizens to monitor hazardous pollution.

Those who argue that government shouldn't intervene in the marketplace should recognize it has been intervening in the energy and climate sectors for centuries, and that action is needed if only to counteract decades and sometimes centuries of previous support for conventional infrastructure. We no longer sell human beings in the free market, even though some were considered legitimate market-based commodities 150 years ago. Government did not wait to abolish the injustice of slavery until the market regulated itself, for the simple reason that the government could not wait.[20] Government can no longer wait to address climate change and challenges to energy security, either.

One does not have to break into coal-fired power plants in the United Kingdom, impede traffic at coal mines in West Virginia, or chain themselves to coal export terminals in Australia to make a difference. We can all undertake simple changes in our home and workplaces; by making intelligent decisions about what we purchase, where we invest, and who we elect; and by educating ourselves about technologies and policies that can quickly and effectively shift our communities to more sustainable forms of energy supply and energy use. Electricity, transport, agriculture and forestry, waste and water, are intersecting climate challenges, but one hidden aspect of that interconnection is that solutions can have multi-faceted and diffuse benefits if strategically designed. In the end, the question is not "How can we improve energy security?" or "How can we protect the climate?" In view of how inextricably linked we are to both energy security and climate change the question is "How can we save ourselves?"

Appendix A: Experts Contacted and Interviewed for Case Studies

Case study	Name	Institution	Location	Date
Chinese cookstoves	Kirk R. Smith	University of California	Berkeley	June 30, 2009
	Peter Catania	International Energy Foundation	Okotoks, Saskatchewan	July 6, 2009
	Junfeng (Jim) Zhang	School of Public Health, Rutgers University	Piscataway, New Jersey	July 16, 2009
	Bina Agarwal	Institute of Economic Growth, Delhi University	Delhi	July 17, 2009
	Melissa Chan	John F. Kennedy School of Government, Harvard University	Cambridge, Massachusetts	July 21, 2009
Danish wind energy	Steffen Nielsen	Danish Energy Authority	Copenhagen	October 29, 2007
	Ole Oddgaard	Danish Energy Authority	Copenhagen	October 29, 2007
	Anders Hasselager	Danish Energy Authority	Copenhagen	October 29, 2007
	Lene Nielsen	Danish Energy Authority	Copenhagen	October 29, 2007
	Simon B. Leonhard	Orbicon	Copenhagen	October 29, 2007
	Hans Henrik Lindboe	EA Energlananalyse	Copenhagen	October 29, 2007
	Charlotte Boesen	Dong Energy	Copenhagen	October 29, 2007

Case study	Name	Institution	Location	Date
	Peter Wenzel Kruse	Vestas	Randers, Denmark	October 30, 2007
	Bo Morup	Vestas	Randers	October 30, 2007
	Martin Molzen	LM Glasfiber	Kolding, Denmark	October 30, 2007
	Henrik Carstens	Ramboll Denmark A/S	Virum, Denmark	October 31, 2007
	Soren Juel Petersen	Ramboll Denmark A/S	Virum	October 31, 2007
	Soren A. Nielsen	MBD Offshore Power A/S	Virum	October 31, 2007
	Tove Feld	Det Norske Veritas	Virum	October 31, 2007
	Claus Gormsen	Niras	Virum	October 31, 2007
	Per Volund	Grontmij / Carl Bro	Virum	October 31, 2007
	Jesper Tornbjerg	El & Energi	Copenhagen	October 31, 2007
	Lise Nielson	EnergiNet DK	Copenhagen	November 1, 2007
	Martin Randrup	EnergiNet DK	Copenhagen	November 1, 2007
	Henrik Bindner	Risø National Laboratory	Roskilde, Denmark	November 1, 2007
	Sten Frandsen	Risø National Laboratory	Roskilde	November 1, 2007
	Jeppe Johansen	Risø National Laboratory	Roskilde	November 1, 2007
	Jyoti Painuly	Risø National Laboratory	Copenhagen	November 2, 2007
Ethanol	Carolina Costa	Brazilian Sugarcane Industry Association (ÚNICA)	São Paulo	April 27, 2009
	Regis Lima Verde Leal	University of Campinas	Campinas, Brazil	May 25, 2009
	Luis Cortez	University of Campinas	Campinas	May 25, 2009
	Arnaldo César Walter	University of Campinas	Campinas	May 25, 2009

Case study	Name	Institution	Location	Date
	Rogerio Cezar de Cerqueira Leite	University of Campinas	Campinas	May 25, 2009
	Antonio M. F. L. J. Bonomi	Centro de Ciência e Tecnologia do Bioetanol (CTBE)	Campinas	May 25, 2009
	Jose Pradella	Centro de Ciência e Tecnologia do Bioetanol (CTBE)	Campinas	May 25, 2009
	Jose Goldemberg	University of São Paulo	São Paulo	May 26, 2009
	Marcelo Galdos	University of São Paulo	São Paulo	May 27, 2009
	Jaime Finguerut	Sugarcane Technology Center (CTC)	São Paulo	June 2, 2009
	Patricia Simoes	Sugarcane Technology Center (CTC)	São Paulo	June 2, 2009
	Jorge Neves	Sugarcane Technology Center (CTC)	São Paulo	June 2, 2009
Germany's Feed-in Tariff	David Elliott	The Open University	London	November 9, 2007
	Till Stenzel	Hazel Capital LLP	London	November 16, 2007
	Hermann Scheer	European Association for Renewable Energy	London	November 16, 2007
	David Toke	University of Birmingham	London	November 17, 2007
	Godfrey Boyle	The Open University	London	November 18, 2007
	Miguel Mendonca	World Future Council	London	November 18, 2007
	Kirsty Hamilton	Chatham House	London	November 18, 2007

Case study	Name	Institution	Location	Date
	Chris Barton	Department for Business, Enterprise & Regulatory Reform	London	November 19, 2007
	Eric Usher	United Nations Environment Programme	London	November 20, 2007
	Clive Mason	World Bank	London	November 20, 2007
	Karsten Neuhoff	Cambridge University	London	November 27, 2007
	Steve Sawyer	Global Wind Energy Council	Brussels	December 1, 2007
	Reinhard Kaiser	Bundesministerium für Umwelt, Naturschutz und Reaktorsicherheit (BMU)	Brussels	April 7, 2008
	Carlos Gasco	Iberdrola	Brussels	April 7, 2008
	Marlene Kratzat	German Aerospace Center	Brussels	April 8, 2008
	Jose Etcheverry	David Suzuki Foundation	Brussels	April 8, 2008
	Ole Langniss	Zentrum fur Sonnenenergie-und Wasserstoff- Forschung Baden- Wurrtemberg (ZSW)	Brussels	April 8, 2008
	Dirk W. Hendricks	World Future Council	Brussels	April 8, 2008
	Bianca Barth	World Future Council	Brussels	April 8, 2008
	Ruth Brand	Enercon	Berlin	July 28, 2008
	Daniel Argyropoulos	Bundesministerium für Umwelt, Naturschutz und Reaktorsicherheit (BMU)	Berlin	July 28, 2008
	Thomas Chrometzka	German Solar Industry Association (BSW-Solar)	Berlin	July 28, 2008

Case study	Name	Institution	Location	Date
	Maike Schmidt	Zentrum fur Sonnenenergie-und Wasserstoff-Forschung Baden-Wurrtemberg (ZSW)	Stuttgart	July 29, 2008
	Peter Bickel	Zentrum fur Sonnenenergie-und Wasserstoff-Forschung Baden-Wurrtemberg (ZSW)	Stuttgart	July 29, 2008
	Karin Schneider	Fraunhofer-Institut für Solare Energiesysteme ISE	Freiburg, Germany	July 31, 2008
	Andrea Ocker	Solar-Fabrik AG	Freiburg, Germany	July 31, 2008
	Matthias Reitzenstein	Solar-Fabrik AG	Freiburg, Germany	July 31, 2008
	Stefan Schurig	World Future Council	Hamburg	August 1, 2008
Grameen Shakti	A. K. M. Sadrul Islam	Bangladesh University of Engineering & Technology	Dhaka, Bangladesh	June 30, 2009
	Fazley Rabbi		Dhaka	July 2, 2009
	Dipal C. Barua		Dhaka	July 8, 2009
	Noim Uddin	Macquarie University	Sydney	July 9, 2009
Oasis Project	Thais Losso	Losso, Tomasetti, and Leonardo	Curitiba, Brazil	May 28, 2009
	Rafael D. Zenni	Fundação O Boticário de Proteção à Natureza	Curitiba	May 29, 2009
	Laurenz Pinder	Fundação O Boticário de Proteção à Natureza	Curitiba	May 29, 2009
	Maisa Guapyassu	Fundação O Boticário de Proteção à Natureza	Curitiba	May 29, 2009
	Roberto Waack	Forest Stewardship Council	São Paulo	June 2, 2009
	Charlotte Streck	Climate Focus	Rotterdam	June 2, 2009

Case study	Name	Institution	Location	Date
Singapore's Urban Transport Policy	Paul Barter	National University of Singapore	Singapore	July 7, 2009
	Sock Yong Phang	Singapore Management University	Singapore	July 7, 2009
	Chin Hoong Choor	National University of Singapore	Singapore	July 7, 2009
	Lee Der-Horng	National University of Singapore	Singapore	July 9, 2009
	Janice Quah	Ministry of Transport	Singapore	July 4, 2009
	Soffy Hariyanti Abdul Rahim	Ministry of Transport	Singapore	July 6, 2009
	Chris Leong	Land Transport Authority	Singapore	July 7, 2009
	Anthony Chin	National University of Singapore	Singapore	July 8, 2009
	Chin Kian Keong	Land Transport Authority	Singapore	July 9, 2009
	Lay Kim Lim	Land Transport Authority	Singapore	July 9, 2009
	Justin How	Land Transport Authority	Singapore	July 9, 2009
	Lily Wong	Office of Sustainability, National University of Singapore	Singapore	July 17, 2009
	Neo Kay Lian	National Environmental Agency	Singapore	July 18, 2009
	Dison Zhang	ecoWise Incorporated	Singapore	July 18, 2009
	Melvin Chen Kai Ching	ecoWise Incorporated	Singapore	July 18, 2009
	Leon Khew	IUT Singapore	Singapore	July 19, 2009
	S. K. Ashraf	IUT Singapore	Singapore	July 19, 2009
	Evelyn Quek	Ministry of Transport	Singapore	July 21, 2009

Case study	Name	Institution	Location	Date
Toxics Release Inventory	TRI Information Center	US Environmental Protection Agency/ Booz Allen Hamilton	Washington	July 6, 2009
	William C. Reilly	US Environmental Protection Agency/ TRI	Washington	July 7, 2009
	Nichelle McDaniel	Waste Division, Virginia Department of Environmental Quality	Richmond, Virginia	July 8, 2009
	Sanjay Thirunagari	Waste Division, Virginia Department of Environmental Quality	Richmond	July 8, 2009
	James T. Hamilton	Duke University	Durham, North Carolina	July 20, 2009

Appendix B: Methodology for Case Studies

In selecting a methodology for the case studies discussed in chapter 8, we utilized a grounded, ethnographic, semi-structured interview approach.

By *case studies* we mean that we endeavored to provide what the methodological theorists Alexander George and Andrew Bennet call a "detailed examination of an aspect of a historical episode to develop or test historical explanations that may be generalizable to other events."[1] Rather than utilizing laboratory samples or statistical analysis to examine variables, case-study methods involve an in-depth, longitudinal assessment of a single instance or group of instances: a case or cases.[2] Put another way, the case-study method is an investigation of a contemporary phenomenon within its real-life context to explore causation in order to find underlying principles.[3] The intention is to bring a more accurate understanding to a complex issue and provide detailed, contextual insight for students (or policy makers).[4]

After all, we hold that the social science research is the study of human affairs, and that only context-dependent knowledge really exists in this realm. Such knowledge is always related to case studies and specific instances. Elucidating the commonalities and the differences between these cases can enhance student learning, critical thinking, and the expertise of policy makers.[5] Case studies have the "irreducible quality" of being "thick," that is, full of rich detail, or "hard to summarize."[6] Only through an intimate connection with experience and reality can students develop testable, relevant, and valid knowledge and information.[7]

By *grounded* we mean that we commenced our research without preformed hypotheses. This method is sometimes called "grounded theory" because it is an inductive discovery method that starts with no theoretical preconception. Instead, researchers develop a conceptual account "from the ground up," the analysis grounded in the data collected itself.[8] A grounded approach works exceptionally well when few relevant theories yet exist to explain what is being studied, as was the situation with our eight cases.

By *ethnographic* we mean that we took what our participants and experts said at face value; we did not prompt them, suggest answers, or intentionally inject our own values.[9] Though ethnographic techniques are typically employed to study foreign cultures, and intended to minimize researcher bias, we found that our training in the social sciences and in science and technology studies made us see energy issues differently than most of our interviewees.

For our *semi-structured interviews* we relied on a purposive sampling strategy, which means that experts were chosen to represent different aspects of the cases in question, some having expertise in public policy, others having knowledge of engineering, and still others having training in economics or even medicine. We relied on semi-structured questions (sometimes referred to as "intensive interviewing" or "responsive interviewing"[10]), asking a set of standard inquiries but then allowing the conversation to build and deviate to explore new directions and areas. Though this method has some challenges, we relied on it extensively throughout the case studies for its ability to promote interactivity, flexibility, sufficiency, and appropriateness.

Semi-structured interviews enable *interactivity* because they encourage participants to talk, and also allow the conversation to attain suitable momentum that enables truly in-depth exploration of topics. They permit the researchers to ask for clarification, elaboration, redirection, and cross-examination, dynamic processes that are not possible with textual surveys. With no fixed time, interviews can sometimes last hours, delivering large amounts of qualitative data.[11]

Semi-structured interviews are *flexible* in that their unscripted dimension can enable the researcher to develop additional lines of inquiry and explore new topics in the conversation that were not originally anticipated. This can facilitate a more iterative and complete flow of knowledge, as answers are not confined to the responses and categories anticipated by the researcher.[12] It also empowers the investigator to adjust questions during the conversation and follow through to provide more complete answers. Rigorous adherence to a script or survey, by contrast, can often freeze the interview and foreclose the collection of information.[13]

Such interviews are *sufficient* as they allow researchers to collect information difficult or impossible to represent in numerical or statistical form without trivializing them. The contextual information gleaned from them escapes the limits of rigidly defined tools and models.[14] Many of the variables of interest to us, such as political factors pushing particular policies or the interests of stakeholders, are difficult to measure, and to describe them with quantitative methods would amount to "conceptual stretching" by "lumping together dissimilar cases to get a larger sample."[15] Such quantitative methods tend to omit contextual factors, whereas qualitative case studies can look at a large number of intervening variables to observe a particular phenomenon or issue. Analogously, qualitative interviews are one of the only ways to deal with value laden or subjective issues such as which policy mechanism for energy "works best" or which explanation for success of a program was most valid.

Semi-structured interviews are *appropriate* in that they enabled us to use face-to-face interaction, or words and images, rather than text to solicit information. Our case studies, for instance, involved site visits and discussions with rural sugarcane farmers in Brazil, solar home system distributors in Bangladesh, and cookstove program managers in China. Though clearly intelligent, some of these participants were illiterate, which made textual collection of data impossible. Moreover, the visual element of the interviews enabled us to look for nonverbal cues to decide whether a respondent understood a question, and enabled us to occasionally use visual aids, such as graphs or photos.

Notwithstanding these collective benefits, our methodology does have some shortcomings. The qualitative aspect of interview responses makes them difficult to code and answers vary for each participant. Some respondents may provide socially desirable responses, telling us what they think we want to hear. Others could deliberately give us answers that they thought would sway the outcome of the study in their favor. Our sample of experts was purposive and relatively small. Our case studies also were selected on the dependent variable,[16] meaning that they all demonstrate effective examples of polycentrism. To truly test whether polycentrism absolutely engenders successful programs, we would also have to examine cases in which non-polycentric intervention results in success, cases in which non-polycentric intervention results in failure, and cases in which polycentric intervention results in failure. Absent such examination, our case studies do not conclusively demonstrate that polycentrism is a panacea.

Ultimately, however, we believe the strengths of our methodology far surpass its weaknesses. Selecting cases on a dependent variable, or cases that are all similar, can have the advantage of investigating that variable in great depth.[17] Similarity can also matter if it appears to occur across a complex number of cases that differ according to time, geography, and context. Here, with our case studies, it is the commonalities that are striking—that is, the fact the elements of polycentrism appear to have resulted in success in the face of so much diversity. In the end, we hold that our method balances the goal of attaining theoretical parsimony with that of explanatory richness and keeping the number of cases manageable. Our cases exhibit parsimony by distilling commonalities across case studies and their implications for public policy. They demonstrate explanatory richness as they cover different sectors, time periods, and countries; they are manageable in that we let the number of them rest at eight.

Notes

Chapter 1

1. M. King Hubbert, *Two Intellectual Systems: Matter-Energy and the Monetary* (MIT Energy Laboratory, 1981).

2. Bjorn Stigson, "The international climate change landscape," presented at National Academy of Sciences Meeting of the Committee on America's Climate Choices, Washington, 2009.

3. This exercise is borrowed from William McDonough and Michael Braungart, *Cradle to Cradle: Remaking the Way We Make Things* (Farrar, Straus, and Giroux, 2002): 18–19.

4. We are careful here to use the term "development" rather than "growth." Growth implies getting bigger at expanding regardless of the consequences, whereas development means getting better and carefully considering social, political, and environmental consequences. See Herman E. Daly, *Steady-State Economics* (Island, 1991); Paul Hawken, *The Ecology of Commerce: A Declaration of Sustainability* (HarperCollins, 1993); Herman E. Daly, *Beyond Growth: The Economics of Sustainable Development* (Beacon, 1996).

5. Paul C. Stern, "A second environmental science: Human-environment interactions," *Science* 260, 1993: 1897–1899.

6. Daniel M. Kammen and Michael R. Dove, "The virtues of mundane science," *Environment* 39, 1997, no. 6: 10–41. See also E. F. Schumacher, *Small Is Beautiful: Economics As If People Mattered: 25 Years Later . . . With Commentaries* (Hartley & Marks, 1999).

7. For more details about these interviews and what they indicate about energy policy, see Benjamin K. Sovacool, "The importance of comprehensiveness in renewable electricity and energy efficiency policy," *Energy Policy* 37, 2009, no. 4: 1529–1541; Marilyn A. Brown, Jess Chandler, Melissa V. Lapsa, and Benjamin K. Sovacool, *Carbon Lock-In: Barriers to the Deployment of Climate Change Mitigation Technologies*, Oak Ridge National Laboratory Report ORNL/TM-2007/124, 2008); Benjamin K. Sovacool, *The Power Production Paradox: Revealing the Socio-Technical Impediments to Distributed Generation Technologies*, doctoral dissertation, Virginia Polytechnic Institute and State University, 2006.

8. The "threat multiplier" was introduced in a 2007 report by the CNA Military Advisory Board titled *National Security and the Threat of Climate Change*. The report is available at SecurityAndClimate.cna.org.

9. Benjamin K. Sovacool and Marilyn A. Brown, "Competing dimensions of energy security: An international perspective," *Annual Review of Environment and Resources* 35, 2010: 77–108.

10. Edward Chow and Jonathan Elkind, "Hurricane Katrina and energy security," *Survival* 47, 2005, no. 4: 145–160.

11. *Energy Security Issues* (World Bank Group, 2005).

12. Kevin D. Stringer, "Energy security: Applying a portfolio approach," *Baltic Security & Defense Review* 10, 2008: 121–142.

13. For just a small sample of this paradigm, see S. Pacala and R. Socolow, "Stabilization wedges: Solving the climate problem for the next 50 years with current technologies," *Science* 305, 2004: 968–972; M. I. Hoffert et al., "Advanced technology paths to global climate stability: Energy for a greenhouse planet," *Science* 298, 2002: 981–987.

14. Michael T. Klare, "The futile pursuit of energy security by military force," *Brown Journal of World Affairs* 13, 2007, no. 2: 139.

15. Benjamin K. Sovacool, "Solving the oil independence problem: Is it possible?" *Energy Policy* 35, 2007, 11: 5505–5514.

16. Both Iran and North Korea signed the Nuclear Non-Proliferation Treaty several decades ago but have since withdrawn. See Donald L. Zillman and Michael T. Bigos, "Security of supply and control of terrorism: Energy security in the US in the early twenty-first century," in *Energy Security: Managing Risk in a Dynamic Legal and Regulatory Environment*, ed. B. Barton et al. (Oxford University Press, 2004); Jan H. Kalicki and David L. Goldwyn, eds., *Energy & Security: Toward a New Foreign Policy* Strategy (Woodrow Wilson Center, 2005); David A. Deese and Joseph S. Nye, eds., *Energy and Security* (Ballinger, 1981).

17. CNA Military Advisory Board, *Powering America's Defense: Energy and the Risks to National Security*, 2009 (www.cna.org).

18. Masahiro Atsumi, "Japanese energy security revisited," *Asia-Pacific Review* 14, 2007, no. 1: 28–43; Tsutomu Toichi, "Energy security in Asia and Japanese policy," *Asia-Pacific Review* 10, 2003, no. 1: 44–51.

19. William J. Nuttall and Devon L. Manz, "A new energy security paradigm for the twenty-first century," *Technological Forecasting & Social Change* 75, 2008: 1249; Youngho Chang and Jian Liang Lee, "Electricity market deregulation and energy security: A study of the UK and Singapore electricity markets," *International Journal of Global Energy Issues* 29, 2008, no. 1: 109–132.

20. For explorations of the Chinese energy security strategy, see Margret J. Kim and Robert E. Jones, "China's energy security and the climate change conundrum," *National Resources & Environment* 19, 2004–2005: 3–8; Joseph Y. S. Cheng, "A Chinese view of China's energy security," *Journal of Contemporary China* 17, 2008, 55: 297–317; Shebonti Ray Dadwal, "China's search for energy security: Emerging dilemmas," *Strategic Analysis* 31, 2007, no. 6: 889–914; Xu Yi-chong, "China's energy security," *Australian Journal of International Affairs* 60, 2006, no. 2: 265–286. On the specific maritime issues associated with China and energy security, see Andrew Erickson and Gabe Collins, "Beijing's energy security strategy: The significance of a Chinese state-owned tanker fleet," *Orbis*, fall 2007: 665–684; Kevin X. Li and Jin Cheng, "Maritime law and policy for energy security in Asia: A Chinese perspective," *Journal of Maritime Law & Commerce* 37, 2006: 567–587.

21. For recent assessments of Russian energy security policy, see Sebastian Mallaby, "What 'energy security' really means," *Washington Post*, July 3, 2006; Sergey Sevastyanov, "The more assertive and pragmatic new energy policy in Putin's Russia: Security Implications for Northeast Asia," *East Asia* 25, 2008: 35–55.

22. Daniel Yergin, "Ensuring energy security," *Foreign Affairs* 85, 2006, no. 2: 69–82.

23. Richard Leaver, "Factoring energy security into Australian foreign and trade policy: Has luck run out?" *International Journal of Global Energy Issues* 29, 2008, no. 4: 388–399; Richard Leaver, "Australia and Asia-Pacific energy security: The rhymes of history," in *Energy Security in Asia,* ed. M. Wesley (Routledge, 2007); Jiaping Wu, Stephen T. Garnett, and Tony Barnes, "Beyond an energy deal: Impacts of the Sino-Australia uranium agreement," *Energy Policy* 36, 2008: 413–422.

24. Lila Barrera-Hernandez, "The Andes: So much energy, so little security," in *Energy Security: Managing Risk in a Dynamic Legal and Regulatory Environment,* ed. B. Barton et al. (Oxford University Press, 2004).

25. Frank Verrastro and Sarah Ladislaw, "Providing energy security in an interdependent world," *Washington Quarterly* 30, 2007, no. 4: 95–104.

26. Also contributing to the volatility of oil prices is the increasingly limited world oil surplus production capacity. In the 1990s, significant surplus capacity for world oil production existed in Saudi Arabia, in the United Arab Emirates, in Iraq, in Kuwait, in Iran, and in other OPEC countries. Since 2005, surplus oil production in the world has been limited essentially to one or two million barrels per day in Saudi Arabia. See Energy Information Administration, *Short-Term Energy Outlook,* 2006 (http://www.eia.doe.gov/pub/forecasting/steo/oldsteos/sep06.pdf).

27. International Energy Agency, *World Energy Outlook 2008*: 105, figure 3.10.

28. Ibid.: 105, figure 3.10.

29. Thomas Fingar, "National intelligence assessment of the national security implications of global climate change to 2030," testimony before House Permanent Select Committee on Intelligence and House Select Committee on Energy Independence and Global Warming, June 25, 2008 (http://www.dni.gov/testimonies/20080625_testimony.pdf).

Chapter 2

1. William McDonough and Michael Braungart, *Cradle to Cradle: Remaking the Way We Make Things* (Farrar, Straus, and Giroux, 2002): 128.

2. David E. Nye, *America as Second Creation: Technology and Narratives of New Beginnings* (MIT Press, 2003): 147.

3. Vaclav Smil, "Energy in the twentieth century: Resources, conversions, costs, uses, and consequences," *Annual Review of Energy and Environment* 25, 2000: 21–51.

4. Charles Hall, Preadeep Tharakan, John Hallock, Cutler Cleveland, and Michael Jefferson, "Hydrocarbons and the evolution of human culture," *Nature* 426 (November 2003): 318–322.

5. Smil, "Energy in the twentieth century."

6. Ibid.

7. Intergovernmental Panel on Climate Change (IPCC), *Climate Change 2007: Synthesis Report* (Geneva: IPCC, 2008): 36. "Carbon dioxide equivalency" describes the quantity of CO_2 that would have the same global-warming potential as a given amount of other greenhouse gas when measured over a specified time period, generally 100 years. It thus reflects the time-integrated radiative forcing of a quantity of greenhouse-gas emissions.

8. David Archer, *The Long Thaw* (Princeton University Press, 2009).

9. Once emitted, a ton of carbon dioxide takes a very long time to process through the atmosphere. According to the latest estimates, one fourth of all fossil fuel derived carbon dioxide emissions will remain in the atmosphere for several centuries and complete removal could take as long as 35,000 years. See J. Hansen, M. Sato, P. Kharecha, D. Beerling, V. Masson-Delmotte, M. Pagani, M. Raymo, D. Royer, and J. Zachos, "Target atmospheric CO_2: Where should humanity aim?" *Atmospheric Science Journal* 2, 2008: 217–231; D. Archer, "Fate of fossil fuel CO_2 in geologic time," *Journal of Geophysical Research* 110, 2005: 26–31.

10. David Victor, Granger Morgan, John Steinbruner, and Kate Ricke, "The geoengineering option: A last resort against global warming?" *Foreign Affairs* 88, 2009: 65.

11. Energy Information Administration, *An Updated Annual Energy Outlook 2009 Reference Case Reflecting Provisions of the American Recovery and Reinvestment Act and Recent Changes in the Economic Outlook,* US Department of Energy Report SR/OIAF/2009-03, 2009 (http://www.eia.doe.gov/oiaf/servicerpt/stimulus/excel/aeostimtab_18.xls), tables A1 and A8.

12. Benjamin K. Sovacool, *The Dirty Energy Dilemma: What's Blocking Clean Power in the US* (Praeger, 2008): 17.

13. Arno A. Evers, *Worldwide Electricity Deprivation* (http://www.fair-pr.com).

14. Vijay Modi, Susan McDade, Dominique Lallement, and Jamil Saghir, *Energy Services for the Millennium Development Goals* (International Bank for Reconstruction and Development/World Bank and United Nations Development Programme, 2005) (http://www.unmillenniumproject.org/documents/MP_Energy_Low_Res.pdf).

15. John P. Holdren and Kirk R. Smith, "Energy, the environment, and health," in *World Energy Assessment: Energy and the Challenge of Sustainability,* ed. T. Kjellstrom et al. (United Nations Development Programme, 2000).

16. National Research Council (NRC), *Hidden Costs of Energy: Unpriced Consequences of Energy Production and Use* (Washington, D. C. : The National Academies Press, 2009).

17. EIA, *An Updated Annual Energy Outlook 2009 Reference Case,* tables A2 and A18.

18. Energy Information Administration (EIA), *Annual Energy Outlook 2008* (http://www.eia.doe.gov/oiaf/archive/aeo08/index.html), table A18.

19. Ailis Aaron Wolf, "New study: Coal ash water contamination much worse than previously thought," 2010 (http://www.environmentalintegrity.org/news_reports/documents/082610EIPEJSCBCcoalashwaterpollutionnewsrelease.pdf).

20. National Research Council, *Hidden Costs of Energy.*

21. Ibid.

22. A spill of coal fly ash slurry spill occurred at the Tennessee Valley Authority's Kingston Fossil Plant, in Kingston, Roane County, Tennessee, on December 22, 2008, when an earthen

dike broke at a 40-acre waste retention pond. More than a billion gallons of coal fly ash slurry were released. The coal-fired power plant uses three ponds to store fly ash, a by-product of coal combustion, in wet form. The slurry (a mixture of fly ash and water) traveled downhill, covering up to 400 acres of the surrounding land, damaging homes and flowing into the Emory River and the Clinch River (tributaries of the Tennessee River). It was the largest spill of fly ash slurry in US history.

23. In 2007 coal provided about 41 percent of the world's 18,930 terawatt-hours of electricity generation, natural gas 20.1 percent, hydroelectric 16 percent, nuclear 14.8 percent, oil 5.8 percent, and other renewable sources 2.3 percent. Every kilowatt-hourh of coal generation has 19.14 cents in negative externalities; every kilowatt-hourof natural gas and oil 12 cents; nuclear 11.1 cents; hydroelectric 5 cents; other renewable sources 3 cents. Weighing these according to the percentage of the world's supply, every kilowatt-hourof electricity has about 13.46 cents of negative externalities. Multiplying 18,930 terawatt-hours by 13.46 cents yields $2,547,978,000,000. See International Energy Agency, *World Energy Outlook 2008*; Sovacool, *The Dirty Energy Dilemma*.

24. D. G. Hawkins, D. A. Lashof, and R. H. Williams, "What to do about coal," *Scientific American* 195, 2006, no. 3: 68–75; R. Socolow and S. W. Pacala, "A plan to keep carbon in check," *Scientific American* 195, 2006, no. 3: 50–57.

25. Sovacool, *The Dirty Energy Dilemma*: 221.

26. Philip Fairey, "The unruly power grid," *IEEE Spectrum* 41 (8), 2004: 22–27.

27. Ibid.

28. US-Canada Power System Outage Task Force, *Final Report on the August 14, 2003 Blackout in the United States and Canada: Causes and Recommendations*, 2004 (https://reports.energy.gov/BlackoutFinal-Web.pdf).

29. National Research Council Board on Energy and Environmental Systems, *What You Need to Know About Energy*, 2008 (http://www.nap.edu/catalog.php?record_id=12204).

30. Amory B. Lovins, "Energy myth nine—Energy efficiency improvements have already maximized their potential," in *Energy and American Society,* ed. B. Sovacool and M. Brown (Springer, 2007).

31. http://www.imfmetal.org/index. cfm?c=22697&l=2

32. Energy Information Administration (EIA), *Annual Energy Outlook 2009* (http://www.eia.doe.gov/oiaf/archive/aeo09/index.html), table A8.

33. Jon Wellinghoff and David L. Morenoff, "Recognizing the importance of demand response: The second half of the wholesale electric market function," *Energy Law Journal* 28, 2007, no. 2: 393.

34. Lawrence Agbemabiese, Kofi Berko, and Peter du Pont, "Air conditioning in the tropics: Cool comfort or cultural conditioning?" presented at 1996 ACEEE Summer Study on Energy Efficiency in Buildings.

35. W. M. Shields, The Automobile as an Open to Closed Technological System: Theory and Practice in the Study of Technological Systems, Ph.D. dissertation, Virginia Polytechnic Institute and State University, 2007.

36. Benjamin K. Sovacool, "Early modes of transport in the US: Lessons for modern energy policymakers," *Policy & Society* 27, 2009, no. 4: 411–427.

37. http://www.fhwa.dot.gov/ohim/tvtw/tvtpage.cfm

38. The National Railroad Passenger Corporation, also known as Amtrak, began operation in 1971. Amtrak's revenue-passenger miles increased at an average annual rate of 2.9 percent from 1971 to 2006, rising to 5.4 billion passenger-miles in 2005. In 2005, passenger-miles of commuter rail reached about 9.5 billion, and rail transit passenger-miles about 16 billion (Stacy C. Davis, Susan W. Diegel, and Robert G. Boundy, *Transportation Data Book: Edition 27*, Oak Ridge National Laboratory Report ORNL-6081, 2008, tables 9.10–9.12). In total, these three rail transportation modes represented 30.9 billion passenger-miles in 2005. Vehicle-miles per capita reached 10,082 in 2005 (ibid., table 8.2). Calculated on the basis of a US population of 296 million in 2005, this amounts to 3.2 trillion miles. Thus, the US total passenger miles on Amtrak, commuter rail, and rail transit represent less than 1 percent of the total vehicle-miles traveled by US passengers in 2005.

39. EIA, *Annual Energy Outlook 2009*, tables A2 and A17.

40. Daniel Sperling and Deborah Gordon, *Two Billion Cars: Driving Toward Sustainability* (Oxford University Press, 2009).

41. Domic Wilson and Roopa Purushothaman, *Dreaming with BRICs: The path to 2050*, Goldman Sachs Global Economics Paper 99, 2003.

42. Sperling and Gordon, *Two Billion Cars: Driving Toward Sustainability*.

43. IEA, *World Energy Outlook 2008*.

44. Sovacool, "Early modes of transport in the US."

45. Bureau of Transportation Statistics, *National Transportation Statistics 2007* (http://www.bts.gov/publications/national_transportation_statistics/), table 1-32.

46. Davis et al., *Transportation Data Book: Edition 27*, table 8.5; US Census Bureau, "Households, by type: 1940 to present" (www.census.gov/population/www/socdemo/hh-fam.html).

47. Bureau of Transportation Statistics, *National Transportation Statistics 2007*, table 1-32.

48. David Schrank and Tim Lomax, *The 2007 Urban Mobility Report* (Texas Transportation Institute, 2007) (http://mobility.tamu.edu).

49. "Jam tomorrow, tomorrow, and tomorrow," *The Economist*, August 28, 2010: 25.

50. P. Hawken, *The ecology of commerce: A declaration of sustainability* (HarperCollins, 1993).

51. EIA, *Light-Duty Diesel Vehicles: Efficiency and Emissions Attributes and Market Issues* (http://www.eia.doe.gov/oiaf/servicerpt/lightduty/chapter2.html).

52. Chen Gang, "Energy efficiency: High politics in China," presented at Conference on Energy Efficiency, Singapore, 2008.

53. Benjamin K. Sovacool, "A transition to plug-in hybrid electric vehicles (PHEVs): Why public health professionals must care," *Journal of Epidemiology and Community Health* 64, 2010, no. 3: 185–187.

54. Julio Godoy, "Auto Emissions Killing Thousands," Common Dreams news release, June 3, 2004 (http://www.commondreams.org/headlines04/0603-08.htm).

55. Ibid.

56. The government website http://www.fueleconomy.gov explains how a gallon of gasoline, which weighs about 6.3 pounds, can produce 20 pounds of carbon dioxide: most of the weight of the CO_2 comes not from the gasoline but from the oxygen in the air.

57. EIA, *An Updated Annual Energy Outlook 2009 Reference Case.*

58. Davis et al., *Transportation Data Book: Edition 27*, table 4.6.

59. Robert Baer, "The fall of the House of Saud," *Atlantic Monthly*, May 2003: 34–48.

60. IEA, *World Energy Outlook 2008.*

61. Benjamin K. Sovacool, "Sound climate, energy, and transport policy for a carbon constrained world," *Policy & Society* 27, 2009, no. 4: 273–283.

62. Luis Echávarri and Yuri Sokolov, *Uranium Resources: Plenty to Sustain Growth of Nuclear Power* (Vienna: Nuclear Energy Agency and IAEA, 2006): 12.

63. EIA, *Short-Term Energy Outlook, September 2006* (http://www.eia.doe.gov/pub/forecasting/steo/oldsteos/sep06.pdf).

64. Herbert Girardet and Miguel Mendonca, *A Renewable World: Energy, Ecology, Equality* (Green Books, 2009): 187.

65. Albert Howard, *An Agricultural Testament* (Oxford University Press, 1943).

66. Roy A. Rappaport, "The flow of energy in an agricultural society," *Scientific American* 25, 1971: 116–132.

67. A person's ecological footprint is the land needed to supply his or her needs relating to food, housing, energy, transport, and consumer goods and services. L. Walker and W. Rees, "Urban density and ecological footprints: An analysis of Canadian households," in *Eco-City Dimensions: Healthy Communities, Healthy Planet*, ed. M. Roseland (New Society Publishers, 1997).

68. Miranda A. Schreurs, "Divergent paths: Environmental policy in Germany, the US, and Japan," *Environment* 45, 2003, no. 8: 9–17.

69. Vaclav Smil, "Energy in the twentieth century: Resources, conversions, costs, uses, and consequences," *Annual Review of Energy,* 21–51.

70. Girardet and Mendonca, *A Renewable World: Energy, Ecology, Equality*, 188.

71. Richard B. Wilk, "Culture and energy consumption," in *Energy: Science, Policy, and the Pursuit of Sustainability*, ed. R. Bent et al. (Island, 2002): 112.

72. Shahbaz Mushtaq, Tek Narayan Maraseni, Jerry Maroulis, and Mohsin Hafeez, "Energy and water tradeoffs in enhancing food security: A selective international assessment," *Energy Policy* 37, 2009, no. 9: 3635–3644.

73. Michael Pollan, *The Omnivore's Dilemma: A Natural History of Four Meals* (Penguin, 2006): 46.

74. Claudia H. Deutsch, "Trying to connect the dinner plate to climate change," *New York Times*, August 29, 2007.

75. Noam Mohr, *A New Global Warming Strategy: How Environmentalists are Overlooking Vegetarianism as the Most Effective Tool Against Climate Change in Our Lifetimes* (Earthsave International, August, 2005).

76. Anthony J. McMichael, John W, Powles, Colin D, Butler, and Ricardo Uauy, "Food, livestock production, energy, climate change, and health," *Lancet* 370, 2007: 1253–1263.

77. Ibid.

78. Elke Stehfest et al., "Climate benefits of changing diet," *Climatic Change* 95 (1–2), 2009: 83–102.

79. McMichael et al., "Food, livestock production, energy, climate change, and health."

80. Energy efficiency in this sense is defined as the percentage of fossil fuel input energy retrieved as edible energy. The study noted that estimated energy efficiency of protein in animal products varies from 0.5 percent for lamb to five percent for chicken to three percent for beef. These efficiencies are compared to 60 percent for tomatoes, 170 percent for oranges and potatoes, and 500 percent for oats. See Gidon Eshel and Pamela A. Martin, "Diet, energy, and global warming," *Earth Interactions* 10, 2006: 1–17.

81. McMichael et al., "Food, livestock production, energy, climate change, and health."

82. Mohr, *A New Global Warming Strategy*.

83. According to the latest data presented in figure 2.1 above, agriculture-related emissions are responsible for 13.5 percent of the world's GHG emissions, transportation only 13.1 percent. Livestock emissions are responsible for 70 percent of agricultural emissions; passenger vehicles are responsible for only 60–65 percent of transport-related emissions (the rest are composed of medium and heavy duty trucks, aircraft, rail, buses, and marine vessels).

84. David Tilman, Kenneth Cassman, Pamela Matson, Rosamond Naylor, and Stephen Polasky, "Agricultural sustainability and intensive production practices," *Nature* 418 (August 8, 2002): 671–677.

85. McMichael et al., "Food, livestock production, energy, climate change, and health."

86. Tilman et al., "Agricultural sustainability and intensive production practices," p. 671.

87. Ibid.: 672–677.

88. Girardet and Mendonca, *A Renewable World*: 188.

89. Charlotte Streck, "Forests, carbon markets, and avoided deforestation: Legal implications," *Carbon & Climate Law Review* 2, 2008, no. 3: 239–247.

90. Charlotte Streck, Lucio Pedroni, Manuel Porrua, and Michael Dutschke, "Creating incentives for avoiding further deforestation: The nested approach," in *Climate Change and Forests: Emerging Policy and Market Opportunities,* ed. C. Streck et al. (Brookings Institution Press, 2008).

91. Doug Boucher, *Money for Nothing? Principles and Rules for REDD and Their Implications for Protected Areas* (Tropical Forest and Climate Initiative of the Union of Concerned Sciences, 2009): 2.

92. R. A. Houghton, *Emissions (and Sinks) of Carbon from Land-Use Change* (World Resources Institute and the Woods Hole Research Center, 2003).

93. Quoted from Article 6 of the Copenhagen Accord (http://unfccc. int/resource/docs/2009/cop15/eng/l07.pdf).

94. United Nations Food and Agricultural Organization, *Global Forest Resource Assessment*, 2006.

95. Doug Boucher, *Out of the Woods: A Realistic Role for Tropical Forests in Curbing Global Warming* (Union of Concerned Scientists, 2008).

96. Union of Concerned Scientists, *Tropical Deforestation and Global Warming: A Solution* (UCS Tropical Forest and Climate Initiative, 2009).

97. Streck et al., "Creating incentives for avoiding further deforestation."

98. "Paying to save trees: Last gasp for the forest," *The Economist*, September 26, 2009: 81.

99. Girardet and Mendonca, *A Renewable World*: 55.

100. Manuel F. Montes and Francisco A. Magno, "Trade and environmental diplomacy: Strategic options for ASEAN," *Pacific Affairs* 70, 1997, no. 3: 351–372.

101. *Second ASEAN State of the Environment Report*, 2000.

102. Indonesian Working Group on Underlying Causes of Deforestation and Forest Degradation, *The Underlying Causes of Deforestation and Forest Degradation*, 1999.

103. James Gustave Speth, *The Bridge at the End of the World: Capitalism, the Environment, and Crossing from Crisis to Sustainability* (Yale University Press, 2008): 31.

104. United Nations Environmental Program, *Last Stand of the Orangutan: Illegal Logging, Fire and Palm Oil in Indonesia's National Parks*, 2007 (http://www.unep-wcmc.org/resources/publications/LastStand.htm).

105. Ibid.

106. Barry W. Brook, Navjot S. Sodhi, and Peter K. L. Ng, "Catastrophic extinctions follow deforestation in Singapore," *Nature* 424, 2003, July 24: 420–426.

107. Rachel Smolker, Brian Tokar, Anne Petermann, and Eva Hernandez, "Devastated lands, displaced peoples," *Pacific Ecologist*, summer 2009: 36–39.

108. Reinhard Madlener, Carmenza Robledo, Bart Muys, and Javier Freja, "A sustainability framework for enhancing the long-term success of LULUCF projects," *Climatic Change* 75, 2006, no. 1–2: 241–271.

109. B. Holvoet and B. Muys, "Sustainable forest management worldwide: A comparative assessment of standards," *International Forestry Review* 6, 2004, no. 2: 99–122.

110. Fresh water is water that contains less than 1,000 milligrams per liter of dissolved solids; salt water contains more than 1,000 mg/l, although generally more than 500 mg/l of dissolved solids is undesirable for drinking and for many industrial uses.

111. The energy of the sun evaporates water into the atmosphere from oceans and land surfaces. Evaporation is the change of liquid water to a vapor. For an excellent introduction see Jill Boberg, *Liquid Assets: How Demographic Changes and Water Management Policies Affect Freshwater Resources* (Santa Monica: Rand Corporation, 2005).

112. US Geological Survey, *Ground-Water Depletion Across the Nation*, Fact Sheet 103-03, 2003 (http://pubs. usgs.gov/fs/fs-103-03/).

113. Amory B. Lovins, L. Hunter Lovins, and Paul Hawken, "A road map for natural capitalism," *Harvard Business Review*, May/June 1999: 146, 152.

114. Jared Diamond, "What's your consumption factor?" *Pacific Ecologist* 16, 2008, winter: 23–24.

115. Ministry of Environment, *The World in Transition and Japan's Efforts to Establish a Sound Material-Cycle Society* (Tokyo: Government of Japan, 2008): 6.

116. Diana Bocco, "How much garbage does a person create each year?" (http://www.wisegeek.com/how-much-garbage-does-a-person-create-in-one-year.htm).

117. Jennifer Clapp, *Distancing of Waste: Overconsumption in a Global Economy*, Working Paper 01/10, Trent International Political Economy Centre (http://www-rohan.sdsu.edu/faculty/dunnweb/rprnts.2005.10.10Clapp.pdf).

118. Rachel Oliver, "Landfills," CNN International, October 2008.

119. Ibid.

120. Lindsey Hoshaw, "Afloat in the ocean, expanding islands of trash," *New York Times*, November 10, 2009 (http://earthsky.org/water/kara-lavender-law-ocean-garbage-patch-in-atlantic-too).

121. Ocean Conservancy, *A Rising Tide of Ocean Debris and What We Can Do About It* (Ocean Conservancy, 2009).

122. Oliver, "Landfills."

123. Brian T. Horowitz, "How green is your cell phone?" Fox News Online, March 16, 2009.

124. Noah Sachs, "Planning the funeral at the birth: Extended producer responsibility in the European Union and the US," *Harvard Environmental Law Review* 30, 2006, no. 1: 51–98.

125. Ibid.: 51–98.

126. World Health Organization and United Nations Environment Program, "Global Water Conservation and Use Survey," 2010.

127. Mark Schaefer, "Water technologies and the environment: Ramping up by scaling down," *Technology in Society* 30, 2008: 415–422.

128. Carl Pope, "The state of nature: Our roof is caving in," *Foreign Policy* 67, 2005: 66–73.

129. Schaefer, "Water technologies and the environment."

130. Ibid.

131. Kenneth D. Frederick, "America's water supply: Status and prospects for the future," *Consequences* 1, 1995, no. 1: 34–41.

132. US General Accounting Office, *Fresh Water Supply: States' Views of How Federal Agencies Could Help them Meet the Challenges of Expected Shortages*, GAO-03-514, 2003 (http://www.gao.gov/new.items/d03514.pdf).

133. Will Dunham, "Fivefold dust increase chokes the West," Reuters, February 24, 2008 (http://www.reuters.com/article/environmentNews/idUSN2259224520080224).

134. These data are from a site visit, on August 7, 2010, to Regional Environmental Awareness Cameron Highlands, or REACH, an NGO located in the state of Pahang near Bringchang in the Cameron Highlands.

135. Peter H. Gleick, "Water use," *Annual Review of Environment and Resources* 28, 2003: 275–314.

136. WHO and the United Nations Environment Program, *Global Water Conservation and Use Survey*.

137. Josiah Smith and Motomu Ibakari, "Data mining and spatiotemporal analysis of extreme precipitation and Northern Great Plains drought," presented at 2007 meeting of Geological Society of America, Denver.

138. Benjamin K. Sovacool and Kelly E. Sovacool, "Preventing national electricity-water crisis areas in the US," *Columbia Journal of Environmental Law* 34, 2009, no. 2: 333–393.

139. Katharine Hayhoe et al., "Emissions pathways, climate change, and impacts on California," *Proceedings of the National Academy of Sciences* 101, 2004, 34: 12422–12424 (http://www.pnas.org/content/101/34/12422.abstract).

140. Sovacool, "Preventing national electricity-water crisis areas in the US."

141. Ibid.

142. Ibid.

143. Jeffrey S. Levinton and John R. Waldman, *The Hudson River Estuary* (Cambridge University Press, 2006): 198–99.

144. Sovacool, "Preventing national electricity-water crisis areas in the US."

145. Ibid.

146. Associated Press, "Leaks spotlight aging nuclear plants," February 1, 2010.

147. Sovacool, "Preventing national electricity-water crisis areas in the US."

148. US Department of Energy, *Energy Demand on Water Resources*, 2006 (http://www.sandia.gov).

149. Benjamin K. Sovacool, "Running on empty: The electricity-water nexus and the US electric utility sector," *Energy Law Journal* 30, 2009, no. 11: 11–51.

150. National Academy of Sciences, *Advancing the Science of Climate Change* (National Academies Press, 2010): 27; National Academy of Sciences, *Limiting the Magnitude of Future Climate Change* (National Academies Press, 2010): 24.

151. IEA, *World Energy Outlook 2008*.

152. *Human Development Report 2007/2008* (United Nations, 2008).

153. *Inventory of US Greenhouse Gas Emissions and Sinks: 1990–2005*, US Environmental Protection Agency Report 430-R-07-002, 2007 (http://www.epa.gov/climatechange/emissions/ usinventoryreport.htm.

154. J. L. Nazarenko et al., "Earth's Energy Imbalance: Confirmation and Implications," *Science* 308, 2005: 1431–1435; James Hansen and Makiko Sato, "Greenhouse gas growth rates," *Proceedings of the National Academys of Sciences* 101, 2004: 16,109–16,111.

155. Nicholas Stern, *The Economics of Climate Change: The Stern Review* (Cambridge University Press, 2007); Intergovernmental Panel on Climate Change, "Summary for policy-makers," in *Climate Change: 2007: The Physical Science Basis. Contribution of Working Group I to the Fourth Assessment Report of the Intergovernmental Panel on Climate Change*, ed. S. Solomon et al. (Cambridge University Press, 2007).

156. William Collins, Robert Colman, James Haywood, Martin R. Manning and Philip Mote, "The physical science behind climate change," *Scientific American* (August 2007): 64–73.

157. The US carbon footprint is derived from table A18 of the EIA's *Annual Energy Outlook 2008*. The global carbon footprint is derived from the EIA's *Emissions of Greenhouse Gases Report*, DOE/EIA-0573 (http://www.eia.doe.gov/oiaf/1605/ggrpt/), and is computed by dividing global emissions (28.1 billion metric tons of carbon dioxide) by the world population (6.4 billion) and converting from carbon dioxide to carbon by dividing by 3.67. See also Marilyn

A. Brown, Frank Southworth, and Andrea Sarzynski, "The geography of metropolitan carbon footprints," *Policy and Society* 27, 2009: 285–304.

158. Council on Competitiveness, "Competitiveness index: Where America stands," 2007 (http://www.compete.org/images/uploads/File/PDF%20Files/Competitiveness_Index_Where _America_Stands_March_2007.pdf).

159. Severin Carrell, "Ocean acidification rates accelerating," *The Hindu*, December 11, 2009.

160. Andy Haines, Kirk R. Smith, Dennis Anderson, Paul R. Epstein, Anthony J. McMichael, Ian Roberts, Paul Wilkinson, James Woodcock, and Jeremy Woods, "Policies for accelerating access to clean energy, improving health, advancing development, and mitigating climate change," *Lancet* 370, 2007: 1264–1281.

161. Anne E. Prouty, "The clean development mechanisms and its implications for climate justice," *Columbia Journal of Environmental Law* 34, 2009, no. 2: 513–540.

162. IPCC, "Summary for policymakers," in *Climate Change: 2007: The Physical Science Basis*, Contribution of Working Group I to the Fourth Assessment Report of the Intergovernmental Panel on Climate Change.

163. United Nations Development Programme, *Energy After Rio: Prospects and Challenges* (Geneva: United Nations, 1997).

164. Klaus S. Lackner and Jeffrey D. Sachs, "A robust strategy for sustainable energy," *Brookings Papers on Economic Activity* 2, 2004: 215–248.

165. Cynthia Rosenzweig et al., "Attributing physical and biological impacts to anthropogenic climate change," *Nature* 453 (May 15), 2008: 353–357.

166. Edward Josberger, William Bidlake, Rod March, and Shad O'Neel, "Fifty year record of glacier change," *US Geological Survey Fact Sheet 2009–3046* (July 6, 2009); E. G. Josberger, W. R. Bidlake, R. S. March, and B. W. Kennedy, "Glacier mass-balance fluctuations in the Pacific Northwest and Alaska," *USA: Annals of Glaciology* 46, 2007: 291–296.

167. Eileen Claussen and Janet Peace, "Energy myth twelve—Climate policy will bankrupt the US economy," in *Energy and American Society—Thirteen Myths,* ed. B. Sovacool and M. Brown (Springer, 2007).

168. Stern, *The Economics of Climate Change*; IPCC, "Summary for policymakers."

169. William D. Nordhaus, "A review of the Stern Review on the Economics of Climate Change," *Journal of Economic Literature* 55 (September 2007): 686–702.

170. B. Sudhakara Reddy and Gaudenz B. Assenza, "The great climate debate," *Energy Policy* 37, 2009: 2997–3008.

171. Economics of Climate Adaptation Working Group, *Shaping Climate-Resilient Development: A Framework for Decision-Making* (Climate Works Foundation, 2009).

172. Asian Development Bank "The economics of climate change in Southeast Asia: a regional review" (Manila: ADB, 2009).

173. Arief Anshory Yusuf and Herminia A. Francisco, *Climate Change Vulnerability Mapping for Southeast Asia* (Singapore: IDRC, 2009).

174. Kay Weir, "Don't cry for Kiribati, Tuvalu, Marshall Islands, parts of Papua New Guinea, the Caribbean, Bangladesh, Africa . . . ," *Pacific Ecologist*, winter 2008: 2.

175. Alex Smith, "Climate refugees in Maldives buy land," press release, Tree Hugger, November 16, 2008.

176. Bianca Jagger, "The threat of a global climate disaster is no longer up for debate," testimony before House Select Committee for Energy Independence and Global Warming, March 6, 2008.

177. Anne E. Prouty, "The clean development mechanisms and its implications for climate justice," *Columbia Journal of Environmental Law* 34, 2009, no. 2: 513–540.

178. Ibid.

179. Richard A. Matthew, "Climate change and human security," in Joseph F. C. DiMento and Pamela Doughman (eds.), *Climate Change: What It Means for Us, Our Children, and Our Grandchildren* (MIT Press, 2007).

180. "Climate change: No hiding place," *The Economist*, January 9, 2010: 70–71.

181. Joseph F. C. DiMento and Pamela Doughman, "Making climate change understandable," in *Climate Change*, ed. DiMento and Doughman.

182. James Hansen, Reto Ruedy, Makiko Sato, and Ken Lo, "If it's that warm, how come it's so damned cold?" (http://www.realclimate.org/index.php/archives/2010/01/2009-temperatures-by-jim-hansen/); NASA, "Research news," January 21, 2010 (http://www.giss.nasa.gov/research/news/20100121/).

183. Naomi Oreskes, "The scientific consensus on climate change: How do we know we're not wrong?" in *Climate Change*, ed. DiMento and Doughman.

184. Paul N. Edwards, *A Vast Machine: Computer Models, Climate Data, and the Politics of Global Warming* (MIT Press, 2010): 439.

185. A. P. Sokolov, P. H. Stone, C. E. Forest, R. Prinn, M. C. Sarofim, M. Webster, S. Paltsev, C. A. Schlosser, D. Kicklighter, S. Dutkiewicz, J. Reilly, C. Wang, B. Felzer, J. M. Melillo, and H. D. Jacoby, "Probabilistic forecast for twenty-first-century climate based on uncertainties in emissions (without policy) and climate parameters," *Journal of Climate* 22, 2009, 19: 5175–5204.

186. David Chandler, "Revised MIT climate model sounds alarm," *Tech Talk* 53 (26), 2009: 5.

187. See James Gustave Speth, "Time for civic unreasonableness," *Environment: Yale*, spring 2008: 2–3; DiMento and Doughman, "Making climate change understandable"; Hugh Doulton and Katrina Brown, "Ten years to prevent catastrophe? Discourses of climate change and international development in the UK press," *Global Environmental Change* 19, 2009: 191–202.

188. Virgilio M. Viana, Mariano C. Cenamo, Mariana N. Pavan, Gabriel C. Carrero, and Matthew D. Quinlan, "Railroads in the Amazon: A key strategy for reducing deforestation," *Carbon & Climate Law Review* 2, 2008, no. 3: 292–299.

189. Adam Simpson, "The environment-energy-security nexus: Critical analysis of an energy 'love triangle' in Southeast Asia," *Third World Quarterly* 28, 2007, no. 3: 539–554.

190. Benjamin K. Sovacool, "Solving the oil independence problem," *Energy Policy* 35, 2007, no. 11: 5512.

191. MIT Energy Initiative, *The Future of Natural Gas: An Interdisciplinary MIT Study*, Interim Report, 2010; Benjamin K. Sovacool, "The problem with the 'portfolio approach' in American energy policy," *Policy Sciences* 41, 2008, no. 3: 245–261.

192. According to data compiled by the World Resources Institute, these ten entities emitted 27.1 billion tons of carbon dioxide equivalent, relative to a world total of 33.3 billion tons in 2000.

Chapter 3

1. Kurt Vonnegut, *Slaughterhouse-Five, or the Children's Crusade; A Duty-Dance with Death* (Doubleday, 1966).

2. An assessment of basic science R&D needs to improve the performance of climate-change technologies was sponsored by the US Climate Change Technology Program. The results are published in Marilyn A. Brown, Matt Antes, Charlotte Franchuk, Burton H. Koske, Gordon Michaels, and Joan Pellegrino, *Results of a Technical Review of the US Climate Change Technology Program's R&D Portfolio*, ORNL-6976 (Oak Ridge National Laboratory, 2006). The results are also discussed in chapter 9 of the US Climate Change Technology Program, *Strategic Plan*, DOE/PI-0005 (Department of Energy, 2006).

3. Daniel M. Kammen and Gregory F. Nemet, "Energy myth eleven—Energy R&D investment takes decades to reach the market," in *Energy and American Society—Thirteen Myths*, ed. B. Sovacool and M. Brown (Springer, 2007); Stephen Alberth, "Forecasting technology cost via the experience curve—Myth or magic?" *Technological Forecasting & Social Change* 75, 2008: 952–983.

4. For residential electricity rates in the United States, see http://tonto.eia.doe.gov/state/ state_energy_rankings.cfm?keyid=18&orderid=1; Michael Rufo and Fred Coito, *California's Secret Energy Surplus: The Potential for Energy Efficiency* (Energy Foundation, 2002); New York State Energy Research and Development Authority, *Energy Efficiency and Renewable Energy Resource Development Potential in New York State, Final Report*, 2003.

5. Cedric Philibert, *Technology Penetration and Capital Stock Turnover: Lessons from IEA Scenario Analysis* (OECD and International Energy Agency, 2008); International Energy Agency, *World Energy Outlook 2008*: 73–75.

6. Marilyn A. Brown and Sharon (Jess) Chandler, "Governing confusion: How statutes, fiscal policy, and regulations impede clean energy technologies," *Stanford Law and Policy Review* 19, 2008, no. 3: 472–509; G. Unruh, "Escaping carbon lock-in," *Energy Policy* 30, 2002: 317–325.

7. CCTP, *Strategic Plan*.

8. Similar categories of technologies have been suggested in previous reports. One typology is described in Y. Kaya, "Impact of carbon dioxide emission control on gnp growth: Interpretation of proposed scenarios," presented to IPCC Energy and Industry Subgroup of Response Strategies Working Group, Paris, 1990. A second typology is offered in National Laboratory Directors, *Technology Opportunities to Reduce US Greenhouse Gas Emissions*, 1997 (http:// www.ornl.gov/~webworks/cppr/y2003/rpt/110512.pdf).

9. Much of the material for this chapter is drawn from the Climate Change Committee on Science and Technology Initiative (CCCSTI), *Strategies for the Commercialization and*

Deployment of Greenhouse Gas-Intensity Reducing Technologies and Practices (US Department of Energy, 2009, DOE/PI-0007).

10. Marilyn A. Brown and Benjamin K. Sovacool, "A Source of Energy Hiding in Plain Sight," *Yale Global Online*, February 18, 2009.

11. Marilyn A. Brown et al., *Energy Efficiency in Appalachia* (Appalachian Regional Council, revised May 2009) (http://www.arc.gov/images/energy/index.html); Marilyn A. Brown, Etan Gumerman, Xiaojing Sun, Youngsun Baek, Joy Wang, Rodrigo Cortes, and Diran Soumonni, *Energy Efficiency in the South* (Southeast Energy Efficiency Alliance, 2010) (http://www.seealliance.org/programs/se-efficiency-study.php).

12. P. Enkvist, T. Nauclér, and J. Rosander, "A cost curve for greenhouse gas reduction," *McKinsey Quarterly*, 2007 (http://www.mckinseyquarterly.com/A_cost_curve_for_greenhouse_gas_reduction_1911).

13. H. C. Granade, J. Creyt, A. Kerkach, P. Farese, S. Nyquist, and K. Ostrowski, *Unlocking Energy Efficiency in the United States* (McKinsey & Company, 2009) (http://www.mckinsey.com/clientservice/electricpowernaturalgas/downloads/US_energy_efficiency_full_report.pdf).

14. Marilyn A. Brown, Mark D. Levine, Walter Short, and Jonathan G. Koomey, "Scenarios for a clean energy future," *Energy Policy* 29, 2001: 1179–1196.

15. Steven Nadel, Anna Shipley, and R. Neal Elliott, "The technical, economic and achievable potential for energy-efficiency in the US—A meta-analysis of recent studies," presented at ACEEE Summer Study on Energy Efficiency in Buildings, 2004 (http://www.aceee.org/conf/04ss/rnemeta.pdf).

16. Leadership Group, *National Action Plan for Energy Efficiency* (Department of Energy and US Environmental Protection Agency, 2006) (www.epa.gov/eeactionplan); National Academy of Sciences, *Real Prospects for Energy Efficiency in the United States* (National Academies Press, 2009).

17. Vlado Vivoda, "Evaluating energy security in the Asia-Pacific Region: A novel methodological approach," *Energy Policy* 38, 2010, no. 9: 5258–5263.

18. Council on Competitiveness, *Competitiveness Index: Where America Stands*, 2006.

19. International Energy Agency (IEA), *World Energy Outlook 2008* (Paris, France: IEA, 2008): 506.

20. Intergovernmental Panel on Climate Change (IPCC), *Climate Change 2007: Mitigation of Climate Change* (Cambridge University Press, 2007): 325.

21. Energy Information Administration (EIA), *International Energy Outlook*, DOE/EIA-0484(2006) (DOE, 2006).

22. Stanton W. Hadley, *Impact of Plug-in Hybrid Vehicles on the Electric Grid*, ORNL/TM-2006/554 (Oak Ridge National Laboratory, 2008).

23. National Academies, *Transitions to Alternative Transportation Technologies- A Focus on Hydrogen* (National Academies, 2009).

24. IEA, *World Energy Outlook 2008*, 506.

25. IPCC, *Climate Change 2007: Mitigation of Climate Change*, 389.

26. Intergovernmental Panel on Climate Change, *Working Group III: Fourth Assessment Report*, 2007 (http://www.mnp.nl/ipcc/pages_media/AR4-chapters.html): chapter 6.

27. Ibid.: 11–12; Meredith Gray and Jay Zarnikau. "Getting to Zero: Green Building and Net Zero Energy Homes."

28. Jinfeng Xu, "Local implementation of building energy policies in China's Jiangsu Province," presented at ACEEE Summer Study on Energy Efficiency in Buildings, Pacific Grove, California, 2010.

29. *Annual Energy Outlook 2006*, Energy Information Administration Report DOE/EIA-0383, 2006; *Annual Energy Outlook 2007 with Projections to 2030—Overview*, Energy Information Administration Report DOE/EIA-0383, 2007 (http://www.eia.doe.gov/oiaf/archive/aeo07/index.html).

30. http://www.energystar.gov/index. cfm?fuseaction=find_a_product. showProductGroup& pgw_code=LB

31. M. G. Craford, "High power LEDs for solid state lighting: Status, trends, and challenges," *Journal of Light and Visual Environment* 32, 2008, no. 2: 58–62.

32. Anna Shipley, Anne Hampson, Bruce Hedman, Patti Garland, and Paul Bautista, *Combined Heat and Power: Effective Energy Solutions for a Sustainable Future* (Oak Ridge National Laboratory, 2008), ORNL/TM-2008/224.

33. http://www.energybiz.com/article/10/09/generating-electricity-excess-heat

34. John Cymbalsky, "The outlook for buildings energy efficiency and electricity use," presented to National Academies Board on Energy and Environmental Systems, Washington, 2010.

35. H. Akbari, P. Berdahl, R. Levinson, R. Wiel, A. Desjarlais, W. Miller, N. Jenkins, A. Rosenfeld, and C. Scruton, "Cool colored materials for roofs," presented at ACEEE Summer Study on Energy Efficiency in Buildings, Pacific Grove, California, 2004.

36. G. Tulley, "Bringing big builders to efficiency," *Home Energy Magazine*, March/April 2000 (http://homeenergy.org/archive/hem.dis.anl.gov/eehem/00/000307.html); H. Wee, "Buildings with built-in energy savings," *BusinessWeek Online*, August 27, 2001 (http://www.businessweek.com/magazine/content/01_35/b3746614.htm).

37. Wee, "Buildings with built-in energy savings."

38. Nicholas B. Rajkovich, William C. Miller, and Anna M. LaRue, "Zeroing in on zero net energy," in *Energy Sustainability and the Environment*, ed. F. Sioshansi (Elsevier, 2011); National Association of Home Builders, *Zero Energy Homes* (http://www.toolbase.org/PDF/CaseStudies/ZEH_Brochure-final-screen.pdf).

39. *Basic Research Needs to Assure a Secure Energy Future* (Oak Ridge National Laboratory, 2003) (http://www.sc.doe.gov/bes/reports/files/SEF_rpt.pdf).

40. Ryoichi Komiyama and Chris Marnay, *Japan's Residential Energy Demand Outlook to 2030 Considering Energy Efficiency Standards "Top Runner Approach,"* Lawrence Berkeley National Laboratory Report LBNL-292E, 2008 (http://eetd.lbl.gov/ea/EMS/reports/lbnl-292e.pdf).

41. IEA, *World Energy Outlook 2008*, 506.

42. IPCC, *Climate Change 2007*, 449.

43. Energy Information Administration, *Annual Energy Outlook 2009* (http://www.eia.doe.gov/oiaf/aeo/), table F1.

44. C. Weber, *Energy Use and Flows in the US Economy, 1997–2002*, report prepared for the National Academy of Sciences by Carnegie Mellon University, 2008.

45. Process-related emissions (including both CO_2 and other gases) account for approximately 5 percent of GHG emissions from all sources in the United States (Inventory of US Greenhouse Gas Emissions and Sinks: 1990–2006, Environmental Protection Agency Report 430-R-08-005, 2008 (http://www.epa.gov/climatechange/emissions/us). CO_2 accounted for about 47 percent of these; the remainder was other gases, including methane and nitrous oxide.

46. Melissa Madgett, personal communication, 2008; Bob Gemmer, "Super boiler update," presented at conference of American Boiler Manufacturers Association, 2007 (http://www1 .eere.energy.gov/industry/combustion/pdfs/abma_2007.pdf).

47. Lee Bruno, "Nano-ceramic boosts fuel efficiency," http://cleantech.com/news/3476/ ceramic-nanotechnology-delivers-efficiency).

48. Distillation of chemicals and petroleum is the largest energy-consuming process in industry. (Oak Ridge National Laboratory and BCA, Inc., Materials for Separation Technologies: Energy and Emission Reduction Opportunities (2005) (http://www1.eere.energy.gov/ industry/imf/pdfs/separationsreport.pdf). Energy efficiency of distillation can be improved by roughly 25 percent by introducing heat exchangers along the column. This has been analyzed and developed over about two decades, and could be put into practice now; see E. Jimenez, P. Salamon, R. Rivero, C. Rendon, K. Hoffmann, M. Schaller, and B. Andresen, "Optimization of a diabatic distillation column with sequential heat exchangers," *Industrial & Engineering Chemistry Research* 43, 2004:7566–7571. Side reboilers are one currently used approximation to the fully optimized approach that would probably only be practical for construction of altogether new columns.

49. http://www.steeldynamics.com/index. php?page_id=118

50. Fiber optic sensors can measure paper basis weight to improve wet-end control in papermaking and make paper of a uniform weight and quality. Laser sensors can measure shear strength and bending stiffness by measuring the rate of propagation of ultrasonic shock waves in the paper. The claim is that this device will save the US paper industry approximately $200 million per year in energy costs. (See "Energy-Saving Paper Sensor Passes Major Milestone" at http://www.physorg.com/news4221.html.)

51. E. Worrell, C. Galitsky, and L. Price, *Energy Efficiency Improvement Opportunities for the Cement Industry* (Lawrence Berkeley National Laboratory, 2008).

52. US Department of Energy, *Energy Technology Solutions: Public-Private Partnerships Transforming Industry* (http://www1.eere.energy.gov/industry/bestpractices/pdfs/itp _successes.pdf).

53. Marilyn A. Brown, Rodrigo Cortes, and Matthew Cox, "Reinventing industrial energy use in a resource-constrained world," in *Energy Sustainability and the Environment*, ed. F. Sioshansi (Elsevier, 2011); J. A. Laitner and M. A. Brown, "Emerging industrial innovations to create new energy efficient technologies," in *Proceedings of the Summer Study on Energy Efficiency in Industry* (American Council for an Energy-Efficient Economy, 2005).

54. Paul Hawken, "Natural capitalism," in *The Energy Reader*, ed. L. Nader (Wiley-Blackwell, 2010).

55. William McDonough and Michael Braungart, *Cradle to Cradle: Remaking the Way We Make Things* (Farrar, Straus, and Giroux, 2002).

56. IEA, *World Energy Outlook 2009.*

57. Energy Information Administration (EIA), *Annual Energy Outlook 2005: with Projections to 2025*, Department of Energy Report DOE/EIA-0383, 2005; Environmental Protection Agency, *Inventory of US Greenhouse Gas Emissions and Sinks: 1990–2005.*

58. US Climate Change Technology Program, *Technology Options for the Near and Long Term* (http://www.climatetechnology.gov/library/2005/tech-options/index.htm).

59. National Academy of Sciences, *Electricity from Renewable Resources: Status, Prospects, and Impediments* (National Academies Press, 2009): 7.

60. Nuclear power is more carbon intensive than electricity produced from renewable resources, but less carbon intensive than fossil fuels, emitting a mean of 66 grams of carbon dioxide equivalency per kWh. See Benjamin K. Sovacool, "Valuing the greenhouse gas emissions from nuclear power: A critical survey," *Energy Policy* 36, 2008, no. 8: 2940–2953.

61. National Academy of Sciences, Liquid Transportation Fuels from Coal and Biomass (National Academies Press, 2009): 1.

62. International Energy Agency (IEA), *Key World Energy Statistics 2008* (http://www.iea .org/Textbase/publications/free_new_Desc. asp?PUBS_ID=1199).

63. EIA, *Annual Energy Outlook 2007.*

64. J. A. Ratafia-Brown, L. Manfredo, J. Hoffmann, M. Ramezan, and G. Stiegel, "An environmental assessment of IGCC power systems," presented at Nineteenth Annual Pittsburgh Coal Conference, 2002 (http://www.netl.doe.gov/technologies/coalpower/gasification/pubs/pdf/18.pdf).

65. The first two of these estimates are from *The Future of Coal* (Massachusetts Institute of Technology, 2007) (http://web.mit.edu/coal/). The estimates for ultracritical pulverized coal plants are from E. Rubin and C. Chen, "Cost and performance of fossil fuel power plants with CO_2 capture and storage," *Energy Policy* 35, 2008: 4444–4454.

66. CCTP, *Strategic Plan.*

67. James J. Markowsky, "Office of Fossil Energy CCS R&D Program Overview," presented to National Academies Board on Energy and Environmental Systems, Washington, 2010.

68. E. Worrell, L. Price, and C. Galitsky, *Emerging Energy-Efficient, Technologies in Industry: Case Studies of Selected Technologies*, Lawrence Berkeley National Laboratory Report LBNL-54828, 2004; S. W. Hadley, J. M. MacDonald, M. Ally, J. Tomlinson, M. Simpson, and W. Miller, *Emerging Energy-Efficient Technologies in Buildings: Technology Characterizations for Energy Modeling*, Oak Ridge National Laboratory Report ORNL/TM-2004/63, 2004; Marilyn A. Brown, Matt Cox, and Rodrigo Cortes, "Transforming industrial energy efficiency," *The Bridge*, fall 2010: 22–30.

69. Marilyn A. Brown, "Energy myth one—Today's energy crisis is 'hype,'" in *Energy and American Society—Thirteen Myths*, ed. Sovacool and Brown.

70. National Research Council, Transitions to Alternative Transportation Technologies—A Focus on Hydrogen (National Academies Press, 2008): 1.

71. Ibid.: 19.

72. J. M. Ogden, "Prospects for building a hydrogen energy infrastructure," *Annual Review of Energy and the Environment* 24, 1999: 227–79.

73. *The Greenhouse Gases, Regulated Emissions, and Energy Use in Transportation (GREET) Model, Version 1.7*, Argonne National Laboratory, 2008.

74. J. M. Ogden, "Hydrogen: The fuel of the future?" *Physics Today* 55, 2002, no. 4: 69–74.

75. National Research Council, *Transitions to Alternative Transportation Technologies—A Focus on Hydrogen* (National Academies Press, 2008): 2.

76. DOE Energy Efficiency and Renewable Energy, Hydrogen, Fuel Cells & Infrastructure Technology Program, *Fuel Cells* (http://www1.eere.energy.gov/hydrogenandfuelcells/fuelcells/).

77. John Teresko, "The hydrogen economy: Lift trucks first?" *Industry Week*, July 1, 2007.

78. CCTP, *Strategic Plan*.

79. One KW is equivalent to 1.34 horsepower.

80. NRC, *A Focus on Hydrogen*, 5.

81. DOE Energy Efficiency and Renewable Energy, Hydrogen, Fuel Cells & Infrastructure Technology Program, *Early Markets: Fuel Cells for Backup Power* (http://www1.eere.energy .gov/hydrogenandfuelcells/education/pdfs/early_markets_backup_power.pdf).

82. Renewable Energy Policy Network for the 21st Century, *Renewables Global Status Report: 2010 Update* (http://www.ren21.net/Portals/97/documents/GSR/REN21_GSR_2010 _full_revised%20Sept2010.pdf).

83. *International Energy Outlook 2009*, Energy Information Administration Report DOE/ EIA-0484, 2009, table 8.

84. Ibid.: 10–11.

85. REN21, *Renewables Global Status Report: 2010 Update*; American Wind Energy Association, "US wind energy industry breaks all records, installs nearly 10,000 MW in 2009," news release, January 26, 2010 (www.awea.org/newsroom/releases/01-26-10_AWEA _Q4_and_Year-End_Report_Release.html).

86. B. Snyder and M. J. Kaiser, *Offshore Wind Power in the US: Regulatory Issues and models for Regulation* (LSU Center for Energy Studies, 2009).

87. W. Musial and S. Butterfield, *Future for Offshore Wind Energy in the United States*, National Renewable Energy Laboratory Report NREL/CP-500-36313, 2004 (http://www .nrel.gov/docs/fy04osti/36313.pdf); W. Musial, S. Butterfield, and B. Ram, *Energy from Offshore Wind*, National Renewable Energy Laboratory Report NREL/CP-500-39450, 2006 (http://www.nrel.gov/docs/fy06osti/39450.pdf).

88. "Salazar Signs Agreement with 10 East Coast Governors to Establish Atlantic Offshore Wind Energy Consortium," US Department of the Interior press release, 2010) (http://www .doi.gov/news/pressreleases/Salazar-Signs-Agreement-with-10-East-Coast-Governors-to -Establish-Atlantic-Offshore-Wind-Energy-Consortium.cfm).

89. Renewable Energy Policy Network for the 21st Century, "Renewables Global Status Report: 2010 Update" (http://www.ren21.net/Portals/97/documents/GSR/REN21_GSR_2010 _full_revised%20Sept2010.pdf).

90. http://www.energybiz.com/article/10/09/china-conquers-renewables

91. http://en.wikipedia.org/wiki/Geothermal_power_in_Iceland

92. Western Governors' Association, *Geothermal Task Force Report*, 2006 (http://www.westgov.org/wga/initiatives/cdeac/Geothermal-full.pdf).

93. B. Green and G. Nix, *Geothermal—The Energy Under Our Feet* (National Renewable Energy Laboratory, 2006) (http://www.nrel.gov/docs/fy07osti/40665.pdf).

94. WGA, *Geothermal Task Force Report*.

95. Jefferson W. Tester et al., *The Future of Geothermal Energy—Impact of Enhanced Geothermal Systems (EGS) on the United States in the 21st Century* (Massachusetts Institute of Technology, 2006) (http://geothermal.inel.gov/publications/future_of_geothermal_energy.pdf).

96. "The Power of Pico Hydro," Asian Trends Monitoring Bulletin 5, August 2010 (http://www.asiantrendsmonitoring.com/subscribe/).

97. Energy Information Administration (EIA). 2009. *An Updated Annual Energy Outlook 2009 Reference Case Reflecting Provisions of the American Recovery and Reinvestment Act and Recent Changes in the Economic Outlook*. SR/OIAF/2009–03, table 16.

98. REN21, *Renewables Global Status Report: 2009 Update*, table R4; EIA, *International Energy Outlook*, tables A8 and A16.

99. D. G. Hall, K. S. Reeves, J. Brizzee, R. D. Lee, G. R. Carroll, and G. L. Sommers, *Feasibility Assessment of the Water Energy Resources of the United States for New Low Power and Small Hydro Classes of Hydroelectric Plants*, US Department of Energy Report DOE-ID-11263, 2006 (http://hydropower.inel.gov/resourceassessment/pdfs/main_report_appendix_a_final.pdf), table B-1.

100. The description of the Wave Energy system is based on the operational specifications of the only commercial wave energy converters in the world: the Pelamis Converter, developed by Pelamis Wave Power Ltd. (formerly Ocean Power Delivery Ltd.).

101. "The RITE Project" (http://www.verdantpower.com/what-initiative).

102. California Independent System Operator, "Integration of energy storage technology in power systems," presented at Northwest Wind Integration Forum's Pumped Hydro Storage Workshop, 2008: 2.

103. *An Updated Annual Energy Outlook 2009 Reference Case Reflecting Provisions of the American Recovery and Reinvestment Act and Recent Changes in the Economic Outlook*, Energy Information Administration Report SR/OIAF/2009-03, 2009 (http://www.eia.doe.gov/oiaf/servicerpt/stimulus/excel/aeostimtab_18.xls), table A9.

104. "Statistics," Renewable Fuels Association (http://www.ethanolrfa.org/industry/statistics/).

105. EIA, *International Energy Outlook*, table 3.

106. According to Wikipedia (http://en. wikipedia.org/wiki/Biodiesel), 5–6 million tonnes of biodiesel was produced in 2006. This is the equivalent of 0.89 million barrels of oil equivalent in 2006 or 0.0024 million barrels of oil equivalent per day. According to table 2-17 in the DOE-EERE Renewable Energy Databook (http://www1.eere.energy.gov/maps_data/pdfs/eere_databook_091208.pdf), almost 400 million gallons of biodiesel were produced in 2002.

107. National Renewable Energy Laboratory, *Learning about Renewable Energy-Biofuels* (http://www.nrel.gov/learning/re_biofuels.html).

108. DOE-EERE, *Renewable Energy Databook*.

109. National Research Council. 2009. *America's Energy Future: Technology and Transformation*. Washington, DC: National Academies Press.

110. National Commission on Energy Policy, *Task Force on Biofuels Infrastructure*, 2009: 8–9.

111. The calculation is based on average offset emissions from 1995 to 2007 divided by 103 plants in operation. Offset emissions calculated by the Nuclear Energy Institute from US EPA and EIA data (*Emissions Avoided by the US Nuclear Industry 1995–2007*, Nuclear Energy Institute, 2008 (http://www.nei.org/filefolder/Emissions_Avoided_by_the_US_Nuclear_Industry_Yearly. xls).

112. IEA, *Key World Energy Statistics 2008*.

113. http://www.euronuclear.org/info/encyclopedia/n/nuclear-power-plant-world-wide.htm; Marvin S. Fertel, "The future of nuclear energy," presented to National Academies Board on Energy and Environmental Systems, Washington, 2010; J. M. Deutsch, D. W. Forsberg, A. C. Kadak, M. S. Kazimi, E. J. Moniz, and J. E. Parsons, *Update of the MIT 2003 Future of Nuclear Power* (http://web. mit.edu).

114. EIA, *International Energy Outlook 2009*: 11–12.

115. Deutsch et al., *Update of the MIT 2003 Future of Nuclear Power*.

116. Ibid.; National Commission on Energy Policy, *Ending the Energy Stalemate*, 2004 (http://www.energycommission.org/).

117. Fertel, "The future of nuclear energy."

118. NECP, *Ending the Energy Stalemate*.

119. Deutsch et al., *Update of the MIT 2003 Future of Nuclear Power*: 4.

120. Energy Information Administration (EIA), *Annual Energy Outlook 2008 with Projections to 2030—Overview*, DOE/EIA-0383(2007) (http://www.eia.doe.gov/oiaf/archive/earlyrelease08/index.html).

121. Nuclear Energy Institute, "Top Ten Reasons Nuclear Power Is Vital to America's Energy Policy" (http://www.nei.org/resourcesandstats/documentlibrary/reliableandaffordableenergy/factsheet/top10reasonstosupportnuclear).

122. CCTP, *Strategic Plan*: 89.

123. E. A. Rosa and R. L. Clark Jr., "Historical routes to technological gridlock: Nuclear technology as prototypical vehicle," *Research in Social Problems and Public Policy* 7, 1999: 21–57; S. C. Whitfield, E. A. Rosa, A. Dan, and T. Dietz, "The future of nuclear power: Value orientations and risk perception," *Risk Analysis* 29, 2009, no. 3: 425–437.

124. Testimony of Michael Corradini before House Energy and Commerce Committee Subcommittee on Oversight and Investigations, April 6, 2011; Nuclear Energy Institute, *Fact Sheet: Comparing Chernobyl and Fukushima*, April 2011.

125. Deutsch et al., *Update of the MIT 2003 Future of Nuclear Power*, 4.

126. International Energy Agency (IEA), *World Energy Outlook 2008* (Paris, France: IEA: 2008); David Hawkins, "Coal capture and storage," in *Climate Change Science and Policy*, ed. S. Schneider et al. (Island, 2010).

127. Intergovernmental Panel on Climate Change, *IPCC Special Report on Carbon Dioxide Capture and Storage* (Cambridge University Press, 2005) (http://www.ipcc.ch/ipccreports/special-reports.htm).

128. Ibid., section 8.3.3.

129. International Energy Agency, *World Energy Outlook 2008*: 74.

130. http://fossil. energy.gov/sequestration/geologic/index.html

131. http://www.euci.com/web_conferences/0909-ccs-mountaineer/index. php?ci=822

132. Electric Power Research Institute (EPRI), *EPRI Post-Combustion CO_2 Capture and Storage Demonstrations: Overview, Value, and Deliverables* (Palo Alto: EPRI, 2009).

133. Department of Energy, *Carbon Sequestration Technology Roadmap and Program Plan 2006* (http://fossil.energy.gov/sequestration/publications/programplans/2006/2006_sequestration_roadmap.pdf): 11–12.

134. Hawkins, "Coal capture and storage"; Intergovernmental Panel on Climate Change, *Carbon Dioxide Capture and Storage* (Cambridge University Press, 2005).

135. In the 1970s, interest in coal gasification was spurred by concerns that the US supply of natural gas was waning. The massive Great Plains Coal Gasification Plant in Beulah, North Dakota, was built with federal government support to use coal gasification to produce methane, the chief constituent of natural gas. When government price controls on natural gas were lifted, however, large quantities of natural gas became available, and no other coal-to-methane gasification plants have been built to date in the United States. Coal gasification, however, returned to the market in the 1980s and the 1990s. Driven primarily by environmental concerns about the traditional burning of coal, gasification emerged as an extremely clean way to generate electric power.

136. Allan Casey, "Carbon cemetery," *Canadian Geographic Magazine*, January/February 2008: 61.

137. The CO_2 content in the natural gas produced from the In Salah project ranges between 4 percent and 9 percent. The target market for the Algerian natural gas is Europe, where the market requires incoming natural gas to contain no more than 0.3% CO_2. As a result, In Solah Gas is required to separate the carbon before exporting the natural gas. The industry's typical practice is to first separate CO_2 from the natural gas (by chemical absorbers in stripper towers) and then vent it into the atmosphere. A commitment by the company shareholders to shrink the emissions footprint of the project and not to employ atmospheric venting of the CO_2 stream resulted in a need to identify an alternative solution. So, rather than venting CO_2, the project compresses it and injects it into a large underground aquifer lower than the gas reservoir at 1,800 meters depth. Source: Redouane Haddadji, "The In-Salah CCS experience: Sonatrach, Algeria," presented at First International Conference on the Clean Development Mechanism, Riyadh, 2006.

138. Markowsky, Office of Fossil Energy CCS RD&D Program Overview.

139. BBC News, "Access all areas: Schwarze Pumpe" (http://news.bbc.co.uk/2/hi/science/nature/7584155.stm).

140. Al-Juared and Whitmore, *Realistic Costs of Carbon Capture.*

141. The first two of these estimates are from the MIT report *The Future of Coal.* The estimates for ultracritical pulverized coal plants are from pp. 4444–4454 of Rubin and Chen, "Cost and performance of fossil fuel power plants with CO_2 capture and storage."

142. Paul W. Parfomak and Peter Folger, *Carbon Dioxide (CO_2) Pipelines for Carbon Sequestration: Emerging Policy Issues* (Congressional Research Service, 2007) (http://assets. openers .com/rpts/RL33971_20080117.pdf).

143. US Department of Transportation, National Pipeline Mapping System database (https:// www.npms.phmsa.dot.gov).

144. Parfomak and Folger, Carbon Dioxide (CO_2) Pipelines for Carbon Sequestration: Emerging Policy Issues.

145. Department of Energy, *Enhanced Oil Recovery/CO_2 Injection* (http://www.fossil.energy .gov/programs/oilgas/eor/index.html).

146. Planners are well aware of the terrible natural disaster that occurred in 1986 at Lake Nyos in Cameroon: carbon dioxide of volcanic origin slowly seeped into the bottom of the lake, which sits in a crater. One night an abrupt overturning of the lake bed let loose between 100,000 and 300,000 tons of CO_2 in a few hours. The gas, which is heavier than air, flowed down through two valleys, asphyxiating 1,700 villagers and thousands of cattle.

147. CCTP, *Strategic Plan.*

148. CCTP, *Technology options for the near and long term.*

149. US Department of Energy, Office of Fossil Energy, *Carbon Sequestration Atlas of the United States and Canada, Second Edition* (National Energy Technology Laboratory, 2008): 134.

150. http://www.fossil. energy.gov/programs/sequestration/terrestrial/index.html

151. EPA, *Inventory of US Greenhouse Gas Emissions and Sinks: 1990–2007,* table ES-4.

152. IPCC, *Working Group III: Fourth Assessment Report.*

153. World Wildlife Fund, "Russia's Boreal Forests" (http://assets.panda.org/downloads/ russia_forest_cc_final_13nov07.pdf).

154. NASA, *Causes of Deforestation: Direct Causes* (http://earthobservatory.nasa.gov/ Features/Deforestation/deforestation_update3).

155. Foundation for Sustainable Development, Environmental Issues in Argentina, http:// www.fsdinternational.org/ntlopps/country/argentina/environment.

156. S. Nilsson et al., "Greenhouse gas balance of Russia," presented at GCP Regional Carbon Budgets Workshop, Beijing, 2004 (http://www.docstoc.com/docs/529629/ Greenhouse-Gas-Balance-of Russia).

157. Lisa Tracy, "The deforestation of Siberia: Economic and environmental problems in Russian forest management," presented in Forest Industry Lecture Series, University of Alberta, 1994 (http://www.uofaweb.ualberta.ca/rr/pdfs/FILS_32.pdf).

158. The estimates are based on summed potentials presented for croplands at 55–164 TgC (R. Lal, J. M. Kimble, R. F. Follett, and C. V. Cole, *The Potential of US Cropland to Sequester Carbon and Mitigate the Greenhouse Effect*, CRC Press, 1998), grazing lands at 29–110 TgC (R. F. Follett, J. M. Kimble, and R. Lal, *The Potential of US Grazing Lands to Sequester*

Carbon and Mitigate the Greenhouse Effect, Lewis, 2001), and forest lands at 210 TgC (L. A. Joyce and R. Birdsey, eds., *The Impact of Climate Change on American's Forests: A Technical Document Supporting the 2000 USDA Forest Service RPA Assessment*, US Department of Agriculture, 2000; http://www.fs.fed.us). Estimates of potential savings from dedicated bioenergy croplands ranging from 91 TgC to 152 TgC (G. A. Tuska and M. E. Walsh, "Short-rotation woody crop systems, atmospheric carbon dioxide and carbon management: A US case study," *Forestry Chronicle* 77, 2001: 259–264) are excluded from this sum.

159. J. Lewandrowski, M. Peters, C. Jones, R. House, M. Sperrow, M. Eve, and K. Paustian, *Economics of Sequestering Carbon in the US Agricultural Sector*, technical bulletin 1909, US Department of Agriculture Economic Research Service, 2004; *Greenhouse Gas Mitigation Potential in US Forestry and Agriculture*, US Environmental Protection Agency Report 430-R-05-006, 2005.

160. IPCC, *Working Group III: Fourth Assessment Report*.

161. CCTP, *Strategic Plan*.

162. EPA, Inventory of US Greenhouse Gas Emissions and Sinks: 1990–2007.

163. Ibid., table ES-2.

164. US Environmental Protection Agency (EPA), *Global Anthropogenic Non-CO_2 GHG Emissions: 1990–2020*, USEPA 430-R-06-005 (http://www.epa.gov/nonco2/econ-inv/international.html#global_anthropogenic).

165. *Inventory of US Greenhouse Gas Emissions and Sinks: 1990–2007*, US Environmental Protection Agency Report ES-9.

166. *Vision and Framework for Strategy and Planning*, US Climate Change Technology Program Report DOE/PI-0003, 2005 (http://www.climatetechnology.gov/vision2005/cctp-vision2005.pdf).

167. Leon Khew, IUT Singaporean Plant Visit: Organic Waste Bio-Methanization Plant, IUT Global, July 19, 2009.

168. EPA, Global Anthropogenic Non-CO_2 GHG Emissions: 1990–2020; EPA, Inventory of US Greenhouse Gas Emissions and Sinks: 1990–2007, table ES-7.

169. *Inventory of US Greenhouse Gas Emissions and Sinks: 1990–2007*, Environmental Protection Agency Report ES-9.

170. US Department of Agriculture (USDA), *Energy Estimator, Energy Consumption Awareness Tool* (http://nfat.sc.egov.usda.gov/Help.aspx).

171. US Department of Agriculture, *Precision Agriculture: Information Technology for Improved Resource Use*, 1998 (http://www.ers.usda.gov/publications/agoutlook/apr1998/ao250f.pdf).

172. EPA, Global Anthropogenic Non-CO_2 GHG Emissions: 1990–2020. EPA, Inventory of US Greenhouse Gas Emissions and Sinks: 1990–2007.

173. EPA, Global Anthropogenic Non-CO_2 GHG Emissions: 1990–2020; EPA, Inventory of US Greenhouse Gas Emissions and Sinks: 1990–2007.

174. *Inventory of US Greenhouse Gas Emissions and Sinks: 1990–2007*, Environmental Protection Agency Report ES-11.

175. CCTP, Strategic Plan.

176. EPA, Global Anthropogenic Non-CO$_2$ GHG Emissions: 1990–2020.

177. *Inventory of US Greenhouse Gas Emissions and Sinks: 1990–2007*, US Environmental Protection Agency Report ES-10.

178. J. H. Jo, J. S. Golden, and S. W. Shin, "Incorporating built environment factors into climate change mitigation strategies for Seoul, South Korea: A sustainable urban systems framework," *Habitat International* 33, 2009: 267–275.

179. Ibid.

180. Juliet Eilperin, "US trying to weaken G8 climate change declaration," *Boston Globe*, May 14, 2007.

181. IEA, *World Energy Outlook 2008*, table 8.2.

182. A. P. Mitra, Chhemendra Sharma, and M. A. Y. Ajeroc, "Energy and emissions in South Asian mega-cities: Study on Kolkata, Delhi and Manila," presented at International Workshop on Policy Integration Toward Sustainable Urban Energy Use for Cities in Asia, East West Center, Honolulu, 2003.

183. Eilperin, "US trying to weaken G8 climate change declaration."

184. Benjamin K. Sovacool and Marilyn A. Brown. 2010. "Twelve Metropolitan Carbon Footprints: A Preliminary Comparative Global Assessment," *Energy Policy* 38, 2010, no. 9: 4856–4869.

185. The largest 100 metropolitan areas in the United States account for only 56 percent of the carbon dioxide emissions from highway transportation and residential buildings, according to Marilyn A. Brown, Frank Southworth, and Andrea Sarzynski, *Shrinking the Carbon Footprint of Metropolitan America* (Brookings Institution, 2008) (http://www.brookings.edu/~/media/Files/rc/reports/2008/05_carbon_footprint_sarzynski/carbonfootprint_report.pdf): 15.

186. IEA, *World Energy Outlook 2008*, 184.

187. Ibid.

188. Brown et al., "Shrinking the carbon footprint of metropolitan America"; Edward L. Glaeser and Matthew Kahn, *The Greenness of Cities*, Policy Brief, Rappaport Institute for Greater Boston, 2008; Sovacool and Brown, "Twelve metropolitan carbon footprints."

189. Reid Ewing, Keith Bartholomew, Steve Winkelman, Jerry Walters, and David Goldberg, *Growing Cooler* (Urban Land Institute, 2007); C. Kennedy, J. Cuddihy, and J. Engel-Yan, "The changing metabolism of cities," *Journal of Industrial Ecology* 11, 2007: 43–59.

190. Avi Friedman, *Sustainable Residential Development: Planning and Design for Green Neighborhoods* (McGraw-Hill, 2007).

191. Phillip R. Berke, "The evolution of green community planning, scholarship, and practice," *Journal of the American Planning Association* 74, 2008, no. 4: 393–407.

192. Eugene L. Birch and Christopher Silver, "One hundred years of city planning's enduring and evolving connections," *Journal of the American Planning Association* 75, 2009, no. 2: 113–122.

193. http://www.wri.org/stories/2008/06/how-we-move-sustainable-transport-around-world

194. Reid Ewing et al., *Growing Cooler*.

195. Marilyn A. Brown, Frank Southworth, and Andrea Sarzynski, "The geography of metropolitan carbon footprints," *Policy and Society* 27, 2009: 285–304.

196. Marilyn A. Brown, Frank Southworth, and Therese Stovall, *Towards a Climate-Friendly Built Environment*, Pew Center on Global Climate Change, 2005 (http://www.pewclimate .org/global-warming-in-depth/all_reports/buildings).

197. http://www.jdhc. or. jp/en/area.html.

198. John Holtzclaw, *A Vision of Energy Efficiency*, American Council for an Energy-Efficient Economy, 2004.

199. Mary Jean Bürer, David Goldstein, and John Holtzclaw, *Location Efficiency as the Missing Piece of the Energy Puzzle: How Smart Growth Can Unlock Trillion Dollar Consumer Cost Savings*, Natural Resources Defense Council, 2004.

200. Thomas F. Golob and David Brownstone, *The Impact of Residential Density on Vehicle Usage and Energy Consumption*, Institute of Transportation Studies, University of California, Irvine, 2005 (http://repositories.cdlib.org/itsirvine/wps/WPS05_01).

201. Marilyn A. Brown et al., "The geography of metropolitan carbon footprints."

202. Marilyn A. Brown et al., *Shrinking the Carbon Footprint of Metropolitan America*.

203. Ewing et al., *Growing Cooler*.

204. G. B. Arrington and Robert Cervero, *Effects of TOD on Housing, Parking, and Travel*, TCRP Report 128, Transportation Research Board, 2008.

205. Randal O'Toole, *The Myth of the Compact City: Why Compact Development Is Not the Way to Reduce Carbon Dioxide Emissions*, Policy Analysis 653, Cato Institute, 2009 (http://www.cato.org/pubs/pas/pa653.pdf).

206. Brian Roberts and Trevor Kanaley, "Overview: Urbanization and sustainable development," in *Urbanization and Sustainability: Case Studies of Good Practice* (Asian Development Bank, 2006).

207. William Harvey, "Renewable energy: Price and policy are key," Environmental Research Web, July 30, 2008 (http://environmentalresearchweb.org/cws/article/opinion/35174).

208. Paul Denholm, "Improving the technical, environmental and social performance of wind energy systems using biomass-based energy storage," *Renewable Energy* 31, 2006: 1356.

209. Federal Ministry of Economics and Technology, *E-Energy: ICT-Based Energy Systems of the Future* (Berlin: BWMi, April, 2008).

210. Those wishing to learn more about the Lewis Center should read David W. Orr, *Design on the Edge: The Making of a High-Performance Building* (MIT Press, 2006).

211. *UNIDO and Renewable Energy: Greening the Industrial Agenda* (United Nations Industrial Development Organization, 2009) (http://www.unido.org/fileadmin/user_media/ Publications/documents/UNIDO_and_renewable_energy.pdf): 20–21.

212. Ibid.

213. Department of Energy/ Energy Efficiency and Renewable Energy, *PV in Hybrid Power Systems* (DOE/EERE, January 5, 2006): 1.

214. "Installed US Wind Capacity and Wind Project Locations," Department of Energy, http://www.windpoweringamerica.gov/wind_installed_capacity.asp.

215. M. Xu, J. C. Crittenden, Y. Chen, V. M. Thomas, D. S. Noonan, R. DesRoches, M. A. Brown, and S. P. French, "Gigaton problems need gigaton solutions," *Environmental Science & Technology* 44, 2010, no. 11: 4037–4041 (http://dx.doi.org/10.1021/es903306e).

216. Stephen W. Pacala and Robert H. Socolow, "Stabilization wedges: Solving the climate problem for the next 50 years with current technologies," *Science* 305, 2004: 968–972.

Chapter 4

1. *Geoengineering the Climate: Science, Governance, and Uncertainty* (Royal Society, 2009).

2. Thomas R. Karl, Jerry M. Melillo, and Thomas C. Peterson. 2009. *Global Climate Change Impacts in the US* (Cambridge University Press): 11.

3. T. J. Wilbanks, P. Leiby, R. Perlack, J. T. Ensminger, and S. B. Wright, "Toward an integrated analysis of migration and adaptation: Some preliminary findings," *Mitigation and Adaptation Strategies for Global Change* 12, 2007, no. 5: 713–725.

4. The phrase "avoiding the unmanageable, managing the unavoidable" was first used in 2007 in the widely circulated report *Confronting Climate Change: Avoiding the Unmanageable and Managing the Unavoidable*, prepared for the United Nations Commission on Sustainable Development (http://www.globalproblems-globalsolutions-files.org/unf_website/PDF/climate%20_change_avoid_unmanagable_manage_unavoidable.pdf).

5. US Global Change Research Program, *Global Climate Change Impacts in the US* (Cambridge University Press, 2009).

6. David Keith, "Engineering: Important questions, state of knowledge, and major uncertainties related to selected geoengineering options," presented at National Academies Workshop on "Geoengineering Options to Respond to Climate Change: Steps to Establish a Research Agenda," Washington, 2009.

7. Intergovernmental Panel on Climate Change (IPCC), *Climate Change 2007: Mitigation of Climate Change* (Cambridge University Press, 2007).

8. M. L. Weitzman, "On modeling and interpreting the economics of catastrophic climate change," *Review of Economics and Statistics* 91, 2009: 1–19.

9. Stephen Schneider, "The Worst-Case Scenario," *Nature* 458, 2009: 1104–1105; Schneider, "Geoengineering: Could we or should we make it work?" *Philosophical Transactions of the Royal Society A* 366, 2008: 3843–3862.

10. David Archer and Stefan Rahmstorf, *The Climate Crisis: An Introductory Guide to Climate Change* (Cambridge University Press, 2010): 225.

11. P. J. Crutzen, "Albedo enhancement by stratospheric sulfur injections: A contribution to resolve a policy dilemma?" *Climatic Change* 77, 2006: 211–219.

12. Granger Morgan, Governance and Geoengineering: Who Decides and How, panel discussion at National Academies Workshop on "Geoeingineering Options to Respond to Climate Change: Steps to Establish a Research Agenda," Washington, 2009.

13. D. W. Keith, "Geoengineering the Climate: History and Prospect," *Annual Review of Energy and Environment* 25, 2000: 245–284.

14. Roger Angel, "Feasibility of cooling the Earth with a cloud of small spacecraft near the inner Lagrange point," *Proceedings of the National Academy of Sciences* 1–3, 2006: 17184–17189.

15. Dan Schrag, "Physical science: Important questions, state of knowledge, and major uncertainties related to selected geoengineering options," presented at National Academies Workshop on "Geoengineering Options to Respond to climate Change: Steps to Establish a Research Agenda," Washington, 2009.

16. Gordon H. Orions and David Policansky, "Scientific bases of macroenvironmental indicators," *Annual Review of Environment and Resources* 34, 2009: 375–404.

17. Daniel G. Boyce, Marlon R. Lewis, and Boris Worm, "Global phytoplankton decline over the past century," *Nature* 466, 2010: 591–596.

18. Ove Hoegh-Guldberg and John F. Bruno, "The impact of climate change on the world's marine ecosystems," *Science* 328, 2010:1523–1528.

19. Crutzen, "Albedo enhancement by stratospheric sulfur injections."

20. J. C. Orr et al., "Anthropogenic ocean acidification over the twenty-first century and its impact on calcifying organisms," *Nature* 437, 2005: 681–686.

21. Crutzen, "Albedo enhancement by stratospheric sulfur injections": 211–219.

22. David Keith, "Engineering the planet," in *Climate Change Science and Policy*, ed. S. Schneider et al. (Island, 2010).

23. For an alternative characterization of geo-engineering options, see Barbara Levi, "Will desperate climates call for desperate geoengineering measures?" *Physics Today*, August 2008: 26–28.

24. T. M. Lenton and N. E. Vaughan, "The Radiative Forcing Potential of Different Climate Geoengineering Options," *Atmospheric Chemistry and Physics* 9, 2009: 2559–2608.

25. Crutzen, "Albedo enhancement by stratospheric sulfur injections": 211–219.

26. Robert Kunzig, "A sunshade for Planet Earth," *Scientific American* 299, 2008, no. 5: 46.

27. Lenton and Vaughan, "The radiative forcing potential of different climate geoengineering options": 2559–2608.

28. Stephen Salter, Graham Sortino, and John Latham, "Sea-going hardware for the cloud albedo method of reversing global warming," *Philosophical Transactions of the Royal Society* 366, 2008: 3989–4006.

29. Bjørn Lomborg, "Engineering the climate: Global warming's cheap, effective solution" (http://issuu.com/copenhagenconsensus/docs/engineeringthe_climate).

30. See, for example, http://www.ghcc.msfc.nasa.gov/urban/urban_heat_island.html.

31. US Climate Change Technology Program, *Technology Options for the Near and Long Term*, Department of Energy Report DOE/PI-0002, 2005 (http://www.climatetechnology.gov/library/2005/tech-options/tor2005-124.pdf).

32. Art Rosenfeld, "Extreme efficiency: Lessons from California," presented to AAAS, Washington, 2005 (http://www.energy.ca.gov/2005publications/CEC-999-2005-003/CEC-999-2005-003.pdf).

33. National Academy of Sciences, *Limiting the Magnitude of Future Climate Change* (National Academies Press, 2010).

34. Mark McHenry, "Agricultural bio-conduction, renewable energy generation and farm carbon sequestration in Western Australia," 1–7.

35. Johannes Lehmann, John Gaunt, and Marco Rondon, "Bio-char sequestration in terrestrial ecosystems—A review," *Mitigation and Adaptation Strategies for Global Change* 11, 2006: 403–427.

36. Ibid, 403–427.

37. McHenry, "Agricultural bio-char production, renewable energy generation and farm carbon sequestration in Western Australia": 1–7.

38. "First successful demonstration of carbon dioxide air capture technology achieved" (http://www.physorg.com/news96732819.html).

39. "Geo-engineering: Every Silver Lining has a Cloud" *The Economist*, February 1, 2009.

40. Lenton and Vaughan, "The radiative forcing potential of different climate geoengineering options": 2559–2608.

41. Levi, "Will desperate climates call for desperate geoengineering measures?"

42. This feedback loop is called the "CLAW" hypothesis. See http://en.wikipedia.org/wiki/CLAW_hypothesis.

43. Kurt Zenz House, Christopher H. House, Daniel P. Schrag, and Michael J. Aziz, "Electrochemical acceleration of chemical weathering as an energetically feasible approach to mitigating anthropogenic climate change," *Environmental Science & Technology* 41, 2007, 24: 8464–8470.

44. "Ocean CO_2 collector could fight global warming and ocean acidification," mongabay .com, November 20, 2007 (http://news.mongabay.com/2007/1120-geoengineering.html).

45. See Levi, "Will desperate climates call for desperate geoengineering measures?"

46. David Victor, Granger Morgan, John Steinbruner, and Kate Ricke, "The geoengineering option: A last resort against global warming?" *Foreign Affairs* 88, 2009: 64–76.

47. John Steinbruner, Governance and Geoengineering: Who Decides and How, panel discussion at National Academies Workshop on "Geoengineering Options to Respond to Climate Change: Steps to Establish a Research Agenda," Washington, 2009.

48. Dale Jamieson, From Research to Field Testing and Deployment: Ethical Issues Raised by Geoengineering, panel discussion at National Academies Workshop on "Geoengineering Options to Respond to Climate Change: Steps to Establish a Research Agenda," Washington, 2009.

49. Michael Oppenheimer, "Concluding remarks," National Academies Workshop on "Geo-engineering Options to Respond to Climate Change: Steps to Establish a Research Agenda," 2009.

50. Thomas R. Karl, Jerry M. Melillo, and Thomas C. Peterson. 2009. *Global Climate Change Impacts in the US* (Cambridge University Press).

51. Intergovernmental Panel on Climate Change (IPCC), *Climate Change 2007: Impacts, Adaptation and Vulnerability* (Cambridge University Press, 2007).

52. Nicholas Stern, *The Economics of Climate Change: The Stern Review* (Cambridge University Press, 2006).

53. Vulnerability refers to "the propensity of human and ecological systems to suffer harm and their ability to respond to stresses imposed as a result of climate change effects." IPCC, *Climate Change 2007: Impacts, Adaptation and Vulnerability*, 720.

54. R. K. Pachauri, "Foreword," *Mitigation and Adaptation Strategies for Global Change* 12, 2007: 643–644.

55. Jessica M. Ayers and Saleemul Huq, "The value of linking mitigation and adaptation: A case study of Bangladesh," *Environmental Management* 43, 2009: 753–764.

56. IPCC, *Climate Change 2007: Impacts, Adaptation and Vulnerability*.

57. Thomas J. Wilbanks. 2007. "Energy Myth Thirteen—Developing Countries Are Not Doing Their Part in Responding to Concerns about Climate Change" in *Energy and American Society—Thirteen Myths*, ed. B. Sovacool and M. Brown (Springer, 2007).

58. Elisabeth M. Hamin, Nicole Gurran, "Urban form and climate change: Balancing adaptation and mitigation in the US and Australia," *Habitat International* 33, 2009: 238–245.

59. Ayers and Huq, "The value of linking mitigation and adaptation": 753–764.

60. Karl et al., *Global Climate Change Impacts in the US*: 11.

61. IPCC, *Climate Change 2007: Impacts, Adaptation and Vulnerability*, 719. National Research Council, *Adapting to the Impact of Climate Change* (National Academies Press, 2010).

62. S. Fankhauser, J. B. Smith, and R. S. J. Tol, "Weathering climate change: Some simple rules to guide adaptation decisions," *Ecological Economics* 30, 1999: 67–78.

63. National Research Council, *Adapting to the Impact of Climate Change*.

64. National Research Council, *The Great Alaska Earthquake of 1964* (National Academies Press, 1974): 978.

65. National Research Council, *Adapting to the Impact of Climate Change*, table S.1.

66. IPCC, *Climate Change 2007: Impacts, Adaptation and Vulnerability*, 722–724.

67. Bruce Stutz, "Adaptation emerges as key part of any climate change plan" (http://e360 .yale.edu/content/feature.msp?id=2156).

68. Ibid.

69. National Research Council, *Adapting to the Impact of Climate Change*.

70. This example of alternative definitions of "adaptation" draws on Ian Noble, "Adaptation: Seeking an international (development) perspective," presented at meeting of National Academy of Sciences, Washington, 2009.

71. Wilbanks et al., "Toward an integrated analysis of migration and adaptation": 713–725.

72. K. de Bruin et al., "Adapting to climate change in the Netherlands: An inventory of climate adaptation options and ranking of alternatives," *Climatic Change* 95, 2009: 23–45.

73. National Research Council, *Adapting to the Impact of Climate Change*, 131–132.

74. Gary Yohe, "Risk assessment and risk management for infrastructure planning and investment," *The Bridge* 40, 2010, no. 3: 14–21.

75. National Research Council, *Adapting to the Impact of Climate Change*; "Climate Change Adaptation in New York City: Building a Risk Management Response," *Annals of the New York Academy of Sciences*, 1196.

76. Rob Swart et al., *Europe Adapts to Climate Change: Comparing National Adaptation Strategies*, Partnership for European Environmental Research Report 1, 2009.

77. *An Investment Framework for Clean Energy and Development: A Progress Report*, World Bank Report DC2006-0012, 2006.

78. Oxfam International, *What's needed in poor countries, and who should pay?* Oxfam Briefing Paper, 2007.

79. S. Fankhauser, "Protection versus retreat: The economic costs of sea-level rise," *Environment and Planning A* 27, 1995: 299–319.

80. R. J. Nichols and R. S. J. Tol, "Impacts and responses to sea-level rise: A global analysis of the SRES scenarios over the 21st century," *Philosophical Transaction of the Royal Society A* 364, 2006: 1073–1095.

81. R. S. J. Tol, "Estimates of the damage costs of climate change. Part 1: Benchmark estimates," *Environmental Resource Economics* 21, 2002: 47–73.

82. Manoj Roy, "Planning for sustainable urbanisation in fast growing cities: Mitigation and adaptation issues addressed in Dhaka, Bangladesh," *Habitat International* 33, 2009: 276–286.

83. A. T. Butt, B. A. McCarl, J. Angerer, P. T. Dyke, and J. W. Stuth, "The economic and food security implications of climate change," *Climate Change* 67, 2005: 355–378.

84. Tol, "Estimates of the damage costs of climate change. Part 1": 47–73.

85. W. Morrison and R. Mendelsohn, "The impact of global warming on US energy expenditures," in *The Impact of Climate Change on the US Economy*, ed. R. Mendelsohn and J. Neumann (Cambridge University Press, 1999); E. T. Mansur, R. Mendelsohn, and W. Morrison, *A Discrete Continuous Choice Model of Climate Change Impacts on Energy*, Working Paper ES-43, Social Science Research Network, Yale School of Management, 2005.

86. M. Dore and I. Burton, *The Costs of Adaptation to Climate Change in Canada: A Stratified Estimate by Sectors and Regions—Social Infrastructure* (St. Catherines, Ontario: Climate Change Laboratory, Brock University, 2001).

87. Ayers and Huq, "The value of linking mitigation and adaptation": 753–764.

88. Wilbanks, "Energy myth thirteen": 341–350.

89. Lisa Friedman and Darren Samuelsohm, "Hillary Clinton pledges $100B for developing countries," *New York Times*, December 17, 2009.

90. http://www.wbcsd.org/web/complus/documents/LDCF_press_release_FINAL.pdf.

Chapter 5

1. Quoted in Frederick Buell, *From Apocalypse to Way of Life: Environmental Crisis in the American Century* (Routledge, 2003): 39.

2. Ibid.: 39.

3. Portions of this chapter are based on extensive reports written for the Oak Ridge National Laboratory, Brown, Marilyn A., Sharon (Jess) Chandler, Melissa V. Lapsa, and Benjamin K. Sovacool, *Carbon Lock-In: Barriers to the Deployment of Climate Change Mitigation Technologies* (Oak Ridge National Laboratory, ORNL/TM-2007/124, November 2008); US Department of Energy, Committee on Climate Change Science and Technology Integration (CCCSTI), *Strategies for the Commercialization and Deployment of Greenhouse Gas-Intensity Reducing Technologies and Practices* (CCCSTI, 2009); as well as two shorter follow-up articles: Marilyn A Brown and Sharon (Jess) Chandler, "Governing confusion: How statutes, fiscal policy, and regulations impede clean energy technologies," *Stanford Law & Policy Review* 19, 2008, no. 3: 472–509; Benjamin K. Sovacool, "Placing a glove on the invisible hand: how intellectual property rights may impede innovation in energy research and development (R&D)," *Albany Law Journal of Science & Technology* 18, 2008, no. 2: 101–161.

4. Interlaboratory Working Group, *Scenarios for a Clean Energy Future*, Oak Ridge National Laboratory Report ORNL/CON-476 and Lawrence Berkeley National Laboratory Report LBNL-44029, 2000; Marilyn A. Brown, Mark D. Levine, Walter Short, and Jonathan G. Koomey, "Scenarios for a clean energy future," *Energy Policy* 29, 2001: 1179–1196.

5. J. P. Painuly, "Barriers to renewable energy penetration; a framework for analysis," *Renewable Energy* 24, 2001: 73–89.

6. F. Beck and E. Martinot, "Renewable energy policies and barriers," in *Encyclopedia of Energy*, volume 4, ed. C. Cleveland (Elsevier, 2004).

7. Benjamin K. Sovacool, "Rejecting renewables: The socio-technical impediments to renewable electricity in the US," *Energy Policy* 37, 2009: 4500–4513; Benjamin K. Sovacool, *The Dirty Energy Dilemma: What's Blocking Clean Power in the US* (Praeger, 2008); Benjamin K. Sovacool and Richard F. Hirsh, "Energy myth six—The barriers to new and innovative energy technologies are primarily technical— The case of distributed generation," in *Energy and American Society—Thirteen Myths*, ed. B. Sovacool and M. Brown (Springer, 2007).

8. Amory B. Lovins, "Energy myth nine—Energy efficiency improvements have already reached their potential," in *Energy and American Society—Thirteen Myths*, ed. B. Sovacool and M. Brown (Springer, 2007).

9. Brown et al., *Carbon Lock-In*.

10. Intergovernmental Panel on Climate Change, *Climate Change 2007: Mitigation of Climate Change* (http://www.ipcc.ch/publications_and_data/publications_ipcc_fourth _assessment_report_wg3_report_mitigation_of_climate_change.htm).

11. A. Myrick Freeman, "The Ethical Basis of the Economic View of the Environment," presented at Morris Colloquium on Ethics and Economics in the Environment, University of Colorado, 1983; Sovacool, *The Dirty Energy Dilemma*.

12. "Canada's Energy Industry: Tarred with the Same Brush," *The Economist*, August 7, 2010: 35–6.

13. Benjamin K. Sovacool, "Renewable energy: Economically sound, politically difficult," *Electricity Journal* 21, 2008, no. 5: 18–29.

14. Elinor Ostrom, *A Polycentric Approach for Coping With Climate Change*, report prepared for WDR2010 Core Team at World Bank, 2009): 4; Elinor Ostrom, "A general framework for analyzing sustainability of socio-ecological systems," *Science* 325, 2009: 419–422.

15. Jerry Taylor and Peter Van Doren, "Energy myth five—Price signals are insufficient to induce efficient energy investments," in *Energy and American Society—Thirteen Myths*, ed. B. Sovacool and M. Brown (Springer, 2007).

16. John D. Donahue, *The Privatization Decision* (Basic Books, 1991): 18.

17. M. A. Brown, M. D. Levine, W. Short, and J. G. Koomey, "Obstacles to Energy Efficiency," *Encyclopedia of Energy* 4 (Academic Press/Elsevier Science, 2004): 465–475; B. Prindle, *Quantifying the Effects of Market Failures in the End-Use of Energy*, American Council for an Energy-Efficient Economy Report E071, 2007 (http://www.aceee.org/energy/IEAmarketbarriers.pdf); Marilyn A. Brown, "Market failures and barriers as a basis for clean energy policies," *Energy Policy* 29, 2001. 14: 1197–1207.

18. George H. W. Bush acceded to the UN Framework Convention on Climate Change treaty produced at the UN Earth Summit in Rio de Janeiro in 1992. The objective of the Convention is to achieve "stabilization of greenhouse gas concentrations in the atmosphere at a level that would prevent dangerous anthropogenic interference with the climate system." In 2002, the George W. Bush administration announced a US goal of reducing GHG intensity (that is, emissions per dollar of real GDP) by 18 percent from 2002 to 2012. Barack Obama pledged in his election campaign to reduce carbon dioxide emissions by 80 percent below 2005 levels in 2050.

19. Brown and Chandler, "Governing confusion."

20. Everett M. Rogers, *Diffusion of Innovations*, fourth edition (Free Press, 1995).

21. G. Unruh, "Understanding carbon lock-in," *Energy Policy* 28, 2000: 817–830.

22. P. DeLaquil III, "Progress commercializing solar-electric power systems," presented at World Renewable Energy Congress, Denver, 1996.

23. Peter Evans, "Clean energy technology innovation and deployment," presented to Council of Foreign Relations, Washington, 2010.

24. Noah Sachs, "Planning the funeral at the birth: Extended producer responsibility in the European Union and the US," *Harvard Environmental Law Review* 30, 2006: 56.

25. Researchers at the University of California at San Francisco identified $7.6 billion in yearly expenses mainly in lost wages and higher health care costs from smoking in California. This amount worked out to the equivalent of $3.43 added on to every pack of cigarettes sold in the state. Paul Hawken, *The Ecology of Commerce: A Declaration of Sustainability* (HarperCollins, 1993): 79.

26. Steven D. Levitt and Stephen J. Dubner, *Freakonomics: A Rogue Economist Explores the Hidden Side of Everything* (Penguin Books, 2006).

27. These "externalities" are benefits or costs resulting from a market transaction that are received or borne by parties not directly involved in the transaction. Externalities can be either positive, when an external benefit is generated, or negative, when an external cost is imposed upon others.

28. IPCC, *Climate Change 2007: Mitigation of Climate Change*, 475.

29. William N. Dunn, *Public Policy Analysis: An Introduction*, fourth edition (Pearson Prentice-Hall, 2008).

30. Mark Z. Jacobson and Gilbert M. Masters, "Exploiting wind versus coal," *Science* 293, 2001: 1438–1439.

31. Hawken, *The Ecology of Commerce: A Declaration of Sustainability*.

32. Charles Wheelan, *Naked Economics: Undressing the Dismal Science* (Norton, 2003): xviii.

33. David W. Orr, *Earth in Mind: On Education, Environment, and the Human Prospect* (Island, 1994): 172.

34. Ulrich Beck, "From industrial society to the risk society: Questions of survival, social structure and ecological enlightenment," *Theory, Culture, & Society* 9, 1992: 97–123.

35. Zachary A. Smith, *The Environmental Policy Paradox*, fifth edition (Prentice-Hall, 2009).

36. R. J. Bulle and D. N. Pellow, "Environmental justice: Human health and environmental inequalities," *Annual Review of Public Health* 27, 2006: 103–124.

37. Cass R. Sunstein, "Social norms and social rules," *Columbia Law Review* 96, 1996: 903–968.

38. These organizations included corporate firms, banks, venture capital investors, start-up companies, and private equity providers covering 15,000 projects and 10,000 transactions involving biomass, geothermal, and wind projects greater than 1 MW, hydro projects between 0.5 and 50 MW, and solar projects more than 300 kW. Sebastian Fritz-Morgenthal, Chris Greenwood, Carola Menzel, Marija Mironjuk, and Virginia Sontag-O'Brien, *The Global Financial Crisis and its Impact on Renewable Energy Finance* (United Nations Environment Program, 2009).

39. DeLaquil III, *Progress commercializing solar-electric power systems*.

40. http://en. wikipedia.org/wiki/Who_Killed_the_Electric_Car%3F

41. Daniel Clery, "A sustainable future, if we pay up front," *Science* 315, 2007: 782–783; Peter H. Kobos, Jo D. Ericson, and Thomas E. Drennen, "Technological learning and renewable energy costs: Implications for us renewable energy policy," *Energy Policy* 34, 2006: 1645–1658; Daniel M. Kammen and Gregory F. Nemet, "Energy myth eleven—Energy R&D investment takes decades to reach the market," in *Energy and American Society—Thirteen Myths*, ed. B. Sovacool and M. Brown (Springer, 2007).

42. Elizabeth J. Wilson, Joseph Plummer, Miriam Fischlein, and Timothy M. Smith, "Implementing energy efficiency: Challenges and opportunities for rural electric cooperatives and small municipal utilities," *Energy Policy* 36, 2008: 3383–3397.

43. Ajeet Rohatgi (Regents' Professor; Georgia Power Distinguished Chair; Director, University Center of Excellence for Photovoltaic Research and Education, Georgia Institute of Technology), telephone interview, November 2006.

44. Jay Braitsch (Director of Strategic Planning, Fossil Energy, US Department of Energy) telephone interview, February 2007.

45. Jim Rushton (Director of Nuclear Science and Technology, Oak Ridge National Laboratory), telephone interview, November 2006.

46. Neal Elliott (Industrial Program Director, American Council for an Energy Efficient Economy) telephone interview October 2006.

47. Bill Prindle, *From Shop Floor to Top Floor: Best Business Practices in Energy Efficiency* (Pew Center, forthcoming).

48. Consider the REScheck tool for assisting building code implementation. This tool incorporates tradeoffs between technologies to meet the code requirements of the state or local code, in the jurisdiction permitting its use; in some cases, these tradeoffs lead to distortions when credits are allowed for practices that have become common, inhibiting further improvements in efficiency. For example, in the Upper Midwest there is upward of 80 percent penetration of condensing gas furnaces. The tradeoffs to meet the code allow savings from this now common high efficiency furnace to be used to offset poor envelopes. As a result, this code specification is no longer promoting improved building practices because it has not adapted to technology advances (Brown and Chandler, "Governing cnfusion"). Perhaps a more current example of a technology barrier is the long retention of the center of glass U-factor criteria rather than the whole window U-factor criteria. The lack of building infiltration criteria, solar heat gain coefficient (SHGC) criteria, and duct testing requirements are also barriers to technological innovation and progress in new construction. (Jean Boulin, US Department of Energy, personal communication, May 14, 2009.).

49. Elliott, telephone interview.

50. Marilyn A. Brown, Jess Chandler, Melissa Lapsa, and Moonis Ally, *Making Homes Part of the Climate Solution*, Oak Ridge National Laboratory Report ORNL/TM-2009/104, 2009 (http://www.ornl.gov/sci/eere/publications. shtml): 77–82.

51. Brown and Chandler, "Governing confusion," 472–509.

52. I. Gritsevich, "Motivation and decision criteria in private households, companies and administration on energy efficiency in Russia," presented at IPCC Expert Meeting on Conceptual Frameworks for Mitigation Assessment from the Perspective of Social Science, Karlsruhe, 2000.

53. International Energy Agency, *World Energy Outlook 2008*.

54. Production Tax Credits are defined in 26 USC § 45(c)(1). EPAct 2005 (P. L. 109–58 § 1301) amended 26 USC § 45. Section 1301 modifies the definition of "qualified energy resources" in section 45(c)(1).

55. American Wind Energy Association, *US Installed Wind Capacity, 1981 to 2006* (http://www.awea.org/projects/,faq/instcap.html).

56. Renewable Energy Policy Network for the 21st Century (REN21), *Renewables Global Status Report: 2009 Update* (http://www.ren21.net/pdf/RE_GSR_2009_Update.pdf).

57. Renewable Energy Policy Network for the 21st Century (REN21), *Renewables 2007: Global Status Report* (REN21, 2008): p. 7; REN21, *Renewables Global Status Report: 2009 Update*.

58. Bruce Tonn, K. C. Healy, Amy Gibson, Ashutosh Ashish, Preston Cody, Drew Beres, Sam Lulla, Jim Mazur, and A. J. Ritter, "Power from perspective: Potential future US energy portfolios," *Energy Policy* 37, 2009: 1432–1443.

59. Benjamin K. Sovacool, "The importance of comprehensiveness in renewable electricity and energy efficiency policy," *Energy Policy* 37, 2009: 1529–1541.

60. Fredric C. Menz and Stephan Vachon, "The effectiveness of different policy regimes for promoting wind power: Experiences from the states," *Energy Policy* 34, 2006: 1786–1796.

61. Brown and Chandler "Governing confusion."

62. Clean Air Act Amendments of 1977 (P. L. 95–95; 91 Stat. 685) and of 1990 (P. L. 101–549).

63. "Standards of Performance for New Stationary Sources" 40 CFR 60.

64. Jay Braitsch (Director of Strategic Planning, Fossil Energy, US Department of Energy) telephone interview February 2007.

65. Robert N. Stavins, "Vintage-differentiated environmental regulation," *Stanford Environmental Law Journal* 25, 2006, no. 1: 29–63.

66. T. Gremillion, "Case comment: *Environmental Defense v. Duke Energy Corporation*," *Harvard Environmental Law Review* 31, 2007: 343.

67. *2007 Commercial Energy Code Compliance Study* (Zing Communications, 2007) (http://www.aboutlightingcontrols.org/education/pdfs/2007%20Commercial%20Energy%20Code%20Compliance%20Study.pdf): 23.

68. Brian Yang, *Residential Energy Code Evaluations: Review and Future Directions*, Building Codes Assistance Project, 2005 (http://www.bcap-energy.org/files/BCAP_RESIDENTIAL_ENERGY_CODE_EVALUATION_STUDY_June2005.pdf).

69. P. L. 95–617, section 202, Interconnection.

70. Steffen Mueller, "Missing the spark: An investigation into the low adoption paradox of combined heat and power technologies," *Energy Policy* 34, 2006: 3153–3164.

71. T. R. Casten and R. U. Ayres, *Energy Myth Eight—Worldwide Power Systems are Economically and Environmentally Optimal* in *Energy and American Society—Thirteen Myths*, ed. B. Sovacool and M. Brown (Springer, 2007).

72. Brown and Chandler, "Governing confusion."

73. Kate Robertson, Jette Findsen, and Steve Messner, *International Carbon Capture and Storage Projects—Overcoming Legal Barriers*, National Energy Technology Laboratory Report DOE/NETL-2006-1236, 2006.

74. Kevin Bliss, 2005. *Carbon Capture and Storage: A Regulatory Framework for States Summary of Recommendations* (Interstate Oil and Gas Compact Commission, 2005): 5–6; Susan Hovorka, telephone interview, November 14, 2006.

75. Ted Sabety, "Nanotechnology innovation and the patent thicket: Which IP policies promote growth?" *Albany Law Journal of Science & Technology* 15, 2005: 477–515.

76. Kurt M. Saunders and Linda Levine, "Better, faster, cheaper-later: What happens when technologies are suppressed," *Michigan Telecommunications and Technology Law Review* 11, 2004: 23–69.

77. Sovacool, "Placing a glove on the invisible hand": 422.

78. Eugene R. Quinn, "Cost of obtaining a patent," IPWatchdog, 2006 (http://www.ipwatchdog.com/patent_cost.html).

79. Jerry R. Potts, What Does It Cost to Obtain a Patent (http://pw1.netcom.com/~patents2/What%20Does%20It%20Cost%20Patent.htm).

80. Sovacool, "Placing a glove on the invisible hand," 101–161.

81. Robert M. Rosenzweig, "Research as intellectual property: Influences within the university," *Science, Technology, & Human Values* 10, 1985: 41–42; Dominique Foray, "Generation and distribution of technological knowledge," in *Systems of Innovation: Technologies,*

Institutions, and Organizations, ed. C. Edquist (Pinter, 1997); Donald Kennedy, "Science and secrecy," *Science* 289, 2000: 724; Nicholas S. Argyres and Julia Porter Liebeskind, "Privatizing the intellectual commons: Universities and the commercialization of biotechnology," *Journal of Economic Behavior & Organization* 35, 1998: 427–454; Aldo Guena and Lionel J. J. Nesta, "University patenting and its effects on academic research: The emerging European evidence," *Research Policy* 35, 2006: 790–807.

82. Walter W. Powell and Jason Owen-Smith, "Universities and the market for intellectual property in the life sciences," *Journal of Policy Analysis and Management* 17, 1998, no. 2: 253–277.

83. P. Biermayer and J. Lin, "Clothes washer standards in china—The problem of water and energy trade-offs in establishing efficiency standards," presented at 2004 ACEEE Summer Study on Energy Efficiency in Buildings.

84. H. Chappells and E. Shove, *Comfort: A Review of Philosophies and Paradigms* (Lancaster University, 2004).

85. George A. Akerlof, "The market for 'lemons': Quality uncertainty and the market mechanism," *Quarterly Journal of Economics* 84, 1970, no. 3: 488–500.

86. P. Bagnoli, J. Château, and Y. Kim, "The incidence of carbon pricing: Norway, Russia and the Middle East," *OECD Journal of Economic Studies* 1, 2008: 239–264.

87. George J. Stigler, "The economics of information," *Journal of Political Economy* 69, 1961, no. 3: 213–225.

88. Herbert A. Simon, "A behavioral model of rational choice," *Quarterly Journal of Economics* 69, 1955, no. 1: 99–118; Herbert A. Simon, "Rational decision making in business organizations," *American Economic Review* 69, 1979, no. 4: 493–513.

89. Wei Shao, Ashley Lye, and Sharyn Rundle-Thiele, "Decisions, decisions, decisions: Multiple pathways to choice," *International Journal of Market Research* 50, 2008, no. 6: 797–816.

90. E. Dawnay and H. Shah, *Behavioural Economics: Seven Principles for Policy-Makers* (New Economics Foundation, 2005); Marilyn A. Brown, Jess Chandler, and Melissa V. Lapsa, "Policy options targeting decision levers: An approach for shrinking the residential energy-efficiency gap," presented at ACEEE Summer Study on Energy Efficiency in Buildings, Pacific Grove, California, 2010.

91. B. van Mierlo and B. Oudshoff, Literature Survey and Analysis of Non-Technical Problems for the Introduction of Building Integrated Photovoltaic Systems (IEA PVPS Task 7-01:1999) (http://www.iea-pvps.org/products/rep7_01.htm); E. Worrell and G. Biermans, "Move over! Stock turnover, retrofit and industrial energy efficiency," Energy Policy 33, 2005, no. 7: 949–962.

92. Neal Elliott, telephone interview, October 2006.

93. David Greene (Corporate Fellow, Center for Transportation Analysis, Oak Ridge National Laboratory) telephone interview, October 2006.

94. Sovacool, *The Dirty Energy Dilemma: What's Blocking Clean Power in the US*, 169.

95. Barbara C. Farhar, "Trends in US public perceptions and preferences on energy and environmental policy," *Annual Review of Energy and Environment* 9, 1994: 211–239.

96. Kentucky Environmental Education Council, *The 2004 Survey of Kentuckians' Environmental Knowledge, Attitudes and Behaviors*, 2005.

97. Suzanne Crofts Shelton, *The Consumer Pulse Survey on Energy Conservation* (Shelton Group, 2006).

98. Glenn Hess, "Bush promotes alternative fuel," *Chemical and Engineering News*, March 6, 2006: 50–58.

99. Amanda R. Carrico, Paul Padgett, Michael P. Vandenbergh, Jonathan Gilligan, and Kenneth A. Wallston, "Costly myths: An analysis of idling beliefs and behavior in personal motor vehicles," *Energy Policy* 39, 2009: 2881–2888.

100. Thomas S. Turrentine and Kenneth S. Kurani, "Car buyers and fuel economy?" *Energy Policy* 35, 2007: 1213–1223.

101. Stephen M. Rosoff, Henry N. Pontell, and Robert Tillman, *Profit Without Honor: White-Collar Crime and the Looting of America* (Prentice-Hall, 1998): viii.

102. Jenny Palm and Erica Lofstrom, "Domestication of new technology in households," presented at annual conference of Society for the Social Studies of Science and European Association for the Study of Science and Technology, Rotterdam, 2008.

103. Gregoire Wallenborn, "The new culture of energy: How to empower energy users?" presented at International Conference on Energy and Culture, Esbjerg, Denmark, 2007.

104. Feng Dianshu, Benjamin K. Sovacool, and Khuong Minh Vu, "The barriers to energy efficiency in China: Assessing household electricity savings and consumer behavior in Liaoning Province," *Energy Policy* 38, 2010, no. 2: 1202–1209.

105. National Geographic/GlobeScan, Consumer Greendex (http://www.nationalgeographic.com/greendex/).

106. Minoru Takada, Ellen Morris, and Sudhir Chella Rajan, *Sustainable Energy Strategies: Materials for Decision-Makers* (United Nations Development Programme, 2000) (http://www.undp.org/energy/publications/2000/2000a.htm), chapter 5, p. 22.

107. Intergovernmental Panel on Climate Change, *Methodological and Technological Issues in Technology Transfer* (Cambridge University Press, 2000).

108. E. Vine, "An international survey of the energy service company (ESCO) industry," *Energy Policy* 33, 2005, no. 5: 691–704; H. Westling, *Performance Contracting: Summary Report from the IEA DSM Task X Within the IEA DSM Implementing Agreement* (IEA, 2003).

109. US Department of Energy (DOE), *National Electric Transmission Congestion Study*, 2006 (http://nietc.anl.gov/documents/docs/Congestion_Study_2006-9MB.pdf).

110. Doug Arent (Center Director, Strategic Energy Analysis Center, National Renewable Energy Laboratory), telephone interview, October 2006; National Commission on Energy Policy, Task Force on Biofuels Infrastructure, 2009.

111. Kevin Bliss, *Carbon Capture and Storage*.

112. The electricity industry has undergone dramatic structural changes over the last 15 years. Before 1990, the US had a system of vertically integrated (generation, transmission, and distribution) monopolies that were highly regulated. The rationale for this industry structure was that electricity production is a natural monopoly: a single firm can produce the total

market output at a lower cost than a collection of individual competitive firms. The industry today, however, is a mixed system that includes some elements of market competition in many states. Privately owned electricity generators have been released from rate-of-return regulation, allowing them to earn market-based rates. Regulations promulgated in the 1980s and 1990s make it possible today for independent power producers and other utilities to compete in the merchant generation sector.

113. B. Sovacool and R. Hirsh, "Energy myth seven—The barriers to new and innovative energy technologies are primarily technical: The case of distributed generation," in *Energy and American Society—Thirteen Myths,* ed. B. Sovacool and M. Brown (Springer, 20007).

114. Aldefer et al., "Making connections."

115. US Department of Energy, Office of Energy Efficiency and Renewable Energy (DOE/EERE), *Buildings Energy Data Book.* (DOE/EERE, 2008), table 2.5.1.

116. *Blueprint for Energy-Efficiency Acceleration Strategies for Buildings in the Western Hemisphere* (Alliance to Save Energy, 2005); Joe Loper, Lowell Ungar, David Weitz, and Harry Misuriello, *Building on Success: Policies to Reduce Energy Waste in Buildings,* report to Alliance to Save Energy, 2005 (http://www.ase.org/images/lib/buildings/Building%20 on%20Success.pdf); Marilyn Brown, Frank Southworth, and Therese K. Stovall, *Toward a Climate-Friendly Built Environment* (Pew Center on Global Climate Change, 2005) (http://www.pewclimate.org/docUploads/Buildings_FINAL.pdf); Benjamin K. Sovacool, *The Dirty Energy Dilemma: What's Blocking Clean Power in the US* (Praeger, 2008).

117. Richard Newell (Gendell Associate Professor of Energy and Environmental Economics, Duke University), telephone interview, November 2006.

118. Katie Southworth, "Corporate voluntary action: A valuable but incomplete solution to climate change and energy security challenges," *Policy and Society* 27, 2009, no. 4: 329–350.

119. Nicholas DiMascio, "Credit where credit is due: The legal treatment of early greenhouse gas emissions reductions," *Duke Law Journal* 56, 2007: 1587.

120. Prindle, *Quantifying the Effects of Market Failures in the End-Use of Energy.*

121. S. Rezessy, K. Dimitrov, D. Urge-Vorsatz, and S. Baruch, "Municipalities and energy efficiency in countries in transition: Review of factors that determine municipal involvement in the markets for energy services and energy efficient equipment, or how to augment the role of municipalities as market players," *Energy Policy* 34, 2006, no. 2: 223–237.

122. Brown et al., *Carbon Lock-In*: 93.

123. *2008 Buildings Energy Data Book* (D&R International) (http://buildingsdatabook.eere .energy.gov/docs/DataBooks/2008_BEDB_Updated.pdf), table 2.2.2.

124. Ibid., table 3.2.3.

125. Brown et al., *Toward a Climate-Friendly Built Environment.*

126. Richard L. Ottinger and Rebecca Williams, "2002 energy law symposium: Renewable energy sources for development," *Environmental Law* 32, 2002: 331–362.

127. D&R International, Ltd. *2008 Buildings Energy Data Book,* November, table 3.2.3.

128. T. Dietz, G. T. Gardner, J. Gilligan, P. C. Stern, and M. P. Vandenbergh, "Household actions can provide a behavioral wedge to rapidly reduce US carbon emissions," *Proceedings*

*of the National Academy of Science*s 106, 2009, 44: 18452–18456 (http://www.pnas.org/content/106/44/18452).

129. Brown and Chandler, "Governing confusion."

Chapter 6

1. A. Denny Ellerman, "A note on tradeable permits," *Environmental & Resource Economics* 31, 2005: 129–130. "Common-pool resources" are natural or human-made resource systems, such as fishing grounds and irrigation systems, from which it is difficult or costly to exclude potential users. Because their benefits are finite, common-pool resources face problems of congestion and overuse.

2. Carolyn Raffensperger and Joel A. Tickner, *Protecting Public Health and the Environment: Implementing the Precautionary Principle* (Island, 1999).

3. Joe Thornton, *Pandora's Poison: Chlorine, Health, and a New Environmental Strategy* (MIT Press, 2001).

4. Ibid.

5. P. F. Ricci, D. Rice, J. Ziagos, and L. A. Cox Jr., "Precaution, uncertainty and causation in environmental decisions," *Environment International* 29, 2003, no. 1: 1–19.

6. Raffensperger and Tickner, *Protecting Public Health*.

7. These following paragraphs borrow heavily from Thornton's book *Pandora's Poison*.

8. Ibid.

9. Scott Douglas Bauer, "The Food Quality Protection Act of 1996: Replacing old impracticalities with new uncertainties in pesticide regulation," *North Carolina Law Review* 75 (April 1997): 1369–1409.

10. National Research Council. 2010. *Advancing the Science of Climate Change* (National Academies Press), chapter 6.

11. Thornton, *Pandora's Poison*.

12. Ibid.

13. Christian Gollier, Bruno Jullien, and Nicolas Treich, "Scientific progress and irreversibility: An economic interpretation of the 'precautionary principle,'" *Journal of Public Economics* 75, 2000, no. 2: 229–253; Ricci et al., "Precaution, uncertainty and causation in environmental decisions."

14. C. R. Sunstein, *Beyond the Precautionary Principle* (Cambridge University Press, 2005); National Research Council, *Informing an Effective Response to Climate Change* (National Academies Press, 2010), chapter 3.

15. Bryan Norton, *Sustainability: A Philosophy of Adaptive Ecosystem Management* (University of Chicago Press, 2005).

16. R. J. Lazarus, "Super wicked problems and climate change: Restraining the present to leverage the future," *Cornell Law Review* 94, 2009: 1153–1234.

17. For a survey of different mechanisms, see the following: Janet Sawin, *The Role of Government in the Development and Diffusion of Renewable Energy Technologies: Wind Power in the United States, California, Denmark, and Germany, 1970–2000*, PhD dissertation, Tufts

University, 2001; Michael P. Vandenbergh, "From smokestack to SUV: The individual as regulated entity in the new era of environmental law," *Vanderbilt Law Review* 57, 2004: 515–610; Carolyn Fischer and Richard G. Newel, "Environmental and technology policies for climate mitigation," *Journal of Environmental Economics and Management* 55, 2008:142–162; Anna-Lisa Linden, Annika Carlsson-Kanyama, and Bjorn Eriksson, "Efficient and inefficient aspects of residential energy behavior: what are the policy instruments for change?" *Energy Policy* 34, 2006: 1918–1927; Carl Blumstein, Betsy Krieg, Lee Schipper, and Carl York, "Overcoming social and institutional barriers to energy conservation," *Energy* 5, 1980: 355–371; Howard Geller, *Energy Revolution* (Island, 2003): 47–92; Michael P. Vandenbergh, "From smokestack to SUV"; Carolyn Fischer and Richard G. Newel, "Environmental and technology policies for climate mitigation," *Journal of Environmental Economics and Management* 55, 2008: 142–162; Anna-Lisa Linden, Annika Carlsson-Kanyama, and Bjorn Eriksson, "Efficient and inefficient aspects of residential energy behavior: What are the policy instruments for change?" *Energy Policy* 34, 2006: 1918–1927; Carl Blumstein, Betsy Krieg, Lee Schipper, and Carl York, "Overcoming social and institutional barriers to energy conservation," *Energy* 5, 1980: 355–371; Howard Geller, *Energy Revolution* (Island, 2003): 47–92.

18. David L. Weimer and Aidan R. Vining, *Policy Analysis: Concepts and Practice*, fifth edition (Prentice-Hall, 2011), chapter 16. William N. Dunn, *Public Policy Analysis: An Introduction*, fourth edition (Prentice-Hall, 2008).

19. Joseph E. Aldy, Alan J. Krupnick, Richard G. Newell, Ian W. H. Parry, and William A. Pizer, *Designing Climate Mitigation Policy*, National Bureau of Economic Research Working Paper 15022, 2009 (www.nber.org/papers/w15022): 2.

20. http://gov. ca.gov/index. php?/press-release/4111/

21. http://www.interaction.org/document/policy-statement-2009-g8-climate-change

22. http://www.climaticoanalysis.org/post/chinas-copenhagen-pledges/

23. R. J. Lempert, M. E. Scheffan, and D. Sprinz, "Methods for long term environmental policy challenges," *Global Environmental Politics* 9, 2009, no. 3: 106–133.

24. Nicholas Kaldor and John R. Hicks, "The valuation of the social income," *Economica* 7, 1940, 26: 105–124.

25. John Rawls, *A Theory of Justice* (Harvard University Press, 1971).

26. Md Rumi Shammin and Clark Bullard, "Impact of cap-and-trade policies for reducing Greenhouse gas emissions on US Households," *Ecological Economics* 68, 2009: 2432–2438.

27. Terry M. Dinan and Diane L. Rogers, "Distributional effects of carbon allowance trading: How government decisions determine winners and losers," *National Tax Journal* 55, 2002: 199–222.

28. T. Marc Jones and Peter Fleming, *Unpacking Complexity Through Critical Stakeholder Analysis: The Case of Globalisation* (Macquarie University, 2003).

29. Paul Dragos Aligica, "Institutional and stakeholder mapping: Frameworks for policy analysis and institutional change," *Public Organizational Review* 6, 2006: 79–90.

30. P. De Leon and D. M. Varda, "Toward a theory of collaborative policy networks: Identifying structural tendencies," *Policy Studies Journal* 37, 2009, no. 1: 59–74; J. S. Dryzek and

A. Tucker, "Deliberative innovation to different effect: Consensus Conferences in Denmark, France and the United States," *Public Administration Review* 68, 2008, 5: 864–876; C. M. Hendriks and L. Carson, "Can the market help the forum? Negotiating the commercialization of deliberative democracy," *Policy Sciences* 41, 2008: 293–313; Jurian Edelenbos and Erik-Hans Klijn, "Managing stakeholder involvement in decision making," *Journal of Public Administration Research and Theory* 16, 2005: 417–446.

31. Susan F. Rockloff and Stewart Lockie, "Participatory tools for coastal zone management: Use of stakeholder analysis and social mapping in Australia," *Journal of Coastal Conservation* 10, 2004: 81–92; Christopher M. Weible, Paul A. Sabatier, and Kelly McQueen, "Themes and variations: Taking stock of the advocacy coalition framework," *Policy Studies Journal* 37, 2009, no. 1: 121–140.

32. Oliver Heidrich, Joan Harvey, and Nicola Tollin, "Stakeholder analysis for industrial waste management systems," *Waste Management* 29, 2009: 965–973.

33. William N. Dunn, *Public Policy Analysis: An Introduction*, fourth edition (Prentice-Hall, 2008).

34. *Limiting Carbon Dioxide Emissions: Prices Versus Caps*, Congressional Budget Office, 2005 (http://www.cbo.gov/doc.cfm?index=6148): 4; *The Economic Effects of Legislation to Reduce Greenhouse Gas Emissions*, Congressional Budget Office, 2009.

35. Earlier bills debated in the US Senate have also had cap-and-trade programs and also received considerable political traction. Perhaps the earliest seriously debated cap-and-trade legislation was the 2005 Bingaman proposal that emerged from the National Commission on Energy Policy's 2004 report *Ending the Energy Stalemate* (http://www.energycommission.org/). That proposal was followed by the Low Carbon Economy Act of 2007 (introduced by Senators Bingaman and Specter) and America's Climate Security Act of 2007 (introduced by Senators Lieberman and Warner).

36. Aldy et al., *Designing Climate Mitigation*; Lawrence H. Goulder, *Carbon Taxes versus Cap-and-Trade*, Working Paper (Department of Economics, Stanford University, 2009).

37. A. Denny Ellerman, Paul L. Joskow, and David Harrison Jr., *Emissions Trading in the US: Experience, Lessons, and Considerations for Greenhouse Gases*, Pew Center on Global Climate Change, 2003 (http://www.pewclimate.org/global-warming-in-depth/all_reports/emissions_trading).

38. National Academy of Sciences, *Limiting the Magnitude of Future Climate Change* (National Academies Press), draft, February, 2010.

39. NCEP, *Ending the Energy Stalemate*; NCEP, "Forging the climate consensus," 2009 (http://www.bipartisanpolicy.org/search/node/type%253Alibrary%2Bcategory%253A18%2BForging%2Bthe%2BClimate%2BConsensus).

40. International Energy Agency, CO_2 *Emissions from Fuel Combustion, 2008 Edition*, 2008 (http://www.iea.org/textbase/publications/free_new_Desc.asp?PUBS_ID=1825), table 1.

41. Emilie Alberola, Julien Chevallier, and Benoît Chèze, "Price drivers and structural breaks in European carbon prices 2005–2007," *Energy Policy* 36, 2008, no. 2: 787–797; T. H. Tietenberg, *Emissions Trading: Principles and Practice* (Resources for the Future, 2006).

42. Goulder, *Carbon Taxes versus Cap-and-Trade*.

43. Aldy et al., *Designing Climate Mitigation Policy*.

44. Daniel Hall, "Mandatory regulation of nontraditional greenhouse gases: Policy options for industrial process emissions and non-CO_2 gases," in *Assessing US Climate Policy Options*, ed. R. Kopp and W. Pizer (Resources for the Future, 2007).

45. NCEP, "Ending the energy stalemate."

46. H. Fell and R. Morgenstern, *Alternative Approaches to Cost Containment in a Cap-and-Trade System*, RFF Discussion Paper DP 09–14 (Resources for the Future, 2009): 23; NCEP, "Forging the climate consensus" (http://www.bipartisanpolicy.org/library/national -commission-energy-policy/forging-climate-consensus-domestic-and-international-offse).

47. NCEP, *Forging the Climate Consensus*.

48. Aldy et al., *Designing Climate Mitigation Policy*.

49. NCEP, *Forging the Climate Consensus*.

50. NCEP, *Forging the Climate Consensus*.

51. "Can Airlines Actually Reduce Their Emissions?" ClimateBiz, January 21, 2008.

52. Air Cairo, Egypt Air Group, Jordan Aviation, Kuwait Airways, Libyan Airlines, Middle East Airlines, Oman Air, Royal Jordanian, Saudi Arabian Airlines, Syrian Arab Airways and Yemen Airways comprise the AACO.

53. National Academy of Sciences, *Limiting the Magnitude of Future Climate Change*.

54. National Academy of Engineering, *Workshop on Assessing Economic Impacts of Greenhouse Gas Emissions*, Washington, October 2–3, 2008; Luis Mundaca et al., "Evaluating Energy Efficiency Policies with Energy-Economy Models," *Annual Review of Environment and Resources* 35, 2010: 305-344; Aldy et al., *Designing Climate Mitigation Policy*.

55. Marilyn Brown and Susan M. Macey, "Residential energy conservation through repetitive household behaviors," *Environment and Behavior* 15, 1983: 123–141.

56. Daniel Kahneman, Jack L. Knetsch, and Richard H. Thaler, "Anomalies: The endowment effect, loss aversion, and status quo bias," *Journal of Economic Perspectives* 5, 1991, no. 1: 193–206.

57. Ahmad Faruqui and Sanem Sergici, *Household Response to Dynamic Pricing of Electricity—A Survey of The Experimental Evidence* (Brattle Group, 2009). (http://www.hks.harvard .edu/hepg/Papers/2009/The%20Power%20of%20Experimentation%20_01-11-09_.pdf).

58. P. Stern, "Changing behavior in households and communities: What have we learned?" in *New Tools for Environmental Protection*, ed. T. Dietz and P. Stern (National Academies Press, 2002).

59. Intergovernmental Panel on Climate Change, *Climate Change 2007: Mitigation of Climate Change* (Cambridge University Press, 2007), table SPM.7.

60. REN 21, *Renewables 2010 Global Status Report*.

61. Ibid.

62. Herbert Girardet and Miguel Mendonca, *A Renewable World: Energy, Ecology, Equality* (Green Books, 2009): 162–169.

63. Jason Coughlin and Karlynn Cory, *Solar Photovoltaic Financing: Residential Sector Deployment* (National Renewable Energy Laboratory, March 2009, NREL/TP-6A2–44853).

64. Ren 2010, ibid.

65. Elisabeth Rosenthal, "German suburb aims to blaze a green trail by giving up the car," *International Herald Tribune*, May 12, 2009.

66. Ibid.

67. Many drivers purchased multiple vehicles or increased the amount they drove on the other days to compensate. See Lucas W. Davis, "The effect of driving restrictions on air quality in Mexico City," *Journal of Political Economy* 116, 2008, no. 1: 38–81.

68. Steven Erlanger, "Israel is set to promote the use of electric cars," *New York Times*, January 21, 2008.

69. Daniela Feldman, "Israelis have their eyes set on electric vehicles," *Jerusalem Post*, July 14, 2009.

70. Jinfeng Xu, "Local implementation of building energy policies in China's Jiangsu Province," presented at ACEEE Summer Study on Energy Efficiency in Buildings, 2010; *Renewables: Global Status Report* (http://www.ren21.net/pdf/RE_GSR_2009_Update.pdf).

71. Merrilee Harrigan, "Can we transform the market without transforming the consumer?" *Home Energy* 11, 1994, no. 1: 17–23.

72. See Howard Geller, Philip Harrington, Arthur H. Rosenfeld, Satoshi Tanishima, and Fridtjof Unande, "Policies for increasing energy efficiency: Thirty years of experience in OECD countries," *Energy Policy* 34, 2006: 556–573; Yukiko Fukasaku, "Energy and environment policy integration: The case of energy conservation policies and technologies in Japan," *Energy Policy* 23, 1995: 1063–1076.

73. Marilyn A. Brown, Jess Chandler, and Melissa V. Lapsa, "Policy options targeting decision levers: an approach for shrinking the residential energy-efficiency gap," presented at ACEEE Summer Study on Energy Efficiency in Buildings, 2010.

74. Merrian Fuller, *Enabling Investments in Energy Efficiency* (California Institute for Energy and Environment, 2009) (http://uc-ciee.org/energyeff/documents/resfinancing.pdf).

75. Wil Nuijen and Meindert Booij, *Experiences with Long Term Agreements on Energy Efficiency and an Outlook to Policy for the Next 10 Years* (Netherlands Agency from Energy and the Environment, 2002); Dian Phylipsen, Kornelis Blok, Ernst Worrell, and Jeroen de Beer, "Benchmarking the Energy Efficiency of Dutch Industry: An Assessment of the Expected Effect on Energy Consumption and CO_2 Emissions," *Energy Policy* 8, 2002: 663-679.

76. James Lamont, "India to launch energy-efficiency trading," *Financial Times*, September 27, 2009; World Bank, "India: Energy efficiency lending focuses on industry clusters," 2008 (http://go.worldbank.org/RWBUAW0RM0); E. Worrell and G. Biermans, "Move over! Stock turnover, retrofit and industrial energy efficiency," *Energy Policy* 33, 2005: 949–962.

77. J. Lin, Nan Zhou, Mark D Levine, David Fridley 2006. *Achieving China's Target for Energy Intensity Reduction in 2010: An Exploration of Recent Trends and Possible Future Scenarios* (Lawrence Berkeley National Laboratory, 2006); J. Sinton, Mark Levine, David Fridley, Fuqiang Yang, Jiang Lin, *Status Report on Energy Efficiency Policy and Programs in China* (Lawrence Berkeley National Laboratory, 1999); W. Wang, Jonathon E Sinton, Mark D Levine 1995. *China's Energy Conservation Policies and Their Implementation* (Lawrence Berkeley National Laboratory, 1995); Z. Zhang, "Why did the energy intensity fall in China's industrial sector in the 1990s? The relative importance of structural change and intensity change," *Energy Economics* 25, 2003, no. 6: 625–638.

78. Dan Bennack, George Brown, Sally Bunning, and Mariangela Hungria da Cunha, "Soil biodiversity management for sustainable and productive agriculture: Lessons from case studies," in *Biodiversity and the Ecosystem Approach in Agriculture, Forestry, and Fisheries* (United Nations Food and Agricultural Organization, 2003).

79. See Susan Subak, "The case of Costa Rica's 'carbon commodity,'" presented at Forest Trends Workshop: New Market Mechanisms for Managing Forests, Victoria, BC, 1999; Lyès Ferroukhi and Alejandra Aguilar Schramm, "Progress and challenges of municipal forest management in costa rica," in *Municipal Forest Management in Latin America*, ed. L. Ferroukhi (International Development Research Center, 2003).

80. Government of Malaysia, "Malaysia's Forest Management with Reference to Ramin," presented at fourteenth meeting of Conference of the Parties to CITES, The Hague, 2007.

81. Allen, D. T., D. J. Bauer, B. Bras, T. G. Gutowski, C. F. Murphy, T. S. Piwonka, P. S. Sheng, J. W. Sutherland, D. L. Th urston, and E. E. Wolff, "Environmentally benign manufacturing: Trends in Europe, Japan and the USA," *ASME Journal of Manufacturing Science* 124, 2002, no. 4: 908–920.

82. Bert Bras, "Product design issues," in *Closed Loop Supply Chains*, ed. M. Ferguson and G. Souza (Taylor and Francis, 2010).

83. EPR is sometimes called "product stewardship" or "take-back" legislation. It was initially discussed as the EU's "Waste Electrical and Electronic Equipment" directive in 1998, or WEEE, before it was implemented as EPR in 2001.

84. See Noah Sachs, "Planning the funeral at the birth: Extended producer responsibility in the European Union and the US," *Harvard Environmental Law Review* 30, 2006: 51–98; Alice Castell, Roland Clift, and Chris France, "Extended producer responsibility policy in the European Union: A horse or a camel?" *Journal of Industrial Ecology* 8, 2004, 1/2: 4–7; Megan Short, "Taking back the trash: Comparing European extended producer responsibility and take-back liability to US environmental policy and attitudes," *Vanderbilt Journal of Transnational Law* 37, 2004: 1217–1301.

85. Chris van Rossem, Naoko Tojo, Thomas Lindhqvist, *Extended Producer Responsibility: An Examination of its Impact on Innovation and Greening Products*, report commissioned by Greenpeace International, Friends of the Earth, and the European Environmental Bureau, 2006.

Chapter 7

1. Bibb Latene, *The Unresponsive Bystander: Why Doesn't He Help?* (Prentice-Hall, 1970).

2. Michael W. Eysenck, *Psychology: An International Perspective* (Psychology Press, 2004): 674–675; T. F. Pettijohn, *Psychology: A Concise Introduction 3rd Edition* (Dushkin, 1992); Lenneth O. Doyle, *Interaction: Readings in Human Psychology* (University of Michigan, 1973): 90–98; F. J. McGuigan and Paul J. Woods, *Contemporary Studies in Psychology* (Appleton-Century, 1972).

3. Berryl Crowe, "The tragedy of the commons revisited," *Science* 166, 1969: 1103–1107.

4. Elinor Ostrom, *A Polycentric Approach for Coping with Climate Change*, report prepared for WDR2010 Core Team at World Bank, 2009.

5. Portions of this chapter are drawn from Benjamin K. Sovacool and Marilyn A. Brown, "Addressing climate change: Global vs. local scales of jurisdiction?" in *Carbon Constrained: Future of Electricity*, ed. F. Sioshansi (Elsevier, 2009): 109–124; Benjamin K. Sovacool and Marilyn A. Brown, "Scaling the response to climate change," *Policy & Society* 27, 2009, no. 4: 317–328; Benjamin K. Sovacool, "The best of both worlds: Environmental federalism and the need for federal action on renewable energy and climate change," *Stanford Environmental Law Journal* 27, 2008, no. 2: 397–476.

6. Richard B. Stewart, "Pyramids of sacrifice? Problems of federalism in mandating state implementation of national environmental policy," *Yale Law Journal* 86, 1976–1977: 1196–1272.

7. Steven G. Calabresi, "A government of limited and enumerated powers": In defense of *United States v. Lopez*," *Michigan Law Review* 94, 1995: 752, 759.

8. Daniel C. Esty, "Revitalizing Environmental Federalism," *Michigan Law Review* 95, 1996: 560.

9. Sovacool, "The best of both worlds."

10. Krister P. Andersson and Elinor Ostrom, "Analyzing decentralized resource regimes from a polycentric perspective," *Policy Sciences* 41, 2008: 71–93.

11. Sovacool, "The best of both worlds."

12. Stewart, "Pyramids of sacrifice": 1212.

13. Kirsten Engel and Susan Rose-Ackerman, "Environmental federalism in the United States: The risks of devolution," in *Regulatory Competition and Economic Integration: Comparative Perspectives*, ed. D. Esty and D. Geradin (Oxford University Press, 2001).

14. Phil Rene Oyono, Jesse C. Ribot, and Anne M. Larson. 2006. "Green and black gold in rural Cameroon: Natural resources for local governance, justice and sustainability," Environmental Governance in Africa Working Paper, World Resources Institute, 2006.

15. Peter Veit, World Resources Institute, personal communication, 2008.

16. Kirsten H. Engel, "State environmental standard-setting: is there a 'race' and is it 'to the bottom'?" *Hastings Law Journal* 48, 1997, no. 271: 317–318.

17. Rena I. Steinzor, "Devolution and the public health," *Harvard Environmental Law Review* 24, 2000: 351.

18. Mancur Olson, *The Logic of Collective Action: Public Goods and the Theory of Groups* (Harvard University Press, 1965).

19. Lee Schipper, *Energy Conservation: Its Nature, Hidden Benefits, and Hidden Barriers*, Lawrence Berkeley National Laboratory Report UCID 3725 ERG 2, 1975.

20. Jonathan B. Weiner, "Think globally, act globally: The limits of local climate policies," *University of Pennsylvania Law Review* 155, 2007: 1961–1979.

21. Jennette Gayer and Jeff Kerr, *Proposed Coal Power in Georgia: A Pollutant Summary* (Environment Georgia Research & Policy Center, 2007).

22. Sovacool, "The best of both worlds."

23. See generally Thomas W. Merrill, "Golden rules for transboundary pollution," *Duke Law Journal* 46, 1997: 931.

24. Ibid.: 985.

25. Christina C. Caplan, "The failure of current legal and regulatory mechanisms to control interstate ozone transport: The need for new national legislation," *Ecology Law Quarterly* 28, 2001, no. 1: 169, 201–202.

26. Steinzor, "Devolution and the Public Health": 462–263.

27. Richard L. Revesz, "Rehabilitating Interstate Competition: Rethinking the "Race-To-The-Bottom" Rationale for Federal Environmental Regulation," New York University Law Review 67, 1992: 1210–1254; Richard L. Revesz, "A race to the bottom and federal environmental regulation: A response to critics," *Minnesota Law Review* 82, 1997: 583–626.

28. Jonathan H. Adler, "Wetlands, waterfowl, and the menace of Mr. Wilson: Commerce clause jurisprudence and the limits of federal wetland regulation," *Environmental Law* 29, 1999: 41–54.

29. Robert E. Roberts, "Debunking the 'race to the bottom' myth," *Ecostates,* November 1997: 13–14.

30. Jonathan H. Adler, "Jurisdictional mismatch in environmental federalism," *New York University Environmental Law Journal* 14, 2005: 130–135.

31. Sovacool, "The best of both worlds."

32. See generally Robert W. Kates and Thomas J. Wilbanks, "Making the global local: Responding to climate change concerns from the ground up," *Environment* 45, 2003: 12–23.

33. Mark Z. Jacobson, "Effects of local CO_2 domes and of global CO_2 changes on California's air pollution and health," testimony before Environmental Protection Agency hearing on California waiver, Washington, March 5, 2009.

34. Charles Tiebout, "Exports and regional economic growth," *Journal of Political Economy* 64, 1956, April: 160–164; William A. Fischel, "Fiscal and environmental considerations in the location of firms in suburban communities," in *Fiscal Zoning and Land Use Controls,* ed. E. Mills and W. Oates (Heath, 1975); William A. Fischel, "Property taxation and the Tiebout model: Evidence for the benefit view from zoning and voting," *Journal of Economic Literature* 30, 1992, no. 1: 171–177; Wallace E. Oates and R. Schwab, "Economic competition among jurisdictions: Efficiency-enhancing or distortion-inducing?" *Journal of Public Economics* 35, 1988, April: 333–354; Wallace E. Oates and R. Schwab, "Community composition and the provision of local public goods: A normative analysis," *Journal of Public Economics* 44, 1991, March: 217–237.

35. David W. Orr, *Earth in Mind: On Education, Environment, and the Human Prospect* (Island, 1994): 161.

36. Brian G. Norton, and Robert E. Ulanowicz, "Scale and biodiversity policy: A hierarchical approach," *Ambio* 21, 1992: 244–249.

37. Orr, *"Earth in Mind."*

38. Sovacool, "The best of both worlds."

39. Ibid.

40. Robert A. Schapiro, "Toward a theory of interactive federalism," *Iowa Law Review* 91, 2005, 243–288.

41. Ibid.

42. Ostrom, *A Polycentric Approach for Coping with Climate Change.*

Chapter 8

1. David A. Stockman, "The political process and energy," in *Future American Energy Policy*, ed. M. Crist and A. Laffer (Heath, 1982).

2. Portions of this case study are based on Benjamin K. Sovacool, Hans Henrik Lindboe, and Ole Odgaard, "Is the Danish wind energy model replicable for other countries?" *Electricity Journal* 21, 2008, no. 2: 27–38.

3. On the Danish approach to wind research and development, see Benjamin K. Sovacool and Janet Sawin, "Rethinking energy innovation and the research process: Lessons from the American and Danish wind energy industries," unpublished manuscript; Raghu Garud and Peter Karnoe, "Bricolage versus breakthrough: Distributed and embedded agency in technology entrepreneurship," *Research Policy* 32, 2003: 277–300; Matthias Heymann, "Signs of hubris: The shaping of wind technology styles in Germany, Denmark, and the United States, 1940–1990," *Technology & Culture* 39, 1998, no. 4: 641–670; Ulrik Jorgensen and Peter Karnoe, "The Danish wind-turbine story: Technical solutions to political visions?" in *Managing Technology in Society*, ed. A. Rip, T. Misa, and J. Schot (Pinter, 1995); David Toke, Sylvia Breukers, and Maarten Wolsink, "Wind power deployment outcomes: How can we account for the differences?" *Renewable and Sustainable Energy Reviews* 12, 2008: 1129–1147; Miguel Mendonça, Stephen Lacey, and Frede Hvelplund, "Stability, participation and transparency in renewable energy policy: Lessons from Denmark and the United States," *Policy & Society* 27, 2009, no. 4: 379–398.

4. P. E. Morthorst, "The development of a green certificate market," *Energy Policy* 28, 2000: 1085–1094.

5. Ole Odgaard, *Energy Policy in Denmark* (Danish Energy Authority, 2007): 9.

6. Joanna Lewis and Ryan Wiser, *Fostering a Renewable Energy Technology Industry: An International Comparison of Wind Industry Policy Support Mechanisms*, Lawrence Berkeley National Laboratory Report LBNL-59116, 2005.

7. US Energy Information Administration, Denmark Energy Profile (http://tonto.eia.doe.gov/country/country_energy_data.cfm?fips=DA).

8. Danish Energy Association, *Danish Electricity Supply Statistical Survey*, 2008 (http://www.danishenergyassociation.com/Statistics.aspx): 1–3.

9. Constantine Hadjilambrinos, "Understanding technology choice in electricity industries: A comparative study of France and Denmark," *Energy Policy* 28, 2000: 1111–1126.

10. David Toke, "Are green electricity certificates the way forward for renewable energy? An evaluation of the United Kingdom's renewables obligation in the context of international comparisons," *Environment and Planning* C 23, 2005: 361–374.

11. Portions of this case study are based on Miguel Mendonça, David Jacobs, and Benjamin K. Sovacool, *Powering the Green Economy: The Feed-in Tariff Handbook* (Earthscan, 2010) and Benjamin K. Sovacool, "Correcting a market failure: Fast tracking renewable energy with the wondrous feed-in tariff," *Pacific Ecologist* 18, 2009, winter: 49–54.

12. Bundesministerium für Umwelt, Naturschutz und Reaktorsicherheit (BMU) [German Federal Ministry of Environment, Nature Conservation, and Nuclear Safety], "Background information on the EEG Progress Report 2007," 2007 (http://www.bmu.de/files/pdfs/allgemein/application/pdf/eeg_kosten_nutzen_hintergrund_en.pdf.lfne);

Beschaffungsmehrkosten der Stromlieferanten durch das Erneuerbare-Energien-Gesetz 2008 (Differenzkosten nach § 15 EEG), Ingenieurbüro für Neue Energie (IfnE), Gutachten im Auftrag des Bundesministeriums für Umwelt, Naturschutz und Reaktorsicherheit, Teltow, März 2009.

13. BMU, "Background information."

14. According to data compiled by the World Resources Institute, these countries had a combined emissions rate of 78 million tons. Source of data: World Resources Institute, Earth Trends Data Tables: Climate and Atmosphere (http://earthtrends.wri.org/datatables/index .php?theme=3).

15. Marc Ringel, "Fostering the use of renewable energies in the european union: The race between feed-in tariffs and green certificates," *Renewable Energy* 31, 2006: 1–17.

16. Mark Bolinger and Ryan Wiser, "Wind power price trends in the United States: Struggling to remain competitive in the face of strong growth," *Energy Policy* 37, 2009: 1061–1071.

17. C. Fischer and B. Praetorius, *Carbon Capture and Storage: Settling the German Coal vs. Climate Change Dispute?* TIPS Discussion Paper 7, 2006 (http://www.fona.de/en/4834).

18. Roy L. Nersesian, *Energy for the 21st century* (Sharpe, 2006): 95.

19. Louisa Schaefer, "Despite climate concerns, Germany plans coal power plants" (http://www.dw-world. de/dw/article/0,,2396828,00.html).

20. Sebastian Knauer and Michael Fröhlingsdorf, "German energy policy at the crossroads," *Spiegel International*, July 26, 2007.

21. International Energy Agency, *Deploying Renewables: Principles for Effective Policies* (OECD, 2008); Doerte Fouquet and Thomas B. Johansson, "European renewable energy policy at a crossroads—focus on electricity support mechanisms," *Energy Policy* 36, 2008, 11: 4079–4092.

22. BMU, "Background information."

23. These numbers come from two studies in California and Arizona. See L. Stoddard, J. Abiecunas, and R. O'Connell, *Economic, Energy, and Environmental Benefits of Concentrating Solar Power in California*, Subcontract Report NREL/SR-550-39291, National Renewable Energy Laboratory, 2006; Arizona Department of Commerce Energy Office, *Energy Dollar Flow Analysis for the State of Arizona*, 2004.

24. Jose Goldemberg, "The ethanol program in Brazil," *Environmental Research Letters* 1, 2006, no. 1: 1–5 (http://www.iop.org/EJ/toc/1748-9326/1/1).

25. Jose Goldemberg, "The Brazilian biofuels industry," *Biotechnology for Biofuels* 1, 2008, no. 6: 1–7.

26. Britt Childs and Rob Bradley, *Plants at the Pump: Biofuels, Climate Change, and Sustainability* (World Resources Institute, 2007): 26.

27. Vanessa M. Cordonnier, "Ethanol's roots: How Brazilian legislation created the international ethanol boom," *Environmental Law & Policy Review* 33, 2008, no. 1: 287–318.

28. Goldemberg, "The ethanol program in Brazil."

29. Haroldo Machado-Filho, "Climate change and the international trade of biofuels," *Carbon and Climate Law Review* 2, 2008, no. 1: 67–77.

30. Ethanol has a lower vapor pressure, and thus results in lower evaporative emissions and pollution, and its flammability in air is lower than that of gasoline, so it reduces the number and severity of explosions and fires in vehicle accidents. The low toxicity and high biodegradability of ethanol relative to oil and gasoline also means that accidental spills and leaks become easier to manage, and its high octane rating means there is no need for additional (and often expensive) lead-based additives. One drawback to ethanol as a fuel is that it has less energy content per gallon than gasoline (usually about 67 percent less), but this is partially offset by its higher-octane burning. Source: Goldemberg, "The Brazilian biofuels industry."

31. At the extreme end of the estimates, and when done poorly, corn ethanol can have a net energy ratio less than 1 (that is, can consume much more energy in inputs than are available in the final product), whereas ethanol from sugarcane can have a net energy ratio as high as 11 to 1.

32. Childs and Bradley, *Plants at the Pump*, 26–27.

33. Manoel Regis Lima Verde Leal, "The Brazilian experience: Proven benefits from integrating the biomass value chain from processing, biofuels refining and electricity production," presented at Biomass & Residues Asia 2008 Conference, Singapore.

34. Jose Goldemberg and Patricia Guardabassi, "Are biofuels a feasible option?" *Energy Policy* 37, 2009: 10–14.

35. State of São Paulo Research Foundation, "Finding riches in leftovers at the mill," in *Brazil World Leader in Sugarcane and Ethanol Knowledge and Technology: FAPESP's Contribution* (State of São Paulo Research Foundation, 2007).

36. State of São Paulo Research Foundation, "Styrofoam vegetable," in *Brazil World Leader in Sugarcane and Ethanol Knowledge and Technology*.

37. Cibelle Boucas, "Mechanization gains ground in São Paulo," in *Sugarcane Ethanol Energy for the World: Brazil Emerges as a Major Producer of Clean Fuels*, ed. A. Felix (Valor Economico, 2008).

38. Jose Goldemberg, "The challenge of biofuels," *Energy & Environmental Science* 1, 2008: 523–525; Daniela Chiaretti, "Four myths fuel attacks on sugarcane ethanol," in *Sugarcane Ethanol Energy for the World*, ed. A. Felix.

39. Jose Goldemberg, Susani Teixeira Coelho, and Patricia Guardabassi, "The sustainability of ethanol production from sugarcane," *Energy Policy* 36, 2008: 2086–2097.

40. Rogerio Cezar de Cerqueira Leite, Manoel Regis Lima Verde Leal, Luis Augusto Barbosa Cortez, W. Michael Griffin, and Mirna Ivonne Gaya Scandiffio, "Can Brazil Replace 5% of the 2025 Gasoline World Demand with Ethanol?" *Energy* 34, 2009: 655–661.

41. Sugar Cane Industry Union (UNICA), *Production and Use of Fuel Ethanol in Brazil: Answers to the Most Frequently Asked Questions*, 2007.

42. According to the World Resource Institute Earth Trends Climate and Atmosphere Database, the emissions from Oceana (the South Pacific, Australia, and New Zealand) in 2000 amounted to 578 million metric tons.

43. Boucas, "Mechanization Gains Ground in São Paulo."

44. Renewable Energy Policy Network for the 21st Century (REN21), *Renewables Global Status Report: 2009 Update* (http://www.ren21.net/pdf/RE_GSR_2009_Update.pdf): 24.

45. Ministry of Agriculture, Livestock, and Food Supply, *Brazil and Agribusiness*, 2007: 40.

46. For a particularly vitriolic example of such commentary, see Chris McGowan, "Biofuel could eat Brazil's savannas and deforest the Amazon," Huffington Post, September 14, 2007.

47. UNICA, *Production and Use of Fuel Ethanol in Brazil*, 42.

48. Goldemberg, "The challenge of biofuels"; Chiaretti, "Four myths fuel attacks on sugarcane ethanol"; Goldemberg and Guardabassi, "Are biofuels a feasible option?"

49. Leite et al., "Can Brazil Replace 5% of the 2025 Gasoline World Demand with Ethanol?"

50. For examples, see Rosamond L. Naylor, Adam J. Liska, Marshall B. Burke, Walter P. Falcon, Joanne C. Gaskell, Scott D. Rozelle, and Kenneth G. Cassman, "The ripple effect: biofuels, food security, and the environment," *Environment* 49, 2007, no. 9: 30–43; Ford Runge and Benjamin Senauer, "How biofuels could starve the poor," *Foreign Affairs* 86, 2007, no. 3: 41–54.

51. Lauro Veiga Filho, "Sustainable energy ignites Brazil's economy," in *Sugarcane Ethanol Energy for the World*, ed. A. Felix.

52. M. Candida Vieira and Antonio Felix, "Biofuels can help combat hunger," in *Sugarcane Ethanol Energy for the World*, ed. A. Felix.

53. Fernando Lopes, "Crops occupy fields where cattle grazed," in *Sugarcane Ethanol Energy for the World*, ed. A. Felix.

54. UNICA, *Production and Use of Fuel Ethanol in Brazil*: 29.

55 Georgina Santos, Wai Wing Li, and Winston T. H. Koh, "Transport policies in Singapore," in *Road Pricing: Theory and Evidence*, ed. G. Santo (Elsevier, 2004).

56. Leo Tan Wee Hin and R. Subramaniam, "Congestion control of heavy vehicles using electronic road pricing: The Singapore experience," *International Journal of Heavy Vehicle Systems* 13, 2006, 1/2: 37–55.

57. Santos et al., "Transport Policies in Singapore."

58. Paul A. Barter, "Singapore's urban transport: Sustainability by design or necessity," in *Spatial Planning for a Sustainable Singapore*, ed. T.-C. Wong, B. Yuen, and C. Goldblum (Springer, 2008).

59. Ibid.

60. The details in this section draw from the excellent depictions in the following: Jeremy Yap, "Implementing road and congestion pricing lessons from Singapore," presented to Workshop on Implementing Sustainable Urban Travel Policies in Japan and Other Asia-Pacific Countries, Tokyo, 2005; J. Y. K. Luk, "Electronic road pricing in Singapore," *Road & Transport Research* 8, 1999, December: 28–40; Mark Goh, "Congestion management and electronic road pricing in Singapore," *Journal of Transport Geography* 10, 2002: 29–38; Environmental Defense Fund, "Singapore: A pioneer in taming traffic," in *Taking Charge of Traffic Congestion: Lessons from Around the Globe* (http://www.edf.org/documents/6116 _SingaporeTraffic_Factsheet.pdf); Chin Kian Keong, "Road pricing: Singapore's experience," presented at seminar of IMPRINT-EUROPE thematic network, Brussels, 2002; Santos et al., "Transport Policies in Singapore."

61. Pascal Poudenx, "The effect of transportation policies on energy consumption and greenhouse gas emission from urban passenger transportation," *Transportation Research A* 42, 2008: 901–909.

62. Goh, "Congestion management and electronic road pricing in Singapore."

63. Ibid.

64. Piotr Olszewski and Litian Xie, "Modeling the effects of road pricing on traffic in Singapore," *Transportation Research A* 39, 2005: 755–772.

65. Goh, "Congestion management and electronic road pricing in Singapore."

66. Barter, "Singapore's urban transport"; Santos et al., "Transport policies in Singapore."

67. Goh, "Congestion management and electronic road pricing in Singapore"; Muhammad Faishal Ibrahim, "Improvements and integration of a public transport system: the case of Singapore," *Cities* 20, 2003, no. 3: 205–216.

68. Chin, "Road pricing."

69. Yap, "Implementing road and congestion pricing lessons from Singapore."

70. "Percentage congestion free" is defined as the percentage of expressways with average speed above 45 kph or percentage of CBD/Arterial Roads with average speed above 20 kph.

71. Barter, "Singapore's Urban Transport."

72. Environmental Defense Fund, "Singapore: A Pioneer In Taming Traffic."

73. Land Transport Authority, *Land Transport Master Plan* (Singapore: Ministry of Transport, 2008) (http://app.lta.gov.sg/ltmp/pdf/LTMP_Report.pdf).

74. Ibid.

75. Poudenx, "The effect of transportation policies."

76. Land Transport Authority, *Land Transport Master Plan*.

77. Yap, "Implementing Road and Congestion Pricing Lessons from Singapore."

78. Ibid.

79. Barter, "Singapore's urban transport."

80. Ibid.

81. Ibid.

82. Ibid.

83. Santos et al., "Transport Policies in Singapore."

84. Sock-Yong Phang and Rex Toh, "Curbing urban traffic congestion in Singapore: A comprehensive review," *Transportation Journal* 37, 1997, no. 2: 24–33.

85. Goh, "Congestion management and electronic road pricing in Singapore."

86. Yap, "Implementing Road and Congestion Pricing Lessons from Singapore."

87. *Energy Security and Sustainable Development in Asia and the Pacific*, UN Economic and Social Commission for Asia and the Pacific Report ST/ESCAP/2494, 2008.

88. Marc Gunther, "Grameen Shakti brings sustainable development closer to reality in Bangladesh," *GreenBiz Magazine*, January 21, 2009.

89. Dewan A. H. Alamgir, *Adaptive Research and Dissemination for Development of PV Technology in Bangladesh* (Grameen Shakti, 1999).

90. A. K. M. Sadrul Islam, Mazharul Islam, and Tazmilur Rahman, "Effective renewable energy activities in Bangladesh," *Renewable Energy* 31, 2006: 677–688.

91. Lisa Schroeder, "Better lives in Bangladesh—through Green Power," *Christian Science Monitor*, June 24, 2009.

92. Danesh Miah, Harun Al Rashid, and Man Yong Shin, "Wood fuel use in the traditional cooking stoves in the rural floodplain areas of Bangladesh: A socio-environmental perspective," *Biomass and Bioenergy* 33, 2009: 70–78.

93. Ibid.

94. Joanna Peios, "Fighting deforestation in Bangladesh," *Geographical*, March 14, 2004: 14.

95. K. R. Islam and R. R. Weil, "Land use effects on soil quality in a tropical forest ecosystem of Bangladesh," *Agriculture, Ecosystems and Environment* 79, 2000: 9–16.

96. M. S. Alam, K. K. Islam, and A. M. Z. Huq, "Simulation of rural household fuel consumption in Bangladesh," *Energy* 24, 1999: 743–752.

97. Wahidul K. Biswas, Paul Bryce, and Mark Diesendorf, "Model for empowering rural poor through renewable energy technologies in Bangladesh," *Environmental Science & Policy* 4, 2001: 333–344.

98. Miah et al., "Wood fuel use."

99. Ibid.

100. Biswas et al., "Model for empowering rural poor."

101. Sk Noim Uddin and Ros Taplin, "Toward sustainable energy development in Bangladesh," *Journal of Environment & Development* 17, 2008, no. 3: 292–315.

102. Sk Noim Uddin and Ros Taplin, "Trends in renewable energy strategy development and the role of CDM in Bangladesh," *Energy Policy* 37, 2009: 281–289.

103. Dipal C. Barua, *Strategy for Promotions and Development of Renewable Technologies in Bangladesh: Experience from Grameen Shakti* (Grameen Shakti, 2001): 5–7.

104. Gunther, "Grameen Shakti."

105. Junfeng Zhang and Kirk R. Smith, "Household air pollution from coal and biomass fuels in China: Measurements, health impacts, and interventions," *Environmental Health Perspectives* 115, 2007, no. 6: 848–855.

106. Peter Catania, "China's rural energy system and management," *Applied Energy* 64, 1999: 238–240.

107. For excellent histories of the program, see Jonathan E. Sinton, Kirk R. Smith, John W. Peabody, Lui Yaping, Zhang Xiliang, Rufus Edwards, and Gan Quan, "An assessment of programs to promote improved household stoves in China," *Energy for Sustainable Development* 8, 2004, no. 3: 33–52; Jonathan E. Sinton, Kirk R. Smith, John W. Peabody, Yaping Liu, Rufus Edwards, Meredith Milet, Quan Gan, and Zheng Yin, *Improved Household Stoves in China: An Assessment of the National Improved Stove Program*, revised edition (Institute for Global Health at the University of California San Francisco and School of Public Health at the University of California Berkeley, 2004); Daxiong Qiu, Shuhua Gu, Peter Catania, and Kun Huang, "Diffusion of improved biomass stoves in China," *Energy Policy* 24, 1996, no. 5: 463–469; Kirk R. Smith, Gu Shuhua, Huang Kun, and Qiu Daxiong, "One hundred million improved cookstoves in China: How was it done?" *World Development* 21, 1993, no. 6: 941–961.

108. Smith et al., "One hundred million improved cookstoves in China."

109. Owing to the number of households involved, the lack of synchronization between national and provisional programs, and unclear definitions as to what constituted an "improved" stove, reaching a definitive definition of stove penetration is difficult. Kirk Smith and his colleagues have noted that their household surveys may suggest that the figures quoted from the Ministry of Agriculture may be overstated by as much as 20 percent. However, other sources cite an upper range of penetration of 210 million households.

110. Melissa Chan, *Air Pollution from Cookstoves: Energy Alternatives and Policy in Rural China* (Carnegie Mellon University, 2000): 10.

111. Majid Ezzati, Robert Bailis, Daniel M. Kammen, Tracey Holloway, Lynn Price, Luis A. Cifuentes, Brendon Barnes, Akanksha Chaurey, and Kiran N. Dhanapala, "Energy management and global health," *Annual Review of Environment and Resources* 29, 2004: 383–419; R. B. Finkelman, H. E. Belkin, and B. Zheng, "Health impacts of domestic coal use in China," *Proceedings of the National Academy of Sciences* 96, 1999: 3427–3431; Kirk R. Smith, "Fuel combustion, air pollution exposure, and health: Situation in developing countries," *Annual Review of Energy and Environment* 18, 1993: 529–566.

112. John W. Peabody, Travis J. Riddell, Kirk R. Smith, Yaping Liu, Yanyun Zhao, Jianghui Gong, Meredith Milet, and Jonathan E. Sinton, "Indoor air pollution in rural China: Cooking fuels, stoves, and health status," *Archives of Environmental and Occupational Health* 60, 2005, no. 2: 86–95.

113. Zhang and Smith, "Household air pollution from coal and biomass fuels in China."

114. Xiadong Wang and Kirk R. Smith, "Secondary benefits of greenhouse gas control: Health benefits in China," *Environmental Science & Technology* 33, 1999: 3056–3061.

115. Catania, "China's rural energy system and management": 240.

116. Ibid.: 229–240.

117. World Resources Institute, "Earth Trends Climate and Atmosphere" (http://earthtrends .wri.org/searchable_db/index. php?theme=3).

118. Zhang and Smith, "Household air pollution."

119. Peabody et al., "Indoor air pollution."

120. R. D. Edwards, Y. Liu, G. He, Z. Zin, J. Sinton, J. Peabody, and K. R. Smith, "Household CO and PM measured as part of a review of China's National Improved Stove Program," *Indoor Air* 17, 2007: 189–203.

121. Yuxing Zhang and Conghe Song, "Impacts of afforestation, deforestation, and reforestation on forest cover in China from 1949 to 2003," *Journal of Forestry* 104, 2006, no. 7: 383–387.

122. Jintao Xu, Ran Tao, and Gregory S. Amacher, "An empirical analysis of China's state owned forests," *Forest Policy and Economics* 6, 2004: 379–390.

123. Justin Zackey, "Peasant perspectives on deforestation in southwest China," *Mountain Research and Development* 27, 2007, no. 2: 153–161.

124. Chan, *Air Pollution from Cookstoves.*

125. Sinton et al., "An assessment of programs."

126. Smith et al., "One hundred million improved cookstoves."

127. Catania, "China's Rural Energy System and Management."

128. This point is made articulately and passionately in the following: Chan, *Air Pollution from Cookstoves*; Bina Agarwal, *Cold Hearths and Barren Slopes: The Woodfuel Crisis in the Third World* (Zed, 1986); D. F. Barnes, K. Openshaw, K. R. Smith, and R. van der Plas, *What Makes People Cook with Improved Biomass Stoves? A Comparative International Review of Cookstove Programs*, Technical Paper 242, World Bank, 1994; Bina Agarwal, "Diffusion of rural innovations: Some analytical issues and the case of wood burning stoves," *World Development* 11, 1983, no. 4: 359–376.

129. On the Oasis Project, see Thais Cercal Dalmina Losso, "The adoption of private natural areas and payment for ecosystem services in Brazil: Analysis of the Oasis Project legal scheme," *Carbon & Climate Law Review* 2, 2008, no. 3: 306–311. This case study is also partially based on Benjamin K. Sovacool, "Using ecosystem valuation to protect the Atlantic Rainforest: The case of the Oasis Project," *Society & Natural Resources* XX, 20XX: XXX– XXX.

130. Haroldo Torres, Humberto Alves, and Maria Aparecida De Oliveira, "São Paulo peri-urban dynamics: Some social causes and environmental consequences," *Environment and Urbanization* 19, 2007, no. 1: 207–223.

131. Mônica Monteiro Schroeder, *São Paulo Metropolitan Region Water Sources* (SocioAmbiental, 2008).

132. Mitsubishi International Corporation Foundation, *Mitsubishi International Corporation Foundation Participates in the Oasis Project, a Watershed Protection Project in São Paulo, Brazil*, 2006.

133. Roberto S. Waack, *Latin American Regional Tropical Forest Investment Forum ITTO* (AMATA, 2008).

134. Two caveats must be mentioned. First, pollutants were not tracked from 1988 to 2007 because the size of the TRI essentially doubled in the 1990s after new classes of chemicals were introduced into the database. Thus, the emissions rates above are for only those chemicals initially listed; total amounts of listed chemicals from 2007 are understandably higher than in 1988 because more chemicals are now reported. Second, the numbers above are for on-site releases; off-site releases for these chemicals have actually increased about 12 percent from 1988 to 2001, although the total amount of off-site releases is much less than those on site. Data taken from US Environmental Protection Agency TRI Explorer On-site and Off-site Releases, 1988, 1998, and 2000–2001 (http://www.epa.gov/triexplorer/).

135. Mary Beth Arnett, "Risky business: OSHA's Hazard Communication Standard, EPA's Toxics Release Inventory, and environmental safety," *Environmental Law Reporter* 22, 1992, no. 7: 10440–10491.

136. David Sarokin, "The Toxics Release Inventory," *Environmental Science & Technology* 22, 1988, no. 6: 616–618.

137. EPCRA is also commonly referred to as the Superfund Amendments and Reauthorization Act (SARA) Title III.

138. In 1988, the first year of the TRI, the threshold amounts were 75,000 pounds released per year for chemicals manufactured, imported, or processed, and 10,000 pounds per year for chemicals "used in any other manner." Source: Sarokin, "The Toxics Release Inventory."

139. Arnett, "Risky business."

140. Virginia W. Gerde and Jeanne M. Logsdon, "Measuring environmental performance: Use of the Toxics Release Inventory and other US environmental databases," *Business Strategy & the Environment* 10, 2001: 269–285; Bradley C. Karkkainen, "Information as environmental regulation: TRI and performance benchmarking, precursor to a new paradigm?" *Georgetown Law Review* 89, 2000–2001: 258–370.

141. Karkkainen, "Information as Environmental Regulation."

142. The TRI is sometimes described as having multiple phases. In the first phase, the EPA added an initial list of 286 chemicals to the TRI list in 1988. A second phase occurred in 1990 when the TRI was expanded by the Pollution Prevention Act of 1990, which added toxic chemicals to its list and also modified reporting requirements for certain long-lived and persistent pollutants, such as mercury and dioxins). During a third phase, the EPA added seven more industrial sectors and increased the number of pollutants to more than 600 in 1994. In a fourth phase, in the late 1990s, the EPA added a new requirement that facilities also report "materials accounting" data.

143. The EPA reported 3,153,729,673 pounds of on-site and off-site releases in 1988, meaning that a 61 percent reduction results in an annual saving of 1.924 pounds for 2007.

144. The *Titanic* had a gross weight of about 46,328 tons, or 103,774,720 pounds; thus, it weighed about 1/18.5 of the 1.924 billion pounds of TRI pollution averted.

145. Shameek Konar and Mark A. Cohen, "Information as regulation: The effect of community right to know laws on toxic emissions," *Journal of Environmental Economics and Management* 32, 1997: 109–124.

146. James T. Hamilton, "Pollution as news: Media and stock market reactions to the Toxics Release Inventory," *Journal of Environmental Economics and Management* 28, 1995: 98–113; James T. Hamilton, *Regulation through Revelation: The Origin, Politics, and Impacts of the Toxics Release Inventory* (Cambridge University Press, 2005).

147. Margaret M. Jobe, "The power of information: The example of the US Toxics Release Inventory," *Journal of Government Information* 26, 1999, no. 3: 287–295.

148. Madhu Khanna, Wilma Rose H. Quimio, and Dora Bojilova, "Toxics release information: A policy tool for environmental protection," *Journal of Environmental Economics & Management* 36, 1998: 243–266.

149. *Drinking Water Infrastructure Needs Survey and Assessment*, Environmental Protection Agency Report EPA 816-R-09-001, 2009.

150. Barbara Ann Clay, "The EPA's proposed Phase-III expansion of the Toxic Release Inventory reporting requirements: Everything *and* the kitchen sink," *Pace Environmental Law Review* 15, 1997–1998: 293–328.

151. Ibid.

152. Ibid.

153. Arnett, "Risky business."

154. According to the EPA, these countries are Australia, Austria, Belgium, Canada, Cyprus, the Czech Republic, Denmark, Estonia, Finland, France, Germany, Greece, Hungary, Ireland, Italy, Latvia, Lithuania, Luxembourg, Malta, Mexico, Netherlands, Norway, Poland, Portugal, Slovakia, Slovenia, Spain, Sweden, the United Kingdom, and the United States.

155. Jeffrey C. Terry and Bruce Yandle, "EPA's Toxic Release Inventory: Stimulus and response," *Managerial and Decision Economics* 18, 1997, no. 6: 433–441.

156. Clay, "The EPA's proposed Phase-III expansion."

157. These groups even went so far as to argue that exposure of TRI data would lower American competitiveness with respect to foreign firms that could decipher the data to uncover trade secrets. The result, the chemical industry ingeniously argued, would be greater pollution and less innovative pollution technology. By compromising intellectual property, the industry also claimed that the TRI would compromise an incentive to innovate, meaning less people would invest in environmental technology. Interestingly, many of the claims from this campaign while clever were flat out false: facilities required to report under the TRI were able to protect the specific identity of reportable chemicals related to trade secrets. The EPCRA explicitly allowed TRI reporting facilities to withhold the identity of a release if the disclosure would "cause substantial harm to its competitive position." This exception is probably why the industry eventually abandoned their campaign, and the TRI amendments proceeded. See Brian Gregg, "Setting priorities for Phase Three of the Toxic Release Inventory: Trade secrets or community right-to-know?" *Environmental Law* 4, 1997–1998: 943–956.

158. Arnett, "Risky business."

159. These changes, known as the "TRI Burden Reduction Rule," were almost uniformly opposed by environmental groups, public health organizations, schools, financial investment firms, labor unions, and many states and municipalities. The rule was changed despite the EPA receiving more than 100,000 comments opposing it. Source: Marie Miranda, Martha Keating, and Sharon Edwards, "Environmental justice implications of reduced reporting requirements of the Toxics Release Inventory Burden Reduction Rule," *Environmental Science & Technology* 42, 2008: 5407–5414.

160. This change was announced on April 21, 2009 as the Toxic Release Inventory A Eligibility Revisions Implementing the 2009 Omnibus Appropriations Act.

161. Arnett, "Risky business."

162. Gerde and Logsdon, "Measuring environmental performance."

163. Ibid.

164. Arnett, "Risky business."

165. Ibid.

166. Gerde and Logsdon, "Measuring environmental performance."

167. Scott de Marchi and James T. Hamilton, "Assessing the accuracy of self-reported data: An evaluation of the toxics release inventory," *Journal of Risk Uncertainty* 13, 2006: 57–76; Dinah A. Koehler and John D. Spengler, "The toxic release inventory: Fact or fiction? A case study of the primary aluminum industry," *Journal of Environmental Management* 85, 2007: 296–307.

168. Koehler and Spengler, "The toxic release inventory."

169. Jobe, "The Power of Information."

170. Linda T. M. Bui and Christopher J. Mayer, "Regulation and capitalization of environmental amenities: Evidence from the Toxic Release Inventory in Massachusetts," *Review of Economics and Statistics* 85, 2003, no. 3: 693–708.

171. Koehler and Spengler, "The toxic release inventory."

172. Larry S. Monroe, "Future options for generation of electricity from coal," hearing before the House Committee on Energy and Commerce, June 24, 2003: 108.

173. Khanna et al., "Toxics release information."

174. D. Sarokin and J. Schulkin, "Environmentalism and the right-to-know: Expanding the practice of democracy," *Ecological Economics* 4, 1991, no. 3: 175–189.

175. Michael Goldblatt, "Registering pollution: The prospects for a pollution information system," in *The Bottom Line: Industry and the Environment in South Africa* (UCT Press, 1997).

Chapter 9

1. John Vidal, "No new coal—The calling card of the 'Green Banksy' who breached Fortress Kingsnorth," *Guardian*, December 11, 2008 (http://www.guardian.co.uk/environment/2008/dec/11/kingsnorth-green-banksy-saboteur).

2. Associated Press, "Hansen and Hannah arrested in West Virginia mining protest," *Guardian*, June 24, 2009 (http://www.guardian.co.uk/environment/2009/jun/24/james-hansen-daryl-hannah-mining-protest).

3. David Barbeler, "Greenpeace ramps up coal protests," *Courier Mail* (Australia), August 6, 2009.

4. Copenhagen's prostitutes offered their services for free to anyone displaying a delegate's badge. See Andrew Gilligan, "Copenhagen climate change conference: Not the science but the vision," *Daily Telegraph* (London), December 7, 2009.

5. "China defends bottom line, but world agrees deal is a dud," *Sunday Morning Post*, December 20, 2009.

6. "World cannot afford a repeat of this travesty," *Sunday Morning Post*, December 20, 2009.

7. Stephen Peake and Joe Smith, *Climate Change: From Science to Sustainability* (Oxford University Press, 2009): 7.

8. Paul J. Crutzen and E. F. Stoermer, "The Anthropocene," *IGBP Newsletter* 41, May 2000.

9. William Rees, "Contemplating the abyss: The role of environmental degradation in the collapse of human societies," in *The Energy Reader*, ed. L. Nader (Wiley-Blackwell, 2010).

10. James Gustave Speth, *The Bridge at the End of the World: Capitalism, the Environment, and Crossing from Crisis to Sustainability* (Yale University Press, 2008).

11. Wiebe Bijker, "Do not despair: There is life after constructivism," *Science, Technology, & Human* Values 18, 1993, no. 1: 124–125.

12. Horst Rittel and Melvin Webber, "Dilemmas of a general theory of planning," *Policy Sciences* 4, 1973: 155–169.

13. Michael P. Vandenbergh and Anne C. Steinemann, "The carbon-neutral individual," *New York University Law Review* 82, 2007: 1673–1745.

14. Ibid.

15. Rachel Carson, *Silent Spring* (Houghton Mifflin, 1962): ix.

16. Jared Diamond, "Will big business save the Earth?" *New York Times*, December 6, 2009.

17. Paul Hawken, *The Ecology of Commerce: A Declaration of Sustainability* (HarperCollins, 1993).

18. Vaclav Smil, *Two Prime Movers of Globalization: The History and Impact of Diesel Engines and Gas Turbines* (MIT Press, 2010): 194–195.

19. Some of these examples are from Benjamin K. Sovacool and Charmaine Watts, "Going completely renewable: Is it possible (let alone desirable)?" *Electricity Journal* 2, 2009, no. 4: 95–111.

20. Hawken, *The Ecology of Commerce*.

Appendix B

1. Alexander L. George and Andrew Bennett, *Case Studies and Theory Development in the Social Sciences* (Harvard University Press, 2004): 5.

2. Bent Flyvbjerg, "Five misunderstandings about case study research," *Qualitative Inquiry* 12, 2006, no. 2: 219–245; Bent Flyvbjerg, *Making Social Science Matter: Why Social Inquiry Fails and How It Can Succeed Again* (Cambridge University Press, 2001).

3. R. K. Yin, *Case Study Research: Design and Methods* (Sage, 2009): 23.

4. Pamela Baxter and Susan Jack, "Qualitative case study methodology: Study design and implementation for novice researchers," *Qualitative Report* 13, 2008, no. 4: 544–559.

5. Flyvbjerg, "Five misunderstandings" and *Making Social Science Matter*.

6. See Gene I. Rochlin and Alexandra von Meier, "Nuclear power operations: A cross-cultural perspective," *Annual Review of Energy and the Environment* 19 (1994): 153–187; Arend Lijphart, "Comparative politics and the comparative method," *American Political Science Review* 65 (1971): 682–693.

7. For more on the benefits and value of case studies, see Bob Hancké, *Intelligent Research Design: A Guide for Beginning Researchers in the Social Sciences* (Oxford University Press, 2009); John Gerring, *Case Study Research* (Cambridge University Press, 2005); Bruce Straits and Royce A. Singleton, *Approaches to Social Research*, fourth edition (Oxford University Press, 2004); Roland W. Scholz and Olaf Tietje, *Embedded Case Study Methods: Integrating Quantitative and Qualitative Knowledge* (Sage, 2002); Robert E. Stake, *The Art of Case Study Research* (Sage, 1995); Charles C. Ragin and Howard S. Becker, eds., *What Is a Case? Exploring the Foundations of Social Inquiry* (Cambridge University Press, 1992); Kathleen M. Eisenhardt, "Building theories from case study research," *Academy of Management Review* 14, 1989, no. 4: 532–550.

8. See A. L. Strauss, *Qualitative Analysis for Social Scientists* (Cambridge University Press, 1987); Patricia Yancey Martin and Barry A. Turner, "Grounded theory and organizational research," *Journal of Applied Behavioral Science* 22, 1986, no. 2: 141–157; B. G. Glaser and A. L. Strauss, *The Discovery of Grounded Theory* (Aldine, 1967).

9. For an introduction to ethnographic methodology, see Paul Atkinson, "Ethnomethodology: A critical review," *Annual Review of Sociology* 14 (1988): 441–465. On its use in energy scholarship, see Willett Kempton and Laura Montgomery, "Folk quantification of energy," *Energy* 7, 1982, 10: 817–827.

10. Elizabeth O'Sullivan, Gary R. Rassel, and Maureen Berner, *Research Methods for Public Administrators* (Pearson Longman, 2010).

11. Eugene Bardach, *A Practical Guide for Policy Analysis: The Eightfold Path to More Effective Problem Solving* (Sage, 2009): 81.

12. Alexander L. George and Andrew Bennett, *Case Studies and Theory Development in the Social Sciences* (Harvard University Press, 2004).

13. Wayne C. Booth, Gregory G. Colomb, and Joseph M. Williams, *The Craft of Research* (University of Chicago Press, 2008): 82.

14. Gary King, Robert O. Keohane, and Sidney Verba, *Designing Social Inquiry: Scientific Inference in Qualitative Research* (Princeton University Press, 1994): 6.

15. Alexander L. George and Andrew Bennett. *Case Studies and Theory Development in the Social Sciences* (Harvard University Press, 2004): 19.

16. B. G. Glaser and A. L. Strauss, *The Discovery of Grounded Theory: Strategies for Qualitative Research* (Aldine de Gruyter, 1997); G. King, R. O. Keohane, and S. Verba, *Designing Social Inquiry: Scientific Inference in Qualitative Research* (Princeton University Press, 1994).

17. Alexander L. George and Andrew Bennett, *Case Studies and Theory Development in the Social Sciences* (Harvard University Press, 2004).

Bibliography

Adler, Jonathan H. Jurisdictional mismatch in environmental federalism. *New York University Environmental Law Journal* 14 (2005): 130–135.

Akerlof, George A. The market for "lemons": Quality uncertainty and the market mechanism. *Quarterly Journal of Economics* 84 (3) (1970): 488–500.

Aligica, Paul Dragos. Institutional and stakeholder mapping: Frameworks for policy analysis and institutional change. *Public Organization Review* 6 (2006): 79–90.

Andersson, Krister P., and Elinor Ostrom. Analyzing decentralized resource regimes from a polycentric perspective. *Policy Sciences* 41 (2008): 71–93.

Archer, David, and Stefan Rahmstorf. *The Climate Crisis: An Introductory Guide to Climate Change*. Cambridge University Press, 2010.

Archer, David. *The Long Thaw*. Princeton University Press, 2009.

Asia Pacific Energy Research Centre. *A Quest for Energy Security in the 21st Century: Resources and Constraints*. 2007.

Ayers, Jessica M., and Saleemul Huq. The value of linking mitigation and adaptation: A Case Study of Bangladesh. *Environmental Management* 43 (2008), no. 5: 753–764.

Barnes, Joe, Mark H. Hayes, Amy M. Jaffe, and David G. Victor, eds. *Natural Gas and Geopolitics: From 1970 to 2040*. Cambridge University Press, 2006.

Barton, Barry, Catherine Redgwell, Anita Ronne, and Donald N. Zillman, eds. *Energy Security: Managing Risk in a Dynamic Legal and Regulatory Environment*. Oxford University Press, 2004.

Berke, Phillip R. The evolution of green community planning, scholarship, and practice. *Journal of the American Planning Association. American Planning Association* 74 (4) (2008): 393–407.

Blumstein, Carl, Betsy Krieg, Lee Schipper, and Carl York. Overcoming social and institutional barriers to energy conservation. *Energy* 5 (1980): 355–371.

Bohi, D. R., and M. A. Toman. *The Economics of Energy Supply Security*. Kluwer, 1996.

Brown, Marilyn A. Market failures and barriers as a basis for clean energy policies. *Energy Policy* 29 (14) (2001): 1197–1207.

Brown, Marilyn A. The multiple dimensions of carbon management: Mitigation, adaptation, and geo-engineering. *Carbon Management* 1 (1) (2010): 27–33.

Brown, Marilyn A., and Sharon (Jess) Chandler. Governing confusion: How statutes, fiscal policy, and regulations impede clean energy technologies. *Stanford Law & Policy Review* 19 (3) (2008): 472–509.

Brown, Marilyn A., Jess Chandler, and Melissa V. Lapsa. Policy options targeting decision levers: An approach for shrinking the residential energy-efficiency gap. In Proceedings of the ACEEE Summer Study on Energy Efficiency in Buildings, August 17, 2010.

Brown, Marilyn A., Jess Chandler, Melissa Lapsa, and Moonis Ally. Making Homes Part of the Climate Solution. Report ORNL/TM-2009/104, Oak Ridge National Laboratory, 2009. Available at http://www. ornl. gov.

Brown, Marilyn A., Jess Chandler, Melissa V. Lapsa, and Benjamin K. Sovacool. Carbon Lock-In: Barriers to the Deployment of Climate Change Mitigation Technologies. Report ORNL/TM-2007/124, Oak Ridge National Laboratory, 2008.

Brown, Marilyn A., Rodrigo Cortes, and Matthew Cox. Reinventing industrial energy use in a resource-constrained world. In *Energy Sustainability and the Environment*, ed. F. Sioshansi. Elsevier, 2011.

Brown, Marilyn A., and Benjamin K. Sovacool. Developing an "energy sustainability index" to evaluate energy policy. *Interdisciplinary Science Reviews* 32 (4) (2007): 335–349.

Bulle, R. J., and D. N. Pellow. Environmental justice: Human health and environmental inequalities. *Annual Review of Public Health* 27 (2006): 103–124.

Butt, A. T., B. A. McCarl, J. Angerer, P. T. Dyke, and J. W. Stuth. The economic and food security implications of climate change. *Climatic Change* 67 (2005): 355–378.

CNA Corporation. *Powering America's Defense: Energy and the Risks to National Security*. 2009.

Collins, William, Robert Colman, James Haywood, Martin R. Manning, and Philip Mote. The physical science behind climate change. *Scientific American* 297 (August 2007): 64–73.

Costantini, V., F. Gracceva, A. Markandya, and G. Vicini. Security of energy supply: Comparing scenarios from a European perspective. *Energy Policy* 35 (1) (2007): 210–226.

de Bruin, K., et al. Adapting to climate change in the Netherlands: An inventory of climate adaptation options and ranking of alternatives. *Climatic Change* 95 (2009): 23–45.

Dietz, T., and P. C. Stern, eds. *New Tools for Environmental Protection: Education, Information, and Voluntary Measures*. National Academy Press, 2002.

DiMento, Joseph F. C., and Pamela Doughman, eds. *Climate Change: What It Means for Us, Our Children, and Our Grandchildren*. MIT Press, 2007.

Dunn, William N. *Public Policy Analysis: An Introduction*, fourth edition. Prentice-Hall, 2008.

Economics of Climate Adaptation Working Group. *Shaping Climate-Resilient Development: A Framework for Decision-Making*. Climate Works Foundation, 2009.

Edelenbos, Jurian, and Erik-Hans Klijn. Managing stakeholder involvement in decision making. *Journal of Public Administration: Research and Theory* 16 (2005): 417–446.

Ediger, Volkan S., and Enes Hosgor, A. Nesen Surmeli, and Huseyin Tatlidil. Fossil fuel sustainability index: An application of resource management. *Energy Policy* 35 (2007): 2969–2977.

Edwards, Paul N. *A Vast Machine: Computer Models, Climate Data, and the Politics of Global Warming*. MIT Press, 2010.

Ellerman, Denny, Paul L. Joskow, and David Harrison Jr. Emissions Trading in the US: Experience, Lessons, and Considerations for Greenhouse Gases. Pew Center on Global Climate Change, 2003. Available at http://www. pewclimate. org.

Ewing, Reid, Keith Bartholomew, Steve Winkelman, Jerry Walters, and David Goldberg. *Growing Cooler: Evidence on Urban Development and Climate Change*. Urban Land Institute, 2007.

Farrell, Alexander, Hisham Zerriffi, and Hadi Dowlatabadi. Energy infrastructure and security. *Annual Review of Environment and Resources* 29 (2004): 421–422.

Fischer, Carolyn, and Richard G. Newell. Environmental and technology policies for climate mitigation. *Journal of Environmental Economics and Management* 55 (2008): 142–162.

Flyvbjerg, Bent. *Making Social Science Matter: Why Social Inquiry Fails and How It Can Succeed Again*. Cambridge University Press, 2001.

Friedman, Avi. *Sustainable Residential Development: Planning and Design for Green Neighborhoods*. McGraw-Hill, 2007.

Geller, Howard. *Energy Revolution*. Island, 2003.

Geller, Howard, Philip Harrington, Arthur H. Rosenfeld, Satoshi Tanishima, and Fridtjof Unande. Policies for increasing energy efficiency: Thirty years of experience in OECD countries. *Energy Policy* 34 (2006): 556–573.

Girardet, Herbert, and Miguel Mendonca. 2009. *A Renewable World: Energy, Ecology, Equality*. Green Books, 2009.

Gnansounou, Edgard. Assessing the energy vulnerability: Case of industrialised countries. *Energy Policy* 36 (2008): 3734–3744.

Gupta, E. Oil vulnerability index of oil-importing countries. *Energy Policy* 36 (3) (2008): 1195–1211.

Hawkins, D. G., D. A. Lashof, and R. H. Williams. What to do about coal. *Scientific American* 195 (3) (2006): 68–75.

Hirst, Eric, and Marilyn A. Brown, with E. Hirst. Closing the efficiency gap: Barriers to improving energy efficiency. *Resources, Conservation and Recycling* 3 (1990): 267–281.

Holdren, John P., and Kirk R. Smith. Energy, the Environment, and Health. In *World Energy Assessment: Energy and the Challenge of Sustainability*, ed. Tord Kjellstrom, David Streets, and Xiadong Wang. United Nations Development Programme, 2000.

Hughes, Larry. The four R's of energy security. *Energy Policy* 37 (6) (June 2009): 2459–2461.

Intergovernmental Panel on Climate Change. *Climate Change 2007: Impacts, Adaptation and Vulnerability*. Cambridge University Press, 2007.

Intergovernmental Panel on Climate Change. *Climate Change 2007: Mitigation of Climate Change*. Cambridge University Press, 2007.

Intergovernmental Panel on Climate Change. *Climate Change: 2007: The Physical Science Basis. Contribution of Working Group I to the Fourth Assessment Report of the Intergovernmental Panel on Climate Change.* Cambridge University Press, 2007.

Intergovernmental Panel on Climate Change (IPCC). *IPCC Special Report on Carbon Dioxide Capture and Storage.* Cambridge: Cambridge University Press, 2005. Available at http://www.ipcc. ch.

International Atomic Energy Agency. *Energy Indicators for Sustainable Development: Guidelines and Methodologies.* IAEA, 2005.

International Energy Agency. *World Energy Outlook 2009.* IEA, 2009.

International Energy Agency. *Security and Climate Policy—Assessing Interactions.* OECD, 2007.

Jacobson, Mark Z. Review of solutions to global warming, air pollution, and energy security. *Energy & Environmental Science* 2 (2009): 148–173.

Jansen, Jaap C., and Ad J. Seebregts. Long-term energy services security: What is it and how can it be measured and valued? *Energy Policy* 38 (2010): 1654–1664.

Jansen, Jaap C. *Energy Services Security Concepts and Metrics.* Vienna: International Atomic Energy Agency Project on Selecting and Defining Integrated Indicators for Nuclear Energy, October 2009, ECN-E-09-080.

Nazarenko, J. L., et al. Earth's energy imbalance: Confirmation and implications. *Science* 308 (2005): 1431–1435.

Kahneman, Daniel, Jack L. Knetsch, and Richard H. Thaler. Anomalies: The endowment effect, loss aversion, and status quo bias. *Journal of Economic Perspectives* 5 (1) (1991): 193–206.

Kalicki, J. H., and D. L. Goldwyn, eds. *Energy Security—Toward a New Foreign Policy Strategy.* Johns Hopkins University Press, 2005.

Karl, Thomas R., Jerry M. Melillo, and Thomas C. Peterson. *Global Climate Change Impacts in the US.* Cambridge University Press, 2009.

Keppler, Jan Horst. *Energy Supply Security and Nuclear Energy: Concepts, Indicators, Policies.* Paris: Nuclear Energy Agency, October 2007.

Kobos, Peter H., Jo D. Ericson, and Thomas E. Drennen. Technological learning and renewable energy costs: Implications for US renewable energy policy. *Energy Policy* 34 (2006): 1645–1658.

Lackner, Klaus S., and Jeffrey D. Sachs. A robust strategy for sustainable energy. *Brookings Papers on Economic Activity* 2 (2004): 215–248.

Lazarus, R. J. Super wicked problems and climate change: Restraining the present to leverage the future. *Cornell Law Review* 94 (2009): 1153–1234.

Lefevre, Nicolas. Measuring the energy security implications of fossil fuel resource concentration. *Energy Policy* 38 (2010): 1635–1644.

Lehmann, Johannes, John Gaunt, and Marco Rondon. Bio-char sequestration in terrestrial ecosystems—A review. *Mitigation and Adaptation Strategies for Global Change* 11 (2006): 403–427.

Lenton, T. M., and N. E. Vaughan. The radiative forcing potential of different climate geoengineering options. *Atmospheric Chemistry and Physics Discussion* 9 (2009): 2559–2608.

Linden, Anna-Lisa, Annika Carlsson-Kanyama, and Bjorn Eriksson. Efficient and inefficient aspects of residential energy behavior: What are the policy instruments for change? *Energy Policy* 34 (2006): 1918–1927.

Loschel, Andreas, Ulf Moslener, and Dirk Rubbelke. Indicators of energy security in industrialized countries. *Energy Policy* 38 (2010): 1665–1671.

Lovins, Amory B., and L. Hunter Lovins. *Brittle Power: Energy Strategy for National Security*. Brick House, 1982.

Lovins, Amory, E. Kyle Datta, Thomas Feiler, Andre Lehmann, Karl Rabago, Joel Swisher, and Ken Wicker. *Small Is Profitable: The Hidden Benefits of Making Electrical Resources the Right Size*. Rocky Mountain Institute, 2002.

Luft, Gal, and Anne Korin, eds. *Energy Security Challenges for the 21st Century*. Praegar International/ABC-CLIO, 2009.

McDonough, William, and Michael Braungart. *Cradle to Cradle: Remaking the Way We Make Things*. Farrar, Straus, and Giroux, 2002.

McMichael, Anthony J., John W. Powles, Colin D. Butler, and Ricardo Uauy. Food, livestock production, energy, climate change, and health. *Lancet* 370 (9594) (2007): 1253–1263.

Mendelsohn, R., and J. Neumann, eds. *The Impact of Climate Change on the US Economy*. Cambridge University Press, 1999.

Mushtaq, Shahbaz, Tek Narayan Maraseni, Jerry Maroulis, and Mohsin Hafeez. Energy and water tradeoffs in enhancing food security: A selective international assessment. *Energy Policy* 37 (9) (2009): 3635–3644.

National Academy of Sciences. *Electricity from Renewable Resources: Status, Prospects, and Impediments*. National Academies Press, 2009.

National Academy of Sciences. *Liquid Transportation Fuels from Coal and Biomass*. National Academies Press, 2009.

National Academy of Sciences. *Real Prospects for Energy Efficiency in the United States*. National Academies Press, 2009.

National Academy of Sciences. *Advancing the Science of Climate Change*. National Academies Press, 2010.

National Academy of Sciences. *Limiting the Magnitude of Future Climate Change*. National Academies Press, 2010.

National Commission on Energy Policy. *Ending the Energy Stalemate*. December, 2004. Available at http://www. energycommission. org.

National Commission on Energy Policy. *Forging the Climate Consensus*. 2009. Available at http://www. energycommission. org.

National Research Council. *Hidden Costs of Energy: Unpriced Consequences of Energy Production and Use*. National Academies Press, 2009.

National Research Council. *Informing an Effective Response to Climate Change*. National Academies Press, 2010.

National Research Council. *Transitions to Alternative Transportation Technologies—A Focus on Hydrogen*. National Academies Press, 2008.

Nordhaus, William D. A review of the Stern review on the economics of climate change. *Journal of Economic Literature* 55 (September 2007): 686–702.

Norton, Brian G., and Robert E. Ulanowicz. Scale and biodiversity policy: A hierarchical approach. *Ambio* 21 (1992): 244–249.

Nuttall, William J., and Devon L. Manz. A new energy security paradigm for the twenty-first century. *Technological Forecasting and Social Change* 75 (2008): 1249.

Nye, David E. *America as Second Creation: Technology and Narratives of New Beginnings.* MIT Press, 2003.

Ogden, Joan M. Prospects for building a hydrogen energy infrastructure. *Annual Review of Energy and the Environment* 24 (1999): 227–279.

Olson, Mancur. *The Logic of Collective Action: Public Goods and the Theory of Groups.* Harvard University Press, 1965.

Olz, Samantha, Ralph Sims, and Nicolai Kirchner. Contributions of Renewables to Energy Security: International Energy Agency Information Paper. Paris: OECD, April 2007.

Orr, David W. *Design on the Edge: The Making of a High-Performance Building.* MIT Press, 2006.

Ostrom, Elinor. A general framework for analyzing sustainability of socio-ecological systems. *Science* 325 (2009): 419–422.

Ostrom, Elinor. *A Polycentric Approach for Coping with Climate Change.* Report Prepared for the WDR2010 Core Team at the World Bank, W09-4, 2009.

Pacala, S., and R. Socolow. Stabilization wedges: Solving the climate problem for the next 50 years with current technologies. *Science* 305 (2004): 968–972.

Pascual, Carlos, and Jonathan Elkind, eds. *Energy Security: Economics, Politics, Strategies, and Implications.* Brookings Institution Press, 2010.

Prouty, Anne E. The clean development mechanisms and its implications for climate justice. *Columbia Journal of Environmental Law* 34 (2) (2009): 513–540.

Quinn, Eugene R. Cost of Obtaining a Patent. Intellectual Property Watchdog, 2006. Available at http://www. ipwatchdog. com.

Reddy, B. Sudhakara, and G. Assenza. The great climate debate. *Energy Policy* 37 (2009): 2997–3008.

Robertson, Kate, Jette Findsen, and Steve Messner. *International Carbon Capture and Storage Projects—Overcoming Legal Barriers.* National Energy Technology Laboratory, DOE/NETL-2006-1236, 2006.

Rogner, Hans-Holger, Lucille M. Langlois, Alan McDonald, Daniel Weisser, and Mark Howells. *The Costs of Energy Supply Security.* Vienna: International Atomic Energy Agency, December 27, 2006.

Rosenzweig, Cynthia, et al. Attributing physical and biological impacts to anthropogenic climate change. *Nature* 453 (May 15), 2008: 353–357.

Royal Society. *The. Geoengineering the Climate: Science, Governance, and Uncertainty.* Royal Society, 2009.

Sachs, Noah. Planning the funeral at the birth: Extended producer responsibility in the European Union and the US. *Harvard Environmental Law Review* 30 (2006): 56.

Saunders, Kurt M., and Linda Levine. Better, faster, cheaper—later: What happens when technologies are suppressed. *Michigan Telecommunications and Technology Law Review* 11 (2004): 23–69.

Schaefer, Mark. Water technologies and the environment: Ramping up by scaling down. *Technology in Society* 30 (2008): 415–422.

Scheepers, Martin, Ad Seebregts, Jacques de Jong, and Hans Maters. *EU Standards for Energy Security of Supply*. ECN, 2006.

Schneider, S. H., A. Rosencranz, M. D. Mastrandrea, and K. Kuntz-Duriseti, eds. *Climate Change Science and Policy*. Island, 2010.

Simon, Herbert A. A behavioral model of rational choice. *Quarterly Journal of Economics* 69 (1) (February 1955): 99–118.

Simon, Herbert A. Rational decision making in business organizations. *American Economic Review* 69 (4) (September 1979): 493–513.

Smil, Vaclav. Energy in the twentieth century: Resources, conversions, costs, uses, and consequences. *Annual Review of Energy and the Environment* 25 (2000): 21–51.

Smil, Vaclav. *Two Prime Movers of Globalization: The History and Impact of Diesel Engines and Gas Turbines*. MIT Press, 2010.

Southworth, Katie. Corporate voluntary action: A valuable but incomplete solution to climate change and energy security challenges. *Policy and Society* 27 (4) (2009): 329–350.

Sovacool, Benjamin K., ed. *The Routledge Handbook of Energy Security*. Routledge, 2010.

Sovacool, Benjamin K. Reassessing energy security and the Trans-ASEAN natural gas pipeline network in Southeast Asia. *Pacific Affairs* 82 (3) (Fall 2009): 467–486.

Sovacool, Benjamin K. Sound climate, energy, and transport policy for a carbon constrained world. *Policy and Society* 27 (4) (2009): 273–283.

Sovacool, Benjamin K. The political economy of oil and gas in Southeast Asia: Heading towards the natural resource curse? *Pacific Review* 23 (2) (May 2010): 225–259.

Sovacool, B. K., and M. A. Brown. *Energy and American Society—Thirteen Myths*. Springer, 2007.

Sovacool, Benjamin K., and Marilyn A. Brown. Twelve metropolitan carbon footprints: A preliminary comparative global assessment. *Energy Policy* 38 (9) (2010): 4856–4869.

Sovacool, Benjamin K., and Marilyn A. Brown. Competing dimensions of energy security: An international perspective. *Annual Review of Environment and Resources* 35 (2010): 77–108. www.annualreviews.org/doi/pdf/10.1146/annurev-environ-042509-143035.

Sovacool, Benjamin K. Running on empty: The electricity-water nexus and the US electric utility sector. *Energy Law Journal* 30 (11) (2009): 11–51.

Sovacool, Benjamin K. Solving the oil independence problem. *Energy Policy* 35 (11) (2007): 5512.

Sovacool, Benjamin K. The importance of comprehensiveness in renewable electricity and energy efficiency policy. *Energy Policy* 37 (2009): 1529–1541.

Sovacool, Benjamin K. *The Dirty Energy Dilemma: What's Blocking Clean Power in the US.* Praeger, 2008.

Sperling, Daniel, and Deborah Gordon. *Two Billion Cars: Driving Toward Sustainability.* Oxford University Press, 2009.

Speth, James Gustave. *The Bridge at the End of the World: Capitalism, the Environment, and Crossing from Crisis to Sustainability.* Yale University Press, 2008.

Schneider, Stephen H. Geoengineering: Could we or should we make it work? *Phil. Trans. R. Soc. A* 366 (2008): 3843–3862.

Schneider, Stephen. The worst-case scenario. *Nature* 458 (2009): 1104–1105.

Stern, Nicholas. *The Economics of Climate Change: The Stern Review.* Cambridge University Press, 2007.

Stern, Paul C., and Elliot Aronson. *Energy Use: The Human Dimension.* Freeman, 1984.

Stigler, George J. The economics of information. *Journal of Political Economy* 69 (3) (June 1961): 213–225.

Streck, Charlotte, Robert O'Sullivan, Toby Janson-Smith, and Richard G. Tarasofsky, eds. *Climate Change and Forests: Emerging Policy and Market Opportunities.* Brookings Institution Press, 2008.

Stringer, John, and Linda Horton. Basic Research Needs to Assure a Secure Energy Future. Oak Ridge National Laboratory, 2003. Available at http://www. sc. doe. gov/bes/reports/files/SEF_rpt. pdf.

Security, Contesting Energy, ed. Subroto. Indonesian Institute for Energy Economics.

Sunstein, C. R. *Beyond the Precautionary Principle.* Cambridge University Press, 2005.

Suplee, Curt, Allen Bard, Marilyn Brown, Mike Corradini, and Jeremy Mark. What You Need to Know About Energy. National Academy of Sciences, 2008. Available at http://sites.nationalacademies.org/energy/Energy_043338.

Tester, Jefferson W., Elisabeth M. Drake, Michael J. Driscoll, Michael W. Golay, and William A. Peters. *Sustainable Energy: Choosing Among Options.* MIT Press, 2005.

United Nations Economic and Social Commission for Asia and the Pacific [UNESCAP]. Energy Security and Sustainable Development in Asia and the Pacific. Geneva: UNESCAP, April 2008, ST/ESCAP/2494.

Vera, I. A., L. M. Langlois, H. H. Rogner, A. I. Jalal, and F. L. Toth. Indicators for sustainable energy development: An initiative by the International Atomic Energy Agency. *Natural Resources Forum* 29 (2005): 274–283.

Victor, David, Granger Morgan, John Steinbruner, and Kate Ricke. The geoengineering option: A last resort against global warming? *Foreign Affairs (Council on Foreign Relations)* 88 (2009): 65.

Weiner, Jonathan B. Think globally, act globally: The limits of local climate policies. *University of Pennsylvania Law Review* 155 (2007): 1961–1979.

Weitzman, M. L. On modeling and interpreting the economics of catastrophic climate change. *Review of Economics and Statistics* 91 (2009): 1–19.

Wellinghoff, Jon, and David L. Morenoff. Recognizing the importance of demand response: The second half of the wholesale electric market function. *Energy Law Journal* 28 (2) (2007): 393.

Wilbanks, T. J., P. Leiby, R. Perlack, J. T. Ensminger, and S. B. Wright. Toward an integrated analysis of migration and adaptation: Some preliminary findings. *Mitigation and Adaptation Strategies for Global Change* 12 (5) (2007): 713–725.

Wilk, Richard B. Culture and Energy Consumption. In *Energy: Science, Policy, and the Pursuit of Sustainability*, ed. Robert Bent, Lloyd Orr, and Randall Baker. Island, 2002.

Wilson, Elizabeth J., Joseph Plummer, Miriam Fischlein, and Timothy M. Smith. Implementing energy efficiency: Challenges and opportunities for rural electric cooperatives and small municipal utilities. *Energy Policy* 36 (2008): 3383–3397.

World Energy Assessment: Energy and the Challenge of Sustainability. United Nations Development Program, 2000.

Worrell, E., and G. Biermans. Move over! Stock turnover, retrofit and industrial energy efficiency. *Energy Policy* 33 (7) (2005): 949–962.

Yergin, Daniel. Ensuring energy security. *Foreign Affairs (Council on Foreign Relations)* 85 (2) (March/April 2006): 69–82.

Yohe, Gary. Risk assessment and risk management for infrastructure planning and investment. *Bridge* 40 (3) (2010): 14–21.

Index